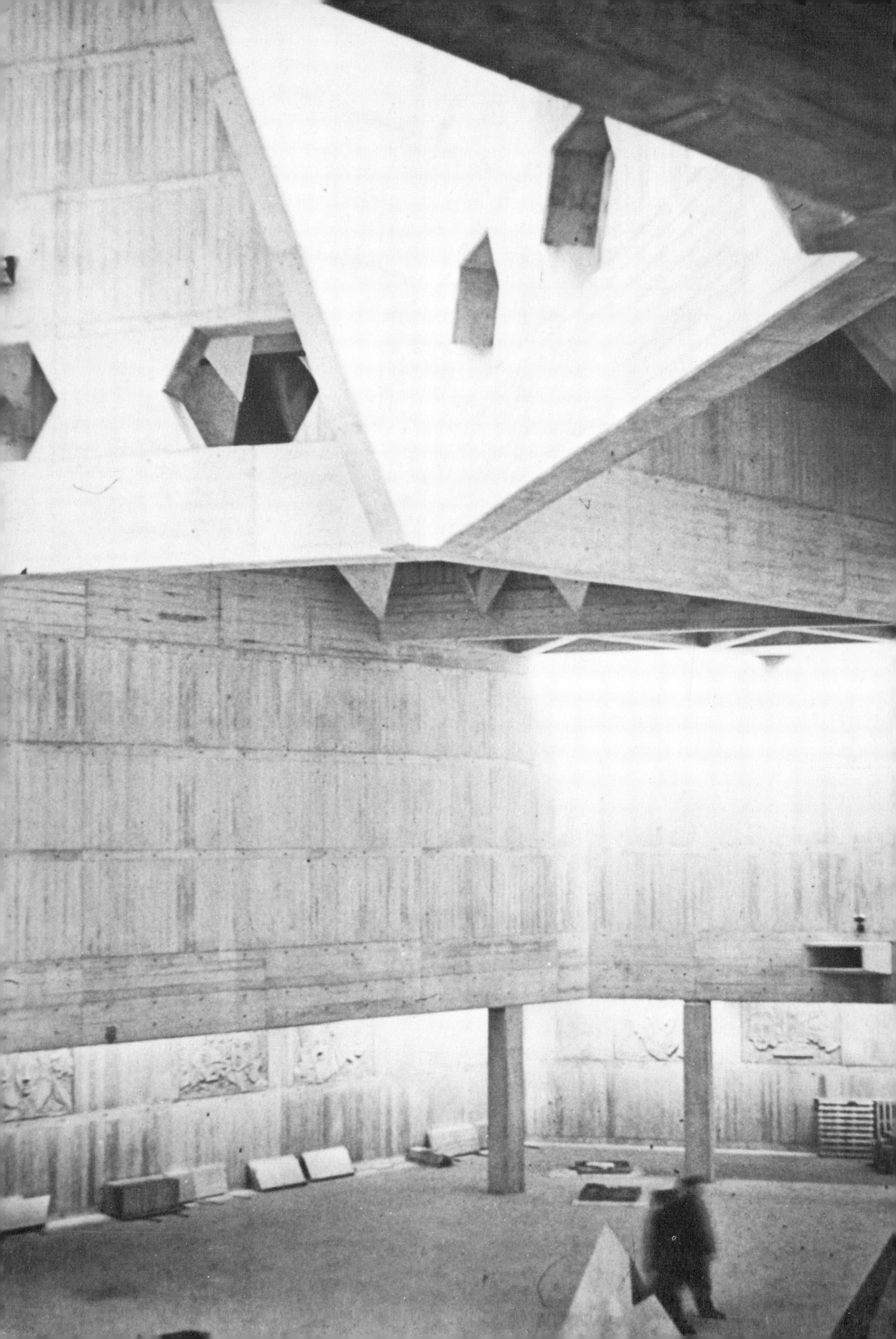

Formwork Construction and Practice

Formwork Construction and Practice

J. G. Richardson FIWM

Lecturer, Construction Department, Training Division, Cement and Concrete Association

A Viewpoint Publication

VIEWPOINT PUBLICATIONS

Books published in the Viewpoint Publications series deal with all practical aspects of concrete, concrete technology and allied subjects in relation to civil and structural engineering, building and architecture.

Contributors to Viewpoint Publications include authors from within the Cement and Concrete Association itself and from the construction industry in general. While the views and opinions expressed in these publications may be in agreement with those of the Association they should be regarded as being independent of Association policy.

13.019 First published 1977

ISBN 0 7210 1058 X

Viewpoint publications are designed and published by the Cement and Concrete Association, Wexham Springs, Slough SL3 6PL

Printed in Great Britain by The Whitefriars Press Ltd, London and Tonbridge

The Author

John Richardson was educated at Christ's College, Finchley, following which he joined the army and underwent training at Mons Officer Cadet Training Establishment and The Royal School of Military Engineering at Ripon and Newark. In 1948, two years after he was commissioned, he left the army and joined a firm of civil engineering contractors where he worked for eight years as a draughtsman/designer of moulds and formwork. During this period he was engaged on formwork design for chimneys, vent shafts, grain silos, elevated road structures and cooling towers. He later joined Holland, Hannen and Cubitts as a mould designer in charge of a mould shop specializing in the manufacture of moulds for all types of precast pre-tensioned and post-tensioned concrete components in timber and composite materials. Further experience in both managerial and supervisory roles was gained from positions held in various precast firms and works. In 1963 he was appointed works manager at Concrete Ltd (Iver Works) and was responsible for all works activity in the production of over 300 tonnes/day of precast and prestressed components.

In 1968 he was awarded the Diploma of The Institute of Works Management and in the same year joined the Training Division of the Cement and Concrete Association as a lecturer. He had, in fact, been a visiting lecturer at the Training Centre since the early 1950s giving lectures on form and mould topics. He organizes many courses at the Training Centre including supervision of precast concrete production; precast concrete production; formwork construction and practice and formwork planning and design.

Mr Richardson also holds the City and Guilds Institute silver medal for formwork design and is very much involved with the Institute's activities on matters relating to formwork construction, planning and design. He was a member of the Institution of Structural Engineers/Concrete Society Joint Committee which recently produced the *Formwork Report.*

John Richardson is the author of *Formwork Notebook, Concrete Notebook, Precast Concrete Production, Practical Formwork and Mould Construction,* and at the time of going to press is preparing another publication, *Concrete Craft Notebook.*

Contents

Introduction

The fascination of concrete construction stems from a number of attributes that the material concrete possesses. Concrete allows expressive moulded forms to be developed in slender members or massive structural components, and it can be designed to exhibit flowing curves or stark functional lines. The plastic nature of the material lends itself to a considerable variety of surface finishes, features and textures, and these in turn can be emphasized by skilled detail combined with the use of colour and choice of exotic aggregates. It can be used for the construction of massive buttresses of gravity dams or, by contrast, for slender prestressed elements such as those employed in the roofs of sports stadia, assembly rooms and concert halls of considerable spans.

Concrete makes various demands on the parties to the contract. The architect and engineer, within the stipulations of governing regulations and Codes of Practice, have complete design freedom to express both aesthetic concept and functional form, while the task of the constructor is that of using the skills of form design and construction to mould and shape the concrete to the designed profiles, and achieving the specified standards of finish and accuracy in a competitive manner.

Formwork is the container into which or against which the concrete is cast to the required designed shape, be it moulded or featured, massive or slender, exposed or hidden within the finished structure. Formwork although only used as a temporary structure has a lasting effect on the final concrete structure thus presenting a commentary on the skills of those employed in its construction.

To achieve a structure that satisfies the demands of structural and architectural designers it is essential that all who are involved in formwork design, construction and supervision should have a sound working knowledge not only of the various formwork operations but also the methods and materials involved.

Concrete technology and the methods of handling, placing and compaction are critical, since they govern the durability of the structure and in many instances its final appearance. Steel reinforcement is critical, its correct positioning, with regard to design, cover and rigidity of the cage that resists the forces imposed on it during the placing operations being essential for sound concrete construction.

However well controlled are the resources deployed in reinforcement design, steelfixing, concrete mixing and placing it is, however, the materials selected for the formwork, its construction, and the fit of the form components that will have the greatest impact in providing a satisfactory structure.

Formwork is the container into which the fresh concrete is placed, and it is in this container that the concrete is compacted by various means so that the reinforcement is totally enclosed and protected. The compactive effort must be such as to ensure dense, void-free concrete capable of developing the designed strength to combat the stresses which develop within the structure. The container must enclose the concrete mass without leakage and without distortion outside that appropriate to the scale of the component. Apart from resisting the pressures that develop during the placing process and the loads imposed by the construction, formwork must also protect the concrete during curing and support the mass until sufficient maturity has been achieved to enable the newly cast concrete itself to contribute structurally. Once this stage has been reached, the container must be such that it can be readily removed for re-use in further construction.

The container is generally a temporary 'structure', often site constructed with the minimum of skilled design assistance. However, for complicated structural or civil engineering work, or where there are special requirements with regard to accuracy and finish, form designers or engineers provide the appropriate mechanical calculations and details. Where systems are used, the development and calculations involved are also inherent parts of the system.

The suitability of the final surface finish, or the accuracy accomplished, is the criteria by which the engineer, the architect and the client evaluate the resulting concrete structure. The ease with which the formwork is used to achieve these results, the number of uses obtained from the equipment and the financial outcome of the whole operation are additional factors by which the contractor assesses the outcome of formwork applications.

These varied interests tend to generate some discussion and even conflict on site, particularly as formwork comprises such a major cost item for the contractor who is responsible for the actual construction. It is essential that the various parties involved on a project are aware of the principles and possibilities of concrete and formwork as constructional 'tools'. Ideally, agreement should be reached at an early stage with respect to suitable standards. Samples of the work should be prepared and agreed by everybody engaged on this particular aspect of a contract.

Specifications lay down certain requirements with regard to surface finish, texture, standards of accuracy and related matters which determine the serviceability of a completed structure. Until recently within the contract,

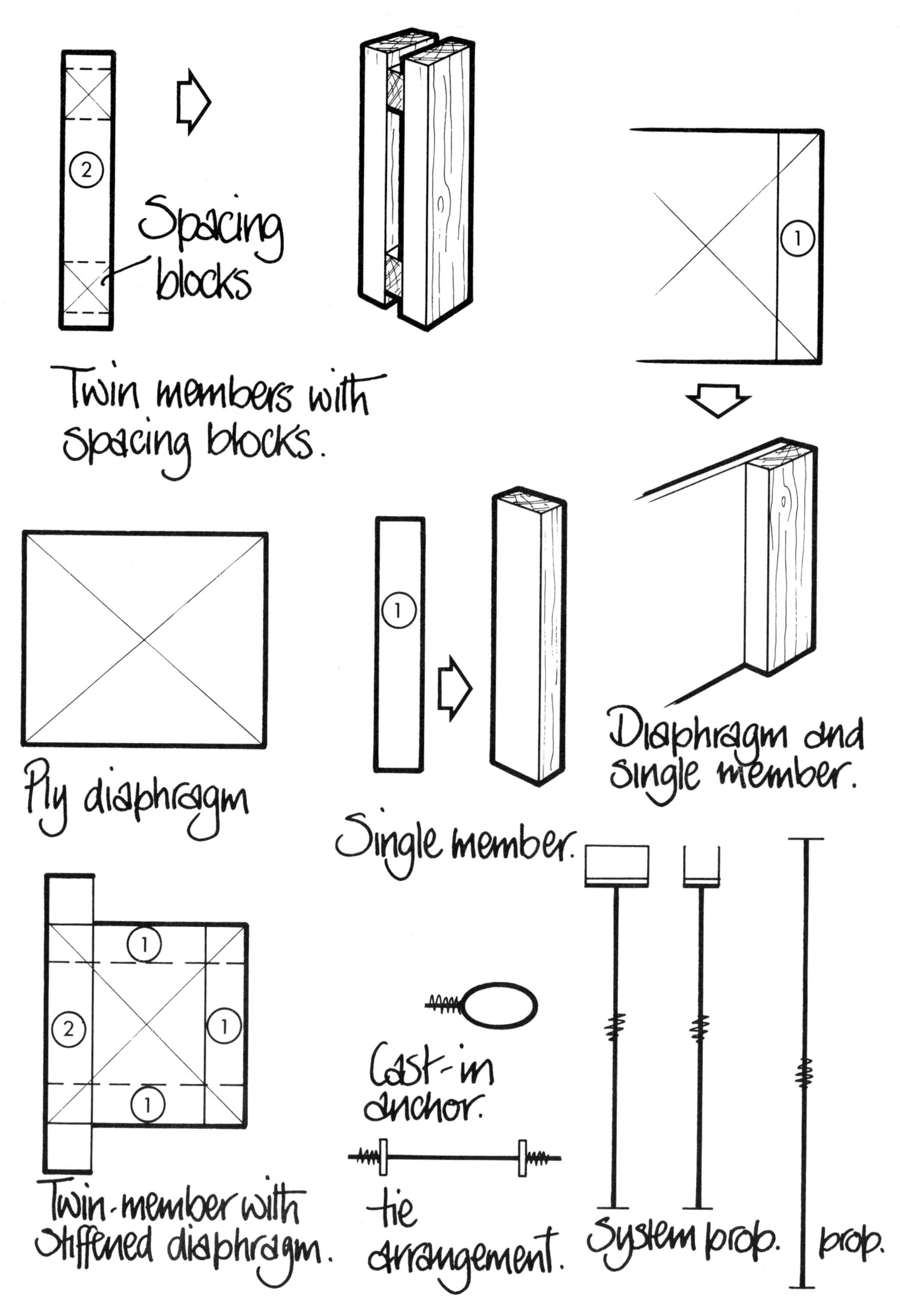

1. Formwork convention

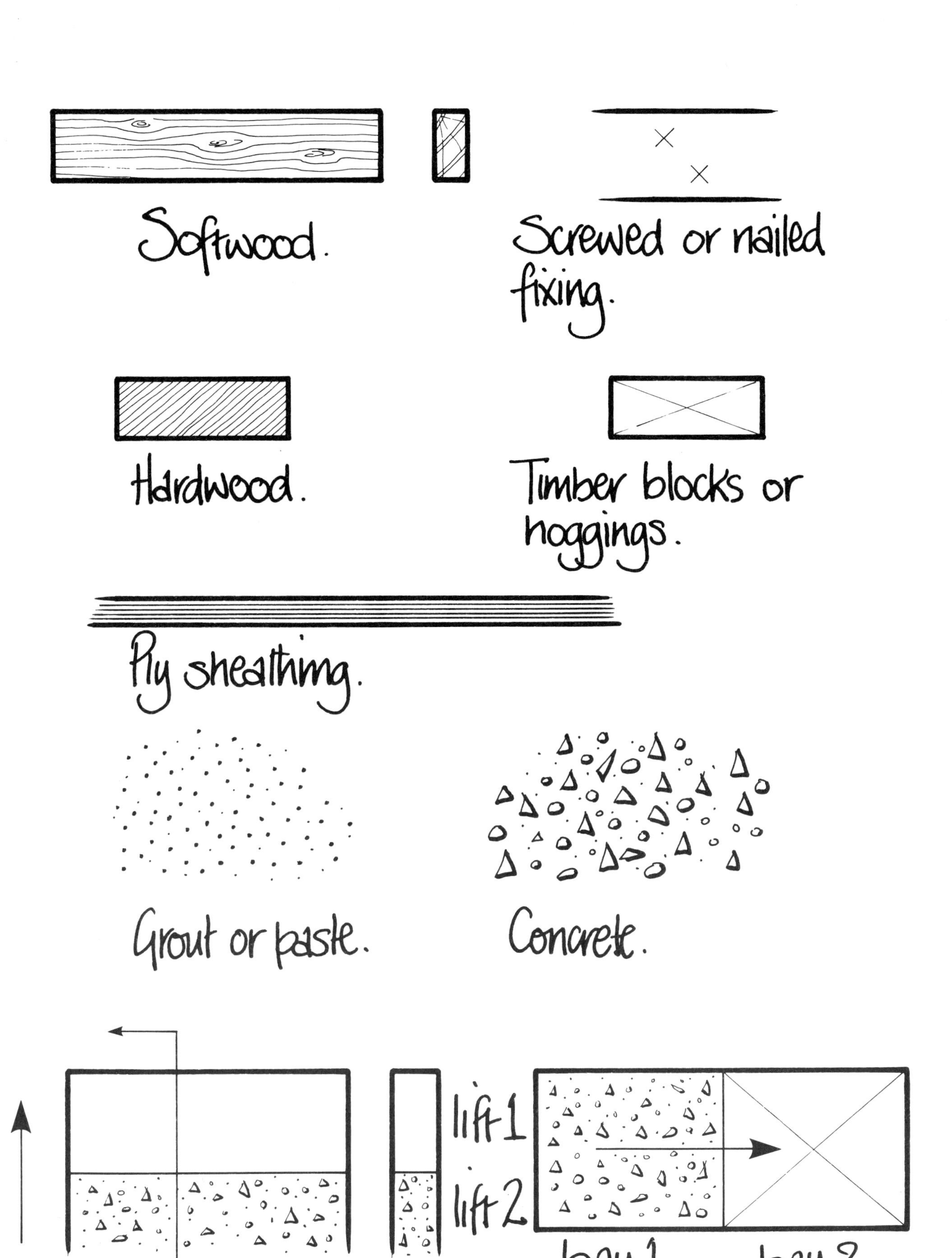

2. Formwork convention

and according to the specification, the selection of method and form materials generally have been considered the prerogative of the contractor. But as architectural and engineering requirements have become more stringent, specifications have developed to the stage where methods and materials may be specified, and such fine points as the type of parting agent, the curing method and other similar details may be described.

As with many aspects of building construction, more and more specialist companies, who produce materials, equipment and techniques for particular formwork applications are now being established. Here, again, there is a need among those responsible for the construction contract to appreciate the factors which govern the suitability of the various arrangements, and to be capable of making objective decisions with regard to the relative values of the facilities offered.

In this book all the major factors are discussed in a way that will be useful to the various people associated on a contract who have to determine a basis for the establishment of a working formwork system. The principles of formwork construction and design, the selection of materials, methods of handling, erection and striking are all covered in depth.

Proprietary arrangements are reviewed, although to avoid this sort of information becoming dated individual systems have not been identified.

Details of handling, placing and compacting concrete are described with particular reference to pressures on formwork, maturity and workability.

The methods of achieving surface finishes direct from formwork, including the associated concrete technology, are covered, and the aspects of management and supervision related to formwork operations are discussed with a view to examining the contribution made by all who are involved in the successful completion of a formwork operation.

The processes of design and manufacture, and the use of formwork, many of which are original, and applicable only to formwork – constitute a complete technology in itself. In addition, the formwork designer and the user need to turn to the technologies of other industries, such as those of plastics and timber to ascertain what techniques and materials can be derived to allow the application to formwork of the most appropriate material and best suited methods. Various aspects of this topic are discussed throughout the book.

The text is illustrated with photographs and line diagrams which do not purport to do more than provide the basis of ideas for development. No solution can be considered to be the ultimate one for any particular formwork problem. All solutions therefore should be regarded as the starting point for the development of future applications; such items as programme, sequence, time, quality standards, budgets and a host of other factors can influence the selection of techniques, materials or solutions to a problem.

It is the author's earnest desire that this book, as with the others he has written, should provide an appreciation of the basics of formwork construction and practice and thus help to eliminate the wasteful process of trial and error. The book attempts to highlight a range of formwork techniques, so that the reader can build a successful system through experience, current knowledge and established criteria.

Again, as with his other books the author has endeavoured to record practical details and to ensure that the skills of formwork practice are fully acknowledged. The techniques which are illustrated are based on many years of practical development, and not solely on research or laboratory developments on which so much technology is necessarily based. The developments outlined are thus results taken from the application of craft skills to fresh problems of construction.

The subject of formwork is wide ranging in that at one end of the scale success is determined by, say, the way two timbers are fastened by nailing or screwing, and at the other by the accuracy of figures that emerge from engineering calculations which relate to pressures, forces and moments of resistance.

The author has dealt with these aspects in an attempt to produce a book which does indeed cover all aspects of 'formwork construction and practice'. The reader, whatever his level of appreciation and knowledge of formwork, having studied this volume should be able to improve on formwork methods and to execute the necessary basic formwork calculations. He should also be able to ensure that his designs are capable of meeting the requirements of the various Codes of Practice and the specification requirements with regard to accuracy and surface finishes.

For timber formwork, the design calculations are quite straightforward and fall within the capabilities of the trade supervisor and junior engineer. However, where substantial loads and pressures are envisaged or where exceptional sections, or patterns of loading are encountered, the help of a suitably qualified engineer *must be sought* in the checking of all calculations and the criteria upon which they are based.

For some considerable time formwork has been considered an area where traditional methods, e.g. empirical methods for the spacing of members and selection of sectional sizes, have sufficed for safe working situations. With the continuing escalating costs of both labour and material, there can be little doubt that any time spent in careful scientific selection and location of members within a planned arrangement for handling, fixing and striking will be amply repaid in improved economic returns both with regard to materials and the expensive labour involved in their use. In the past the author has found that a number of aspects of design almost automatically are considered whenever a new problem arises. The identification of the more critical problems is a facility which the designer acquires through experience on various types of work. Many factors, for example those concerning scale and mass, need to be considered for every aspect of the work. There are, however, certain factors which relate to specific types of work, and which become evident as the formwork design process is developed.

Several chapters include simple checklists generally as a series of questions to which the designer or supervisor should obtain answers during the course of his work. The lists are by no means exhaustive, and the reader would be well advised to add in his own specific questions as they arise during his involvement on various points of construction. To avoid undue repetition those factors which would be considered inherent in the design process have been omitted, although the need for constant assessment of such matters as size, scale, mass, accessibility and quality standards cannot be over-emphasized.

Both designer and constructor are advised to adopt, as part of their normal procedure, a process similar to the 'critical examination' processes of Work Study, which involves the examination of any process by sets of detailed questions. These are intended to establish basic facts and possible alternative solutions to a particular problem. Thus the questions given in the checklists are intended to set the framework for technical decision and to supplement basic questions with respect to purpose, place, sequence, person and means posed by an intelligent designer in connection with all formwork design situations. The reader who makes use of this book should be able to consolidate his basic knowledge of formwork technology and be capable of producing practical design, safe in the knowledge that his details will be based on the established principles of sound formwork construction and practice.

Figure 1.1 When designing formwork every care must be taken to give clear and concise details.

1. Making a start in formwork design

General

The person who wishes to become skilled in formwork design will find that a number of courses are open to him, and depending on what sort of education and training he may have had, his requirements will naturally vary. The designer, be he a skilled mathematician, a practical tradesman, a supervisor or an engineer constantly needs to widen his knowledge of the various facets of the subject. Essentially form designers combine theory and practice in order to prepare a practical, mechanically sound formwork system which is based on a good appreciation of the technologies involved.

Formwork design comprises an area of formwork technology where everyone concerned in the construction of a project has an important contribution to make. The person who is designated as 'formwork designer' is fortunate in that he has the opportunity of combining the various contributions and turning them into a unified, practical, soundly engineered and, hopefully, economic arrangement.

The tradesman

A tradesman or skilled operative with practical experience will already have had a good grounding in basic practicalities, but he will need to apply himself to concrete technology and that which involves timber, plastics and adhesives. He will need to improve his knowledge of geometry and be able to accustom himself with simple mechanical calculations concerning loadings and forces.

For the younger man, and in particular a tradesman with site experience, the best 'way in' to the whole business of formwork design is by asking his contract manager, training or personnel officer whether any vacancies exist within that department of his firm which prepares formwork designs and details. People who have practical knowledge of site operations are often in demand for these vacancies since they are capable of devising practical methods appropriate to the particular site and type of construction. Provided that they possess an aptitude for simple calculations, new recruits to the formwork department should rapidly become useful members of the team, and should be able to contribute initially on practical matters, after which at a later stage in their development they will be able to base designs and methods on practical techniques.

In practice, the keen tradesman will be greatly encouraged by the way in which his improved knowledge of calculation and design, coupled with the growing awareness of formwork technology, gradually endorse the principles and factors that have been laid before him during his trades training.

The formwork engineer

Although it is obvious that the planning engineer or section supervisor who is concerned with a variety of constructional materials and techniques will not need such a complete knowledge of formwork, much of the following applies equally to his training as to that of the designer who is more specifically concerned with formwork proper.

For the engineer who has had formal training, there will be an urgent demand for the skills that he can offer. Often formwork design can be carried out by semi-skilled people who work from charts and manuals, but as structural concrete work becomes more complicated and new methods develop, there is a growing need for designers who are capable of innovating and developing methods and equipment, and who are able to carry out specialized calculations for specific applications.

The engineer, while having a good appreciation of the mechanics of pressures and loads will also need to be fully conversant with the associated technologies of timber, steel and plastics, and he will need to study the practical aspects of materials usage and construction.

The construction supervisor

The supervisor who moves into the formwork aspects of construction from some other trade, or from say the drawing office, will also need to study the various factors mentioned in the previous section in addition to those of site or trade practice. It may be that a person engaged as a draughtsman will find sufficient interest in form design to incite him to specialize. His background knowledge of shape, form, geometry and development will stand him in good stead and, together with an understanding of materials technology, will serve as a sound basis for furthering his interest in formwork design.

The opportunities

The author does not want to suggest that formwork design offers a career for many, although some of the experts who are at present designing formwork with major contractors have found that their jobs have taken them into interesting work often overseas on contracts which have provided outstanding challenges. What is more likely to happen is that formwork will offer an opportunity for studying a fascinating science that creates avenues for the application of engineering skill and practical knowledge. The formwork designer, during the course of a few months, will encounter a whole range of problems, the solutions of which will serve to broaden his experience. He may spend one day working on a design for an intricate mould and perhaps the next engaged on a massive form design. Each problem presents what is in effect a 'mini contract' with its own peculiar factors and considerations. Every facet of construction has a bearing on the formwork method, while every development in the course of a contract impinges on the formwork problem. It is hardly surprising that those involved soon find a need to specialize in some particular section of this work.

The following chapters outline the range of topics of which the formwork designer should have at least a basic knowledge. For those wishing to obtain further information a comprehensive bibliography is given in Appendix 4.

Sources of information

All those who are involved with formwork need to collect data on costs of materials, outputs and various aspects of form and mould construction and usage. It is useful to keep a notebook in order to record times and outputs for various types of tasks, materials prices, labour rates and allowances. Care should always be taken to record dates and particular factors such as the location of works, weather conditions, type of plant available, materials used and such like. This ensures that a valuable collection of basic data will be accumulated which can serve to provide guidelines on subsequent schemes.

A wealth of information is always available from the various proprietary manufacturers of formwork systems and associated plant. It is advisable in any enquiry to state where information is for general use and not related to a particular contract. This is because the supplier will not want to waste valuable time which could otherwise be applied elsewhere in obtaining further work. In this way information can be sent on at some convenient time when pressure of work permits.

System formwork manufacturers, on request, will supply erection manuals and every opportunity should be taken at trade exhibitions to visit the manufacturers' stands and compare the various types of equipment on show. Here, again, there will be certain periods when exhibitors can spare time to explain the finer points of their systems.

Those interested in formwork should always make extensive use of the readers' service pages of trade literature and magazines. Technical journals usually have a considerable number of articles, with illustrations, on the various types of formwork used on contracts. Those journals that are particularly concerned with advertising new products and developments for general engineering purposes often provide interesting information, and many of the products advertised inevitably have some application in the formwork field.

A more direct source of information is offered by the specialist journals such as *Concrete,* the journal of The Concrete Society in the United Kingdom. This monthly journal provides case studies and information on all sorts of contracts, and nearly every issue shows some aspect of concrete construction which depends for its shape or finish on the formwork provided.

The Concrete Society

Membership of The Concrete Society is open to all who have an interest in the applications of concrete in construction. Quite apart from the fact that the Society provides information on technical literature and notes on new publications and new developments, it organizes many site visits which afford the opportunity for participants to study the activities of other companies and the chance to discuss such details as materials selection, system application and method. Such visits are particularly useful because visitors have the chance to meet all the parties involved on a contract, namely the contractor, the consultant and the site authorities, and there is usually ample opportunity to study particular points of interest.

Informal visits

Apart from organized visits, the student of formwork should always make time to study the contract operations on any site by watching closely from outside the hoardings or boundary to the site. Often when something of particular interest is taking place, for example where some special finish is being achieved, or where some unusual type of formwork is being used, the contractor will allow interested parties to visit the site by arrangement. However, care must be taken to apply for such visits through the appropriate channels, and under no circumstances should any visit be made to the site without prior permission.

Training courses

Local Technical colleges or Polytechnics usually run courses on formwork and allied topics. Where a local college is unable to help, the registrar should be able to give advice on where the nearest course or courses are being conducted. The course on Formwork for Concrete Construction, organized under the auspices of the Institute of City and Guilds of London, is a particularly useful one. This one-year course leads to an examination in formwork technology and drawing and the certificate awarded for passing it provides useful evidence of applied study in that a particular standard has been achieved. Other CGLI courses such as those on Concrete Practice and Concrete Technology at Supervisory Level also involve some study of formwork arrangements in the general context of construction. These provide a useful framework for later specialization since they are concerned with concrete technology and general construction methods.

The Training Division of the Cement and Concrete Association have for many years conducted courses on formwork, mainly of one week's duration. Several different courses are run, each being involved with various aspects of formwork technology from the basic introductory course to the design course which is intended for engineers concerned with the mechanical design of formwork arrangements. Details of such courses can be obtained from the address given in Appendix 3.

Reading material

Books on formwork and allied technologies can be obtained from the local library, while the Cement and Concrete Association's library at Wexham Springs is always prepared to obtain books on specific topics. The bibliography in Appendix 4 provides a useful reference for further reading.

The formwork student is advised to purchase at least two or three of these publications so that he can compare the various methods of approach to problems. As discussed elsewhere in this book, solutions to a problem are legion but it is always helpful to have some starting point, or some reference, which can serve as the basis for development in a specific case where a system is to be designed to accommodate special factors and conditions.

It is important here to mention two publications which are extremely valuable to the formwork designer. Both books have been produced by joint committees of the Institution of Structural Engineers and The Concrete Society.

The first is the *Falsework Report* originally published in 1971 and now embodied in the draft British Standard Code of Practice on Falsework, and the second is the *Formwork Report* published in 1977. These two Reports are essential for every formwork designer's reference library and complement the recognized standard works on formwork.

Figure 2.1 The complex nature of the planning arrangements for formwork and concreting operations is evident in this view of construction of a ventilation building. Note transition form (centre).

2. Designing the formwork system

General

No matter who is responsible for the design of formwork, whether it is carried out in the contractor's office or on site, the designer or the design team must work systematically through a logical series of steps to achieve a suitable 'system'. The term 'system' describes the complete arrangement of support, substructure (sometimes known as carcass) and sheathing which comprises formwork that is used in reinforced concrete construction. 'System' implies a fully compatible arrangement of components which when assembled meet the following requirements:

Support and mould the plastic concrete

Contain the complete mix without leakage or distortion caused by concrete pressures, construction loads and external forces

Provide the intended number of re-uses while maintaining a satisfactory standard of accuracy and surface finish

Can be removed from the concrete without sustaining damage or causing damage to the newly cast component

Generate the critical geometry and face profile with the minimum amount of further labour being required to achieve the specified finish

Be capable of being worked by available labour and handled by the equipment available on site

Where manufactured on site, the manufacture must be within the capability of those employed.

It will become apparent from subsequent chapters of this book that many other factors govern the design of the system, each factor being specific to a particular type of work. When considering the various factors the designer must satisfy all the requirements with regard to safety and structural integrity, and those that relate to the economics of the contract. Safety considerations are paramount, and override all but the technical considerations with which they are inevitably interrelated.

During the design process the designer usually has consultations and negotiations with various authorities, including those who are concerned with aesthetics, structural design, cost and price as well as with specialists in the aspects of construction. He must satisfy the criteria which govern the performance of the finished structure and the essential practical requirements of those who during construction are responsible for outputs and standards. He must be aware of the demands of the concrete technologist, and should possess a sound knowledge of general construction techniques including the needs of the various trades that depend on the formwork system for an ensured continuity of work.

He needs to study plant capabilities, crane and hoist capacities and what are now termed the 'logistics' of supply, and as well as possessing a knowledge of the work carried out by electrical, heating, ventilating and mechanical engineers he must be familiar with the requirements of those whose operations are carried out after the completion of the building shell or superstructure. The former work concurrently with those engaged on the operations of forming the concrete while the latter work on fixings and openings provided during these operations.

The logical sequence of formwork design activities is akin to the critical examination technique carried out by the method study engineer. As with the planning of the construction process, the priority or weighting of the various factors may vary throughout the design period as new facts and details emerge. The interdependence of these factors is such that it is difficult to establish hard and fast rules with regard to sequence or method of approach.

Without doubt, the most important activity is that of obtaining and establishing all the information possible of the contract. After the estimator, the formwork designer and planning engineer are often the first of a contractor's staff to study the exact details of the structure.

Preliminary details

Before a designer can start, he has to establish the following basic criteria:

The scale and proportion of the work

His company's policy regarding standards of construction

Details of the specification which govern the work, i.e. performance, accuracy and surface finish

Economic considerations

Availability of labour and supervisory arrangements

Available resources and details of concurrent contracts which could make demands on these resources

Previous information derived from work of a similar nature

An assessment of both the architect's and engineer's abilities.

As a designer makes his preliminary appraisal of these factors he should soon start to get the feel of the job. He will need to interpret all the details as they arrive and be able to visualize the scale and proportion of the structure, including its mass and the location of special sections, and assess all the geometrical aspects. At this stage, a great deal of the designer's work depends on his experience gained from previous contracts, or from his basic training in some trade.

Some formwork design is governed to a large extent by engineering considerations, but the majority of it involves the design of formwork arrangements, and thus combines both the disciplines of an engineering approach and the application of practical experience. On the other hand some design aspects are governed by purely economic considerations, e.g. devising the safest method for an allocated expenditure. Most decisions hinge on details which are either calculated or estimated and demand careful evaluation by the designer with regard to their impact on the total construction performance.

A good designer will be led initially by instinct towards several feasible arrangements each of which at first sight appears to meet the most apparent requirements. The preliminary schemes must then be examined as new information and established criteria emerge. Almost certainly, modification of one of the original speculative schemes will result in a solution that best fits the demand as dictated by the contract.

Matters involving company policy, the preferences of those actually carrying out the work and safety aspects, have a major effect on the more technological decisions. Nobody at this stage can be sure of the best way to carry out the work since every solution must be given equal consideration, being accepted or rejected on the basis of objective study.

During and after the work, costs will inevitably emerge which will validate the decisions on formwork design, but in the initial stages it will only be necessary to rely on forecasts and, when available, established data.

Ideally, the formwork designer acts largely as a co-ordinator, in that he receives information and facts, speculative viewpoints and ideas which, based on his training and experience, he must combine into a firm formwork arrangement. This formwork arrangement, based on consultancy and the injection of the designer's skill, must be supported by method statements, materials, schedules and construction details for the supply and maintenance of an effective formwork system.

Discussion at the pre-tender stage will have established the rate at which the concrete construction is to be carried out, while the estimator and various other people, possibly the agent and certainly the project manager and planning staff, will have introduced some aspects of their practical experience in the planning of the programme.

Often a client's requirements, and factors such as seasonal applications of structures or demands arising from some aspect of the industry or authority for whom the work is being carried out, may determine the actual completion dates to be achieved by the contractor. With the overall construction period determined, and possibly enforced by penalty clauses or the reduction of margins forthcoming from management contracts, the formwork item will invariably feature among the critical series of activities determining the actual structural completion date. Formwork can always be found on the critical path in network analysis.

The formwork designer should always attend any pre-contract discussions, and it is more than likely that during these he will have formulated a provisional scheme, possibly taking into account any new information which he may have received.

Detailed considerations

It is essential that dates be established for critical details. The designer must work in conjunction with the planning engineer to ensure that the profile details and the extent of the concrete structure can be fully visualized. When these are established outline drawings can be of assistance as can all layout details, block plans and contract drawings. Full use should be made of anything that may help the designer to assess the scope of the work, the scale of the components and the location of the various critical sections, such as architectural details or features of the structural design which may in any way affect the formwork system. Models, sketches and similar aids will also prove useful.

Dates must be established, in conjunction with the authorities, for the issue of detail drawings, steel schedules and such like, and in this respect the formwork designer will need to study the structural engineer's details or the civil engineering drawings.

Naturally, architectural drawings are considerably helpful with respect to the aesthetics of a building. For example, they will need to indicate where provisions are to be made for all fixings, fastenings and openings as well as showing which of the specialist or subcontractors' drawings give the final detailed instructions. It must be emphasized, however, that it is the structural drawings which govern the contract.

The formwork designer will be looking at the way in which the main members of a structure vary at different levels of the construction. In point of fact he will be concerned not only with the first use of the system adopted but also all subsequent applications and the necessary modifications needed to deal with those variations that undoubtably occur during construction. He must be familiar with the general specification so that he clearly understands the required quality of surface finish that has to be exhibited at the various positions and levels.

He will soon build up a general picture by studying the preliminary drawings, specification and details as and when they are received by the contractor. All queries must be resolved before any new aspect of design or selection of formwork material or method is considered.

Both the architect and engineer greatly benefit from the experience of a good formwork designer, since he can pinpoint various aspects of the intended construction which may present problems of, say, appearance or

Figure 2.2 Culvert construction – sequence of construction. Note access platform for concrete placing.

Figure 2.3 Close up of formwork used in Figure 2.2 during concreting.

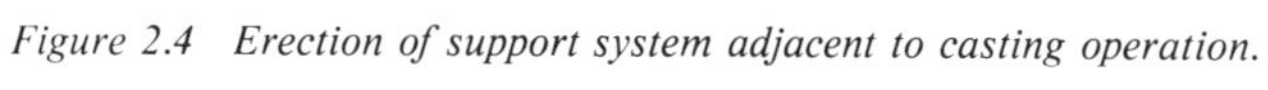

Figure 2.4 Erection of support system adjacent to casting operation.

accuracy. Undoubtedly he will offer useful suggestions for modifications to sections or dimensions which may well improve the performance of the structure, its surface appearance, the accuracy of the concrete components or speed of construction.

Following discussions with the estimator, surveyors and those who are responsible for construction on site, the formwork designer may have to clarify with the engineer and architect any points of detail or specification about which he is in doubt. He may thus be able to propose a modification of the details which in the end, could simplify the work carried out by the site personnel and also eliminate impracticalities. This is not meant to imply that the architectural or structural scheme may be impractical, but rather to convey that a person who possesses the particular skills of the formwork designer can visualize the intricacy of construction as dictated by the design engineer's or architect's stipulations, and thus is able to make a major contribution to rationalization.

An example of this might be where a designer has not realized the full implications on the overall construction process of the importance of geometry, or the interpretation of lines and details especially where he might have detailed a structural component. The formwork designer may well suggest that the design be made simple, or modified, to allow easier construction, and improved profitability, thus leading to a more satisfactory structure. Reference to Chapter 11 will show where small adjustments to detail help to simplify construction. Even minor adjustments to angles and splays can make a considerable difference to the problems of forming a particular profile.

Figure 2.5 and Figure 2.6 Formwork or falsework? Support complex for a massive cantilever slab on a major contract. Considerable liaison is required between the temporary works designer, site manager and sub-contractors to ensure successful operations. (Bryants Ltd)

Standards and construction detail

During preparatory work, the designer must look at the engineer's interpretation of the specification clauses which relate to line and level, construction and day joint arrangements and striking times. Every specification for concrete work contains a clause which states 'that all structural concrete shall be formed accurately as regards line and level, and plumb,' and every structural designer, contractor and manufacturer understands that except in structurally critical positions, or where visual aspects govern, the line and level and degree of plumb may in fact prove to be other than true. Although this may be generally acceptable, some misunderstandings may occur on site where the resident engineer or clerk of works puts a different interpretation on detail from that of the contractor. Ideally a specification will state reasonable deviations which are related to the scale and the function of the concrete component. However if this is not the case, then appropriate standards must be established so that the contractor can maintain them throughout the contract.

The location and formation of day and construction joints is discussed in some detail elsewhere in the book, but it should be stated that there can be little that is more critical in the design of formwork than the decisions which relate to lift, height, bay size, joint detail and such like. The ramifications of these decisions are apparent not only in the structural behaviour of the completed work but also in the quality of appearance. The decisions will to a great extent determine the formwork design and construction details, and will have a major effect on the selection of form materials and on construction methods generally.

The engineer and the architect will both be concerned in the establishment of construction joint positions and details. Of course, the design of reinforcement, joints, laps and location are very much interrelated with the joint considerations.

Once these details have been worked out in design procedure, firm details of structural profile and reinforcement arrangements become available to the contractor.

The designer needs to scrutinize all drawings which relate to such items and he must query any discrepancy of detail, or changes in profile, with those responsible for the costing and accounting of the contract. This check of the drawings is often the first detailed one to be made by any member of the construction team. During the checking process the designer should establish a sound working relationship with the designer/detailer engaged by the structural and civil engineers. To some extent he can provide an informal assurance against errors that may creep into the actual construction. Discussions over the details during the checking process will help the formwork designer to establish a clear picture of the function of the various concrete sections, and improve his understanding of all the critical details.

Timing

The reader by now may be wondering about the time involved in this whole process. Obviously a lot of time is spent on the activities, while the expense of the meetings and discussions described earlier, together with the travelling involved, is often quite considerable. However these activities are spread over a considerable period and the work is generally phased-in with that which relates to other contracts.

The designer will almost certainly be involved in a considerable amount of detailed work on every contract, but it is essential that he maintains the same degree of interest on each one. It is unusual for a contractor to start more than two major contracts in the ground at the same time. Where the design office comprises more than one formwork designer, the work can be so allocated that each member of staff will have several contracts to deal with, all at various stages of construction.

On a large contract, more than one designer or detailer may be involved. Indeed the demands of a contract may be such that the whole office will be involved in providing a design service at some stage of the work.

Once the basic details and timing have been established further discussions usually follow with the site agent and trades supervisor over the proposed system. During these discussions the formwork designer must establish his standing as an important member of the construction team so that he understands the interests and requirements of the site staff. If he has progressed from the contract team to formwork designer, this should be relatively easy to achieve, although in all cases he may well have to deal with people with whom he has previously had little contact. Site staff will have to accept that he will regard himself as an extension of their team in that he is prepared to take considerable trouble over all matters of detail, and the supply and use of the formwork which they are to use.

Preferences of the site supervisors regarding particular types of equipment, or methods of working, soon become apparent, and many ideas about how particular sections of the work should be carried out will be put forward by the site staff. The designer should note all suggestions for consideration, and use them where appropriate whatever the source from which they originate, and he should also discuss them so that he fully understands and appreciates the originator's intention. Obviously not every idea or innovation can be incorporated into a design, although some may well be used elsewhere or retained for consideration at a subsequent stage of the contract. All suggestions should be acknowledged and where one cannot be incorporated, a reason should be given as to why it has not been used. Many instances have occurred where an apparently wild scheme has been converted by a skilled designer into a useful working idea.

Associated interests

When details regarding supply lifts, plant and equipment, heating, ventilating, plumbing and electrical services are received further discussions will take place between the various interested parties, especially as many of the activities will have some affect on the formwork operations.

The planning engineer is of course the main channel of communication with regard to these aspects, but the formwork designer must work closely with the planning department, since he will often be in direct contact with suppliers and sub-contractors.

In addition to the considerations mentioned earlier, a planning department provides programmes and targets thus completing the information needed for the first steps of the formwork design.

During the early stages of design, the formwork designer will need to make contact with the concrete technologist so that an assessment can be made of the relevant details of the concrete to be used during the construction process. Topics discussed would include surface finishes, placing and compacting of the concrete, workability, admixtures, rate of hardening, rate of gain in strength and the final appearance of the structure, since all these have a bearing on the mechanics of the formwork and its subsequent removal once the concrete has been cast.

An example of the way in which the designer and the technologist can work together exists when in a certain situation the formwork designer, after discussions with the structural engineer, may require some help regarding the addition of an integral retarder to ensure uniformity of camber and prevent distortion. On the other hand it may be necessary to discuss the use of admixtures and heating arrangements to accelerate the rate of gain of strength to maintain outputs and obtain early re-use of formwork. One can visualize in the near future the use of concrete of very high workability, but with a minimum of applied compaction, being used in deep lifts. In this respect the technologist would need to be a party to the formwork design.

There have been occasions when concrete technologists have only been consulted as a result of difficulties which have arisen during construction. The trend now is to involve the technologist during the early stages of work to ensure that the requirements of the concrete are incorporated into the designed formwork arrangements. This will help to reduce the occurrence of defects due to pumping, bleeding, aggregate transparency and such like.

The designer, technologist and concrete plant manager will need to remember that there is considerable interaction between the method of placing the concrete and the pressures and loads that develop on the formwork. The planning engineer and those involved on site management in their decisions on the method of placing the concrete, (i.e. rate and volume of concrete placed in any given period), have to study how these will affect the formwork design. The modern tendency towards rapid placing, deep lifts and large casts places considerable emphasis on the application of mechanical design techniques.

When the designer has achieved a sound appreciation of scale, form, and considerations regarding timing and the interrelation of the construction activities, he can then begin to devise the formwork arrangement.

Major design factors

The design process now develops into a systematic evaluation of the following:

1. Formwork profile and quantity
2. Labour
3. Plant and equipment
4. Materials
5. Ancillary arrangements.

Formwork profile and quantity

As stated earlier it is difficult to establish priorities with regard to formwork design considerations, but high on the list will be the selection of the optimum profile and the quantity of formwork to be supplied for a given contract, especially in relation to practical details and economy. From the various structural drawings, and in particular from the sections at various levels throughout the job, the designer must establish the optimum profile to which the major form components are to be constructed. The identification and isolation of this optimum profile is one of the main skills in formwork design. Once the basic unit of formwork has been decided upon, be it a small modular-type panel or a large traveller-mounted component or crane handled assembly, it can then be used with modification for repetitive casting operations throughout the contract.

A basic unit might comprise the ring of forms for a chimney or shaft, the flight of tables for a multi-storey cellular structure, the cross-wall component in a similar structure or traveller-mounted forms for sea walls and retaining walls.

When considering the basic unit, the designer will need to have full details of the structure so that, in conjunction with the arrangements regarding height of lifts, lengths of bays and position of construction joints, suitable sheathing or facing material can be selected, and

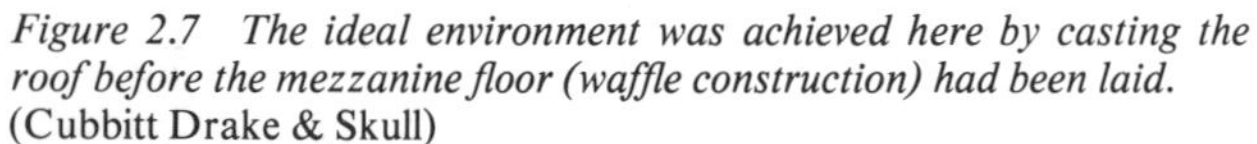

Figure 2.7 The ideal environment was achieved here by casting the roof before the mezzanine floor (waffle construction) had been laid. (Cubbitt Drake & Skull)

Work on several levels ensures continuity of work for all trades.

Delays in stair or service tower casting reduce the amount of work available on adjacent parts of the structure.

Careful planning is essential for continuity of work.

preliminary ideas as to means of support, handling and operating can be formulated. The quantity of formwork required for the selected profile will depend not only on the programme to be achieved, but also on method of handling and degree of mechanization to be incorporated in the system.

It is usual to provide a complete ring of chimney forms, a complete lift of forms for a cooling tower or shaft, a set of two or three sections of tunnel forms, a set of cross-wall forms sufficient to allow the required floor casting cycle to be achieved, one or perhaps two bays of formwork for repetitive bays in wall construction, one traveller mounted form for long runs of retaining wall and so on. In each case the quantity will be a function of the required cycle of construction and reflect the time taken by the associated activities of steelfixing, form handling and concreting, combined with the time stipulated by the specification regarding striking times for formwork which are governed by conditions of weather and temperature.

Labour considerations

The design of a formwork system may well be affected by the quality and availability of labour and trade skills on site. Fortunately, in general, the construction industry contains a pool of skilled tradesmen and supervisors who possess practical knowledge of the principles of formwork practice. However this pool, due to the normal functions of wastage and promotion, is fast diminishing, as fewer tradesmen are entering the industry and wages rapidly increase. The designer is thus faced with the problem of using more effectively the skills he has to hand. Historically, formwork has been oriented to joiners or carpenters and the supervision of formwork construction (this includes steel, glass reinforced plastic or other materials) has been carried out by the carpenter supervisor. It is evident that increased outputs and performance can be achieved by the formwork designer who can clearly define operations and ensure that suitable equipment and methods are made available. He must consider what skills are available, and in discussion with site management, arrange the system to meet those skills since these may determine the place of manufacture of the forms, the material used in the manufacture and the way in which they are handled. He must also assess the efficiency of those who are to supervise the formwork operations, particularly as much will depend on their abilities to achieve outputs, ensure economic re-use and be accurate. In effect he must design his system within their capabilities. Aspects of capability and availability of labour will often recur in the selection of both materials and plant, and in considerations with regard to the formwork quantities to be provided.

The designer must be sure that the system will work effectively and in such a way that all design features which are aimed at providing accuracy and finish are achieved. This will depend to a large extent on the type of labour and supervision that is available on site. There is for example little point in building a sensitive adjustment into the system if nobody on site has the skill required to effect it.

Material selection also depends on this assessment of labour and supervision, just as the rate of re-use depends on the way in which site operatives treat the materials between concreting operations.

The designer is well advised to study the degree of organization of labour and trades, this being a matter of site location, particularly where demarcation is concerned. Demarcation may affect the types of work which the various trades are prepared to carry out and the demands made in respect of access, scaffold, working hours and re-imbursement. Another important item is the degree of innovation and mechanization which operatives and trades unions will accept in a given situation.

Plant and equipment

A lot of formwork design depends on the plant and equipment that is available on site. The designer should study plant considerations and details since the selection of the main plant and equipment is related to the demands of the contract. For example, cranes are gauged to the main loads to be handled over given radii and thus may not be of the type best suited to the formwork and concreting operations. Cranes and associated equipment are used on a variety of operations and although the provision of formwork and the placing of concrete are generally critical to the programme, they may also be needed to install structural steelwork, precast cladding and other items elsewhere while the formwork and concreting operations are being carried out. The designer therefore, in liaison with the planning engineer, needs to set up time schedules for crane demand, and arrange for the critical formwork handling operations to phase in with the other general lifting duties.

It is often necessary to arrange, by the provision of travellers or gantries, for the uncoupling of the general construction operations from the formwork operations in order to achieve continuity of work for the available labour and, in the long term, for all dependent trades.

In building construction, formwork may comprise the largest unit load that has to be handled, so that crane capacity will be related to the weight of a cross-wall panel, or the reach necessary for the extraction of a table form from within the extreme sections of a building at the limit of the crane radius.

For a great deal of construction work it is necessary to supplement the primary lifting arrangement by hoists, pumps, placers or elevators so that the concreting operations can continue uninterrupted within the main construction cycle. Where this occurs the formwork designer must consider the implications of the rate of placing and the volume of concrete on the formwork quantities and the mechanical design of the form arrangement. In-contract changes of plant, or equipment used to place concrete and handle forms, can affect the performance of the forms and the designer must stipulate within his details the design assumptions made regarding method and rate of placing, and he must also be alert to any changes that may arise from any contingency or innovation during construction. He must pay particular attention to lifting and slinging arrangements to ensure that the lifting stresses are transmitted into the substructure of the form with the minimum of distortion and wracking.

Material selection

By now the designer should have established the degree of repetition which he can obtain from the forms and this, coupled with the requirements of surface finish derived from the specification and by discussions with the authorities, should guide him in his initial deliberation with regard to material selection. Materials considerations are governed by the number of re-uses attainable, the specified surface finish and also the policy of the company on material usage.

Many quite complicated projects are carried out using site manufactured or assembled 'traditional formwork', i.e. that which consists of timber and plywood. During the early stages of construction many uses can be obtained and a reasonable finish achieved. However the standards, both of finish and accuracy, tend to fall off fairly rapidly, and unless substantial carcassing or proprietary strongbacks are used framed panels soon deteriorate due to wrack or wind. It may be a company's policy to hire or even buy standard panels which can be either fabricated from ply and timber, or take the form of the commercial steel-framed, ply-faced panel. Provided the panels are systematically used and attention is paid to the treatment of joints, they can ensure considerable re-use with high quality results.

Where these materials are selected, the work of the formwork designer will centre on the calculation or selection of panels in the first instance, with perhaps some preparation of layout drawings for successive uses throughout the contract. It is essential for the designer to keep abreast of new developments in materials technology, especially the implications of coatings and parting agents, new materials coming onto the market, and new techniques evolved for material handling. As an example, the increased availability of plastics to produce decorative surface finishes is a reflection on the recent advances made in the development of this material. Previously, features and profiles were expensive to produce but can now be achieved by the insertion into the form of various formulations of plastics materials with textures prepared from either natural materials or from fabricated plys manufactured to an architect's requirement.

Every piece of material, from the traditional timber sheathing to the relatively sophisticated moulded plastics face, can provide a number of uses depending on the way in which it is fabricated into the form, how it is maintained during and between uses, and its suitability for the work to be done.

It is the designer's responsibility to weigh the factors that govern each material, and to select those that are good enough to meet the demands of the specification with regard to accuracy and finish. The selection of the sheathing material, and consequently the backing, carcassing and framing arrangement, determines, to a large extent, the performance of the form. However it is essential that the requirements of the supporting system receive the same attention that is given to sheathing and carcassing.

With the trend towards exotic liners, for visual effect and the availability of special plies for normal concrete production, there is a tendency for the sheathing and supports to be considered separately. While this may be valid where more exotic concrete finishes are required, it is important, particularly where the larger component or the repetitive form arrangement is concerned, that the form should be considered as a complete system, i.e. sheathing, substructure and supporting arrangement. This is so where table forms are used and where large panel construction is adopted. Where special steel forms are used, not only must formwork be considered in this way, but handling and access arrangements should be integrated within the form design. The adoption of stressed skin steel panels makes it possible to reduce the number of supports and thus allow the design of a complete system.

Figure 2.8 Activities in the area adjacent to stairs and lift walls may determine the length of time taken to construct a frame. Here, there is a lack of planning as evidenced by the ad hoc and dangerous arrangements employed.

Figure 2.9 Perhaps the most difficult of operations to plan and design – little repetition requires heavy cash outlet on materials and makes allocation of labour difficult. (Constain Construction Co. Ltd)

Ancillary arrangements

There are a number of manufacturers who supply the contracting industry with ancillary equipment such as vibrating screeds, vibrators, both internal and external, equipment for use in vacuum treatment of freshly placed concrete, and items for accelerated curing. The designer must always update his knowledge of such equipment and he should maintain constant contact with technical representatives, or at least keep a file of information on equipment which will prove to be of interest on forthcoming work.

Formwork arrangements are often blamed where substandard finishes have occurred or structural components become distorted, where in point of fact the reasons are just as likely to be the result of a badly applied compactive effort, inadequate curing or some other faulty technique. The designer's knowledge of the principles of compaction and the use of the necessary equipment should be sound enough for him to advise site personnel, and be such that he will be able to discuss subsequent results with both the engineer and the concrete technologist.

It is not unusual for the contract staff to increase outputs in order to meet programmes or recover lost time, and it may well be necessary to accelerate the casting cycle. The designer must thus have a working knowledge of all available equipment and also the implications of accelerated curing on concrete technology. At the present time a large number of reinforced concrete projects are programmed on the basis of a rapid turn around which is achieved by accelerated means. This applies particularly where there is a major investment in purpose-made special formwork or some proprietary system. Here it is necessary for the designer to calculate the cost of investing in accelerated curing equipment and the power or energy required to achieve the nominated striking time, after which he must compare the results with those as a result of reduction in equipment required and earlier completion of sections of the work, which in turn, provide earlier access for following trades with subsequent reduction in supervisory time.

Thus far in the consideration of the factors that affect formwork design, the designer will have reached decisions that are based on the details of the particular contract as they become evident from drawings, specification and discussions, and these together with many other factors must be systematically examined. A checklist which accompanies this chapter sets down certain key points, and it can be seen that the designer is really making an attempt to thread his way through somewhat nebulous criteria in order to establish what is apparently the most suitable system.

Of course each stage of construction will introduce further considerations, so that the eventual system that emerges is, in all respects, a compromise. The outcome, however, must be the provision of details for the construction of a mechanically sound, safe method of formwork that is capable of being used to cast concrete to the required standard and to offer all concerned, together with allied trades, a continuity of employment throughout the whole job.

Although so far not discussed, economics have a major influence on design. Rates for the work are usually established on the basis of the accepted tender or bid while the contractor's margins will have been established by the estimator working in conjunction with his board of directors, and it is obviously the responsibility of all engaged on the contract to carry out the construction within the estimates prepared earlier.

There is thus no opportunity to make any new profit. The success of a design can only now be expressed in terms of the opportunity of increasing margins as a result of the careful selection of method, application of labour and skills and use of materials.

At the formwork design stage it is possible to make savings by the skilful application of the factors of production, and indeed, some additional expenditure on formwork material or method may make it possible to achieve considerable savings on the subsequent operations during the service installation, decoration and the activities needed in finishing the structure.

When mentioning economics it must be appreciated that the decisions regarding formwork arrangement will affect the profitability of those sections of the work that are carried out by the major part of the direct labour engaged on site. These decisions govern the rate of progress on the structural components, the progress of the overall construction and thus the rate of cash flow with respect to the contract.

While all this has been taking place some considerable time will have elapsed and the designer will need to recap on any decisions he has made so far, especially with respect to the ordering of materials, bought-in components and hire of equipment. As the work progresses and as contingencies arise, certain changes in method may result, e.g. new techniques for placing and handling may be necessary or changes in site conditions may occur, and these will need accommodating by some modification of design. Certain structural design features will inevitably emerge as being critical, for example, architectural requirements may alter later, and the designer will have to produce a design that is capable of meeting the current or revised requirements.

Formwork quantities

One of the most important decisions which the formwork designer has to make is that regarding the quantity of formwork to be supplied for a given project. The decision is essentially one of economics, although basic construction technology will have some bearing on it. Safety considerations, striking times, location of construction joints, access, type of work and provisions for continuity of work are also factors which will need consideration.

The quantity of formwork supplied may only represent a small part of the concrete surface area, although attempts to minimize the quantity involved may reduce the viability of the total construction by causing peaks in demand for labour, or by failing to provide a continuity of work for the various other trades that depend on formwork for continuity of productive work.

Decisions about formwork quantities are usually made by the contractor. A specification governs the location and format of construction joints, while the design drawings provide details such as the continuity of reinforcement. Although specification clauses dictate the striking times to be adopted these may well be modified as a result of prevailing site conditions, or the development of concrete strength and the early application of loading to a newly constructed component.

Figure 2.10 The designer must be fully alert to possibilities afforded by alternative techniques. Slipform construction has been employed because of the quantity of walling required.

Considerable quantities of formwork are required to ensure continuity of work for all the trades involved, and especially for the continuous construction of suspended floor slabs or cross-wall and slab construction.

The formwork designer must arrange for the work to be continuous and whenever possible provide sufficient formwork material to cater for the demands of the electrical tradesmen, heating and ventilating and mechanical engineers. For example, while the target for concreting work may be two bays of a slab per working day, it may be necessary in order to achieve this target to supply eight bays of complete floor soffit material and a further 16 bays of standing props to support the newly cast concrete. The flooring materials provide work for carpenters during the erection process, for electricians in laying conduits, for steelfixers in placing steel and for joiners in forming day joints, and for edging and positioning openings. The supports supplied depend on the specification, although they must of course be carried through to concrete which has achieved a strength that the engineer considers satisfactory and is capable of supporting the live and dead loads.

The situation can be complicated where a stringent specification exists with regard to the striking times, and in particular during winter where low temperatures extend the striking times even further.

The so-called 'quick strip' systems enable soffit and beam sheathing to be removed and re-used without causing upset to the essential standing support arrangement, thus economizing on the amount of equipment required for casting floors and beams.

When formwork involves expensive construction, or where special geometry is involved, or where special surface finishes have dictated the selection of some particular sheathing material, the designer must be careful with his striking details to ensure an economical number of uses are obtained. Provisions can be made to ensure that the major sheathing areas are removed while the supporting pads or ribands which support subsequent components remain in place. The positions of these pads should be agreed with the engineers, while surface detail should be discussed with the architect.

The cost of materials versus the cost of the labour in recovery, striking and re-use must be carefully balanced. Sometimes it may be possible to cast some of the work in sections, for example a podium soffit on multi-storey structure, the formwork being moved from bay to bay. Alternatively it may be possible, on overbridge construction, to erect formwork to parts of the bridge beams and soffit, and to re-use it on subsequent operations.

In practice however, particularly bearing in mind the implications of concrete provision and distribution, it is more economical to invest in providing sheathing and a complete soffit support, the whole being erected and cast in one operation. Again, for economical reasons, parts of this support system can be removed and re-used while the whole soffit is supported on pads. Commercial systems and bird cage scaffolds are often used thus and can be sent off-site as soon as they are released from the operation.

To a large extent space restrictions and handling problems will dictate the re-use value obtained from the various materials employed, as will changes in section of columns, beams and storey heights.

The designer must always consider the seasonal aspects of construction, for example, delays in striking caused by low temperatures during the winter may upset the re-use calculations. In fact he is bound to provide extra quantities of materials during winter to allow for the rate of construction to be maintained, remembering the rate with which materials can be released and re-used.

Contractors can benefit from the available resources of the proprietary suppliers who maintain large stocks of equipment for hiring purposes. When striking is delayed excessively, equipment supply can be timed to help maintain the required progress.

The beam and slab construction for a roof to a single-storey warehouse of considerable span can be quoted as an example of the economical provision of formwork. Provided that the design engineer agrees, it is convenient to set up a number of beam soffits between the columns, and to erect on these steel reinforcing cages. The beam side formwork members can then be moved along the soffits from one to another. The daily cycle will ensure that one or two beams are cast each day to provide continuous work for a small gang of carpenters. Further teams of operatives can then be employed to erect slab soffits, while edge support for these soffits can be obtained from the suitably supported beams. This technique, quite apart from giving continuous work, reduces the quantity of expensive forms required and avoids peak demands on both labour and equipment. Another advantage is the stability achieved within the system from previously cast components.

In general, the following formwork quantities are required when cyclic construction techniques are employed.

Columns

Sufficient forms to provide continuity of work for two or more teams of carpenters. Columns can and should be cast daily to generate work for the subsequent trades. The number of simple column forms supplied should be geared to the method of beam or slab construction, since there is little value in having a large number of columns cast too far in advance of the general construction.

Where specially designed columns are required for elevated roadways or for podiums of large reinforced concrete structures, it may be sufficient to have only one, or at most, two purpose-made form arrangements to ensure constant programmed re-use of the form. For sloping columns extra faces or parts of face forms may be required which can remain in position as parts of the standing support arrangement.

For circular forms the designer may make effective savings by selecting foil formers which require only skeleton supports and which remain in position as a curing and protection measure. Where groups of four or five columns are encountered forms can be ganged and while it is more expensive to provide speedy erection, the reduced amount of handling involved can provide economies.

Forms for stairs and lift enclosures

The provision of stair forms has a greater significance than is perhaps normally apparent. During normal multi-storey construction, stairs provide means for access for workmen and ease the load on the temporary hoist or passenger lift facilities. A lot of the traffic that passes up and down stairs during the course of construction is caused by service contractors, electricians, plant engineers and others. Good access to the working level is important for the successful progress of a contract. Ideally, sufficient formwork and labour should be made available to keep stair casting in line with the floor construction programme. Any delay will result in some uneconomical arrangements being required to fit the forms that follow on behind the main construction.

Stair enclosures and service tower forms often regulate

the progress of construction. Slipform and lift over can be employed to speed up the erection of the central core or gable service areas on a multi-storey structure. Either a sophisticated slipform system or a carefully designed wall form system is employed, each comprising at least a complete lift of formwork. Nowadays many tower structures are cast using two sets of forms that leapfrog up the building or climb on a landing ring arrangement.

For regular construction, it is usual to supply one complete set or lift of formwork and the designer can be sure that time carefully spent in refining, simplifying and generally speeding this part of construction will effect sound returns for the total time expended.

Figure 2.11 The advantages of composite construction. Timber soffits and forms combined with steel supports used in the casting of cantilever beams at the perimeter of a slab.

Cross-walls

Quantities of formwork required for cross-wall construction, where this construction is based on a dwelling or office module, depends on the shape of the final structure. The walls are normally cast through a daily cycle and the quantity of formwork required has to be geared to this demand.

Most structures are constructed on a 'floor a week' basis, and often 6 or 12 cross-wall forms will be required to maintain continuous floor construction. Many contractors provide a full flight of table on angle forms, so that the cross-wall process becomes a critical aspect of the production cycle.

Floors and structural reinforced concrete slabs

It is normal practice to allocate one complete floor slab soffit, or allow sufficient soffit area, to achieve continuous erection, using form materials as they are stripped from previously cast concrete.

Obviously the process of erection takes longer than the formwork striking operations. The steelfixing of reinforcement and the quantities of materials required must link in with the erection and concreting cycle. Whenever possible cycle times can be reduced by using high, early strength mixes, quickstrip arrangements and sometimes accelerated curing techniques. Recently, contractors have endeavoured to effect earlier release of formwork by working from the results of comparative cube tests or those devised by curing concrete cubes in conditions similar to those experienced when floors are laid. Trough and waffle floor forms which are quite expensive should be released such that the structural floor is fully supported.

Formers for openings and ducts

These can be extremely expensive because they are constantly cast into concrete and seldom are recovered once the formwork has been removed. A designer can effect savings by designing the formers to be integral with the forms, this can be carried out as soon as the lead and draw has been determined. Savings in materials can be made by attaching the formers into the main forms provided that a sound construction is achieved.

Most of the site produced formwork for these openings can be regarded as a 'one-off' exercise, which means that there will be a lot of materials wastage. Perhaps the material that suffers most in this respect is expanded polystyrene.

Retaining walls

Retaining walls for the basements and sub-basements of multi-storey structures are difficult to gauge with regard to formwork quantity. Quite naturally tradesmen can only work as and when work becomes available, and since formwork and shoring or support operations are particularly interdependent, formwork erection must be geared to the removal and 're-insertion' of shoring members. Under such circumstances modular panels are erected and handled, space permitting, for the formwork assembly, the supervisor working closely with those who are concerned with the support works.

Where long runs of retaining wall are used to support embankments or where, for example, continuous walls are needed as water-retaining structures for river and sea defence work, the formwork operations are far more predictable, and lend themselves to normal planning, programming and designs techniques.

Where timber or composite forms are used, it is usual to employ two or three forms as required to maintain a daily cycle. A daily cycle can also be achieved when proprietary special forms are employed if a travelling crane is available to handle the forms.

Outputs, however, depend on a design that incorporates not only stopend arrangements but also a formalized system of bracing from kentledge or cast-in anchors. High outputs have been achieved by adopting in-built accelerated curing equipment, the most popular of which is that where the forms are electrically heated.

Quality of construction

Quality of construction has a direct bearing on the re-use value and quantity of formwork employed. This is particularly difficult to control and requires skilled detailing.

Problems can be created where striated or concrete finishes are required. Contractors have often spent a lot of money in fabricating a featured form only to find that after one use the form has had to be re-made.

Whether in a works or on site, the designer and fabricator must decide and agree on a properly detailed

Figure 2.12 The three stages of construction from kicker to completed wall.

feature which incorporates splay and bevelled ends, and be satisfied that the form will yield the required re-use value.

The initial cost of the formwork materials is another factor which will affect formwork quality. For example, where some exotic finish is required, a contractor may elect to use a casting polyurethane, and make matrices from a master cast. Because the casting of polyurethanes is relatively expensive there may be a temptation to reduce the construction cost by extending the casting programme and increasing the re-use value.

The success of this depends on the achievement of a well made mould and a carefully produced sheathing matrix, since the surface finish will deteriorate as the number of re-uses increases. The formwork designer, materials supplier and all site personnel who are responsible for the use of the forms must co-operate to ensure that the best use is made of the selected material, that the form is properly constructed and that it is suitably handled, and treated with the correct parting agent.

Form deterioration is caused by misapplication of parting agents, bad use of vibrators and careless handling of forms between uses. The designer will need to give full and proper instruction to site staff on the correct applications of all techniques. In his design he will need to give adequate details of all jacking points, strong points and strongbacks, so that full re-use values are achieved.

Height of lift and length of bays

Although high lifts reduce the number of handling operations, they require strongly constructed forms to contain the pressures which result from the concrete placing and compacting operations. Tall lifts require careful design especially where there are limitations on the position of through ties. They are often selected for the construction of basement and retaining walls.

The designer will have to consider whether in some situations shallower lifts do maintain continuity of work with a higher re-use value. The thickness of the concrete wall requires careful consideration because it is difficult to ensure adequate compaction in tall slender sections.

Full height forms, which are crane handled, should be used for bays of the full height; however, where shoring members have to be removed within a sequence it will be necessary to employ smaller forms. Where forms can be mounted on a traveller and cranes do not have to be employed, it is more economical to use full height forms.

Shallow lifts mean that there will be far more construction joints and a greater danger of water penetration. Maintaining the plumb and line of the wall as a complete unit may also be a problem. It is difficult to form accurate openings where these have to extend through more than one lift. Shallow lifts require great quantities of access scaffold because there must be provision at every joint level for the concretors to place

Figure 2.13 As part of the overall approach to the construction problem considerable attention has been given to access for the concretors and compaction equipment.

and compact the concrete. The main advantages of casting shallow sections are that the smallest ties can be used (these being quite cheap) improved access for steelfixing requirements and good continuity of work.

Bay sizes are largely related to the structural design so that in general the larger the walling bay the greater will be the possibility of cracking. Bay lengths of 6 m for walling are taken as the optimum when there is restraint from continuity steel in the kickers or previous lifts. Ideally, the bays should be placed one on top of the other with the minimum of delay so that the concrete mass moves monolithically with the least number of differentials.

For the casting of silos and tanks, a complete ring of concrete is generally acceptable provided that no stresses are set up by beams on slabs which could restrain the main concrete elements.

Generally, bay lengths are given in the specification as are the stipulations with regard to the sequence of construction and the timing of concreting operations to adjacent bays. The formwork designer, where possible, should base his design module on the building module, although of course reinforcement and concrete quantities and the optimum areas of formwork also determine the bay sizes. Day joints, which are determined by these factors are expensive to provide and locate, and the cost of providing them can amount to some 40% of the total formwork cost, especially for something like raft foundations on heavy beam construction. The number of day joints should thus be kept to a minimum, particularly where it is necessary to cut the formwork around the continuity steel. In addition to the construction and location costs, the cost of striking stopends from between steel reinforcement can also prove to be high.

The heavy costs involved in formwork construction and removal, and the possibilities of bad workmanship often persuade a contractor to cast the largest possible areas of concrete that he can negotiate with the design engineer.

Joints should be formed where they are masked by returns, or changes of section, or shadows on the completed work. Where no such features exist the designer will need to discuss the formation of some recess or detail which will help in masking the line of the joint.

The final locations of the joints and the resulting bay sizes are determined by most or all of the previously mentioned factors, in addition to the requirements of the design engineer, architect, contract manager, agent and trades supervisor.

Figure 2.14 The designer, in keeping abreast of new developments and techniques, must constantly add to his technical library.

Checklist

Have the architectural details been checked for unintended geometry?

Have specialist drawings been compared with architectural detail?

Have full details of structure been considered?

What is Company policy and agreed standards of:
- accuracy?
- surface finish?

Have samples been prepared and agreed?

Have proposals regarding modified section and detail been cleared with architect and engineer?

Have lift and bay sizes and construction joint positions been agreed?

What are the recommendations of the:
Estimator
Temporary works engineer?
Plant engineer?
Services engineer?
Heating and ventilating engineer?
Specialist sub-contractor?
Concrete technologist?

What are the preferences regarding method of the
Contract manager
Agent
Project engineer
Planning engineer
Safety officer
Trades supervisor

What plant is available?

What systems of formwork particularly commend themselves?

Has there been a survey of new materials, techniques and equipment?

What are the specific requirements of the contract regarding
Access?
Tidal work?
Falsework?
Ground conditions?
Season of year?
Contract period?

What resources are available in terms of:
Engineering skills?
Supervisor?
Trade skills?

What data is available from previous work of similar nature in form of:
Records of outputs?
Accounts and costings?

What is the sequence of construction?

What are specification requirements regarding lift heights and bay sizes?

What is the proposed rate of placing concrete?

What are implications of concrete mix specification?

What are implications of:
placing method?
compaction technique?
curving?
specification regarding shrinking times?

Can alternative techniques offer savings for say:
precasting?
permanent formwork?
slipforming techniques?
lift slabbing?
extrusion?

Where is formwork to be designed?

Where is formwork to be constructed?

What space and equipment is required for manufacture?

What quality of formwork will be required and when?

What is the optimum size of:
form panel?
form table?

What materials best meet requirements of specification regarding finish and accuracy?

What is the target for re-use?

Can the formwork activities be simplified by adopting:
proprietary formwork?
special forms?
a specialist service for design and manufacture?

Will form operations be labour intensive, or can some degree of mechanization by achieved?

Who will be responsible for formwork operations on site?

What are checks to be made prior to concreting and by whom:
for the contractor?
for the client?

Are the staff from previous contract available:
for consultancy?
to work on forthcoming contracts?

What are the local TV considerations regarding:
demarcation?
consultation?

Do special demands of contract call for discussion with H.M. Factory Inspectorate regarding:
process?
location?
materials?

Are there exceptional lifts or bays or heavy foundations which require special attention?

What formwork drawings are required?

What is drawing distribution?

What calculations have to be submitted?

Can models be used to advantage?

Are handbooks available on method for:
traditional formwork?
proprietary formwork?
special formwork?

What is programme for:
form design?
form fabrication?
erection, fill and shrinking?

Figure 3.1 The success of formwork activities often depends on the simplest skills.

3. Formwork – the people involved

The architect

The architect is often the least informed person on formwork material and method, and although he will have decided on the aesthetics of the structure, he may have little knowledge of the available methods, or the standards that can be achieved. Naturally he is concerned essentially with the outcome of the formwork operations and welcomes being involved in discussions which affect the outcome in terms of finished appearance and accuracy. The architect who specializes in designs in concrete has a sound working knowledge of the principles involved, and it could be that too great an involvement in the practical details might divert him from achieving the results that he originally conceived. Undoubtedly architects are now becoming more interested in the inherent qualities of the materials for which they design, and if there are omissions in their knowledge this is, to some extent, the fault of contractors and producers who previously have tended to discourage architects from any involvement in these matters.

More recently, because of the greater volumes of concrete now being used, architects have gone to great lengths in coming to grips with the problems and related technology. This has been achieved by consultation with advisory bodies, and by attendance at seminars, conferences and training centres. The Cement and Concrete Association, the major cement companies and many contractors employ specialists who deal with matters of durability and the appearance of concrete.

Considerable emphasis is now placed on concrete topics in professional courses at the various Schools of Architecture. As a result, architects are becoming better informed, so that the gap between them and the contractors has narrowed. Contractors appreciate that many of their problems can be resolved by careful detailing and ensure that they become involved where possible in the initial design arrangements. By working in parallel with contractors, multi-disciplinary design teams are now making themselves felt in integrating design with construction, and generally the trend is towards more intelligent detailing with a co-operative approach being made by the contractor.

On a broader scale considerable work has been carried out by the Construction Industry Training Board and the Cement and Concrete Association to focus attention on the roles of those involved with the formwork industry. The Concrete Society through its Technical Reports, its other publications and its meetings (which are open to all who are interested in concrete) has a working party and committee that is devoted to promoting knowledge of formwork and construction. Those interested in anything to do with concrete can join the Concrete Society and participate in its many visits and meetings.

The resident engineer

The resident engineer (RE) on a civil engineering contract is vitally concerned with all aspects of formwork. His task is to ensure the accuracy and suitability of the structure and to maintain standards with respect to the interpretation of the specification. A well qualified man, generally a more senior resident engineer, has a wide knowledge of most aspects of formwork, including the mechanics of design. Generally he will have been involved with a succession of similar contracts, while in some instances he may supervise the construction of a structure which he originally designed.

One of the features of the contract arrangement, where performance specifications are concerned, is that although this well qualified and experienced person may be in a position to assess the suitability of the formwork for a given application, he will only make comments until such time as the formwork is struck, whereupon he will assess the accuracy and surface finish requirements as determined in the specification.

Obviously it would be quite wrong for those responsible for the erection of formwork to disregard any constructive comment that the RE may care to make. Indeed on the question of safe working every regard must be paid to his instructions. Fortunately, many specifications call for the submission of method statements and calculations, and where this is the case communications between the contractor and the resident engineer should be such that any doubtful aspects can be discussed at the planning stage, thus eliminating the possibility of errors or extra expense. A great deal can be achieved towards realizing successful operations by early agreement on samples and standards.

The clerk of works

In some respects the duties of the clerk of works on a building contract are similar to the resident engineer, although on a smaller job, he tends to be an experienced person with an excellent general knowledge of construction, albeit possibly less qualified, academically speaking, than the resident engineer. More often than not he will have undergone trade training, and will have progressed through trade supervision before being appointed a clerk of works. This means that he will possess a very good knowledge of either formwork or some other feature of construction, with possibly a general knowledge of some other aspects. The same considerations of performance and method apply, although the clerk of works often has to consult his employers before being able to make a decision on the acceptability of some method or material.

On larger contracts, or contracts carried out for municipal authorities, clerks of works generally have good technical back-up facilities which make their work and contribution on site extremely effective. Where a clerk of works makes a positive contribution to the effort of those concerned with production, a sound relationship usually develops which results in a reasoned approach being made to the various problems that arise. Possibly the greatest hazard to this relationship is that in which the clerk of works, who has been used to instructing and controlling labour, exceeds his brief, and deals directly with site tradesmen and operatives. This causes problems of control, and can lead to conflict at all levels. The best situations develop where he is respected for ability to resolve problems and speedily interpret details. Prior notice of proposed construction techniques and the contractor's intended use of particular materials will greatly assist him. Advance notice allows time for him to discuss details with his principals, prior to their approval. A contractor often finds it easier to introduce new methods of materials when the clerk of works has been forewarned.

Site engineer

The site engineer spends a considerable amount of time early on in a contract in providing lines and levels for the construction of the superstructure. Together with the formwork tradesman and steelfixers he becomes involved in details and queries which may arise from drawings. This initiates a close co-operation throughout the construction of the structure, although, often when pressure of work is heavy, there may be disputes as to the accuracy of levels and the accuracy to which the formwork has to be fixed. The formwork supervisor who will always be alert to keeping his gangs of operatives fully occupied, may often complain that he has to wait for lines and levels, although it should be admitted that this situation may well arise where, on preparing his basic levels, the engineer has encountered some inaccuracy that demands telephone calls to, or some discussion, with the designer.

For complicated construction, there must always be a close working relationship between the various parties involved at all stages in the erection and use of the formwork, especially when important adjustments and checks have to be made.

The general foreman

The general foreman is always an experienced person who appreciates the importance of the formwork operations. If he has progressed into foremanship through the carpentry trade, then he may tend to over-concentrate on the carpentry side thus making it difficult for the foreman carpenter to 'run his own show'. Often, when a general foreman has been a carpenter, or carpenter supervisor, he will make excessive demands for outputs, or tend to saddle the carpenter foreman with other tasks beyond his normal responsibility. This is less likely to occur where both men have worked together in other roles and have found good agreement on method and technique.

Personnel manager

In many firms the personnel manager is the man responsible for staffing the trades on site. Unfortunately, much of his work is conducted at head office or at the contractor's yard and he tends to become detached from many essential matters. Sometimes it would appear, where form trades are concerned, that those responsible are not really aware of the requirements with regard to skills and experience. At present it is difficult to obtain sufficient numbers of experienced operatives who can cope with the demands being experienced in the construction industry; this is evident from the way things are being run on most construction sites. The site supervisor however may well be dismayed when, in answer to a request for say a carpenter to work on formwork for excavations, he is sent a man who has had only limited experience on multi-storey frame construction. It is therefore essential that those responsible for personnel should understand the special requirements of the various types of construction, and be able to liaise with site staff in providing, or redirecting suitable operatives on site.

Where demand exceeds supply it may be that personnel departments send partially qualified people onto site in a well intentioned attempt to meet site demands. This could result in a floating population of poorly skilled personnel who are constantly being moved from site to site; obviously strained relationships between the personnel department and the production supervisor could ensue. Undoubtedly, the best solution is first-hand contact between all parties. It is usually possible to identify a number of matters which can cause friction and detract from performance, provided an intelligent approach is made by those employed on a particular process. An experienced or trained supervisor can soon avoid or overcome the problems involved. It cannot be over-emphasized, however, that much of the delay and upset which results from difficult relationships can be overcome by intelligent design and sound planning. Site conflicts are generally the result of incomplete or inaccurate detail, poor appreciation of the practical problems of construction, a lack of follow-up on the part of designers and badly prepared production plans.

In recent years – largely prompted by the high increase of sub-letting and sub-contract work – most contractors have adopted the procedure of appointing a services co-ordinator, who is often a skilled engineer with proper training in one or other of the construction regimes, and who also acts as communicator or co-ordinator between the various tradesmen and sub-contractors with respect to information and detail. The recent emphasis on planning and programming, now evident in the construction industry, has resulted in the greater employment of engineers and experienced construction men in the pre-planning stages of the contract. Often agents and general foremen spend their time between contracts planning for the next venture; this results in a more unified approach, with greater trades co-ordination.

Safety officer

The safety officer plays a very important part in connection with the formwork operations. It cannot be over-stressed just how important are the safe methods of formwork erection, or indeed, what a massive affect formwork has on the safety of the whole construction operation. Where building works are concerned, formwork tradesmen are responsible not only for the formwork operations but also for, in civil engineering terms, a considerable amount of falsework.

Formwork tradesmen become involved in matters of access, shoring movement, support of permanent linings and formwork, in each, questions of safety are paramount. It is more than likely that there will be a need for co-operation between the formwork supervisor and the safety officer on the maintenance of safe working arrangements for small tools, sawbenches and even the running of a small site workshop. The safety officer's role in advising the supervisors on the safety arrangements required should pave the way to excellent relationships. Generally, the man who has come up through a trade may well treat matters which involve records and registers with some trepidation, and it is a wise safety officer who provides help on these points.

The training officer

The training officer deals with the personnel department on all matters of staffing on a long term basis. He is particularly concerned with the maintenance of a long term training programme aimed at inducting and training young men within a firm or company. For many years formwork tradesmen have been neglected and indeed there have been very few recognized schemes for recruiting youngsters into the formwork trades.

Unfortunately formwork is very much tied to productivity, and the labour involved tends to be extremely mobile, so that little opportunity exists for the long-term training of young men. Apprenticeship schemes are generally more appropriate to joiners or shop-based personnel, so that a different training system on various basic topics is needed. The training officer who introduces beginners into the formwork industry is to be congratulated; it is essential that some means of encouraging or promoting entry into this sector should be established.

The ganger or section foreman

The formwork tradesman's best ally is the more experienced ganger who, during the course of his varied work, has learnt to appreciate the value of formwork and form components, and is qualified to discipline his gang such that it handles items reasonably and stacks carefully. A good relationship must be established with the ganger to ensure that careful assistance with offloading and movement of equipment is obtained when needed. Many contractors employ gangers as regulators of movement of equipment and goods which come onto site. In practice, the ganger often controls site progress by his use or misuse of cranage; this is particularly the case where form handling is involved.

The tradesman and site operative

The carpenter tradesman, or operative, generally has a clear understanding of the role which formwork provides during the series of construction activities. He is aware that formwork is necessary not only to contain fresh concrete but also to mould it to a required shape. Unfortunately many other tradesmen regard formwork tradesmen as poor relations and often think that the carpenters involved tend to be something less than skilled tradesmen. This idea is perhaps traditional, although with the increased earning power of those who deal with formwork, together with the emphasis which has recently focused good quality formwork, it can be said that the formwork tradesman has gained some status.

The site operative however is often slow to realise the importance of the formworker's contribution to productivity, although the impact of formwork outputs on the earnings of most people on site should be readily apparent. The concreter will often blame displacement or distortion of formwork and other discrepancies on the formwork carpenter, although he may well cause difficulties himself by his treatment of the forms when it comes to placing concrete or by his mishandling of skips and vibrators.

One of the main problems is the contempt with which the general labourer treats formwork items, either in the way he offloads or handles, or when he applies excessive forces during any handling or striking operations. They are often unable to appreciate that a rather dirty or concrete-stained component, be it a piece of timber, ply or proprietary piece of form equipment, can have some sort of value. With the general run of construction workers, these men do not have a feel for the finer points of construction, and are certainly unaware of the economics of re-use of formwork. This is an area where training can improve relationships through a better understanding of the construction process as a corporate effort.

The crane driver is an important operative, because his crane driving skill has a major bearing on the formwork operations. While it can be argued that he is only as good as his banksman, there is a lot more to crane driving than the simple interpretation of signals from the ground. As a skilled operator he plays an important role in two aspects of the formwork operations: handling and filling. The driver's skill lies in interpreting the signals and carrying out all operations while making allowances for mass and scale of load and weather conditions. His skill therefore has a real impact on the number of uses obtained from a form, from handling the panels to the way in which he carries out the allied work of manouvering skips, reinforcement loads, cages of steel and a whole host of other materials which could possibly scour or damage the formwork, sometimes beyond repair. A skilled crane driver can save considerable time and effort by the way in which he places the crane and the way in which he anticipates movements.

The steelfixer

The steelfixer is constantly concerned with formwork considerations, and just as the formwork tradesman is involved with accuracy of the structural profile, so the steelfixer is vitally involved when it comes to positioning steel reinforcement.

The two skills, and here it may be noted that steelfixing does not constitute a 'trade' as such, are concerned with accuracy. Both the steelfixer and carpenter have to interpret drawings, sometimes under the most difficult weather conditions. Often, due to omissions by the drawing office, both will need to check one another's drawings, with the result that a good working relationship develops. This sort of co-operation is necessary where there are numerous through-holes, ducts or slab and wall openings when tradesman may need assistance, either in the installation of steel or the installation of formers.

The scaffolder

The scaffolder's interest in the formwork operations is viewed from quite a different angle. His main interest is provision of access for the formwork trades and, more specifically, provision of supports of a 'birdcage' nature. Good access is a critical factor in any formwork operation and formwork tradesmen and scaffolders are quite used to the existence of detached conditions. It is usual for an excellent working relationship to develop between the leading hands and supervisors, but this is often more difficult to achieve where standing supports or special birdcages have to be provided. The scaffolder works to quite a different set of standards of accuracy to those required for formwork and this, coupled with a reluctance to cut or use special lengths, tends to generate problems at the interface between the scaffold and the formwork.

Trade unions

The trades union representative features on matters which relate to formwork and safety. Safety constitutes a major area where agreement is readily reached, the whole process being devoted to the welfare of the tradesman and operative. It is imperative that the union representatives should be allowed to participate in anything which involves safety and method, and it must be stated that many representatives now have a sound understanding of method study, and planning by review techniques, and are thus able to make valuable contributions to productivity. There are certain regional differences in formwork method, access arrangements and areas of activity of the various tradesmen, but union participation during the planning stages will help to promote clear definition of responsibilities and improve communications.

The formwork sub-contractor

General. The emergence of the sub-contractor who becomes involved with formwork operations is a notable feature of late. He undertakes to provide labour, or labour and materials, for the various operations and has thrived, particularly in those areas where there have been shortages of skilled carpenters or carpenter supervisors.

Unlike the past when they were not able to obtain many jobs on the medium sized contracts because of the normal peaks and slumps in demand, formwork sub-contractors can provide continuity of work for their own operatives. By transferring operatives from one site to another the sub-contractor is able to balance his resources to best meet demand when it arises.

He depends on highly skilled, well experienced foremen and supervisors to set the pace on site and thus help with the maintenance of demand. This together with good supervisors who are capable of planning ahead means that the efficient sub-contractor can also become the pace-setter on a contract.

Often in the smaller companies who undertake formwork operations the foreman or supervisor at site may be one of the managers of the company. Obviously there will be a keen interest in the profitability of the activities, and urgent attention will be paid to outputs – particularly where a sub-contractor needs to draw sufficient money from the certificates of payment to meet his wages bill and purchase further material. Most sub-contractors pay their operatives by means of incentive schemes, or as a payment by results scheme, and thus achieve high outputs from the teams employed.

Sub-contractors generally employ a resident supervisor who deals with the day-to-day operations on site. A visiting supervisor or contract manager makes all the arrangements for the supply of materials, wages and welfare, and handled administrative aspects such as planning site meetings with agents, architects and engineers.

Factors that govern sub-letting. There are undoubtedly advantages to be obtained by the main contractor who sublets the formwork operations provided that:

(1) He is selective as to the type of work sublet

(2) He deals mainly in terms of labour *and* materials

(3) He lets out a complete contract.

These stipulations are the result of a series of considerations:

(i) The sub-contractor, due to the system of payment he adopts, is generally in a better position to handle repetitive, commercial quality operations, although he may not always achieve the standard of concrete finish required on many of the prestigious contracts.

(ii) Although every attempt is made to ensure that the sub-contractor should be made liable for the accuracy of his work, it is extremely difficult due, to the interrelation of the formwork and concreting operations, to isolate any particular reason for inaccuracy and displacement; in any event the sub-contractor almost invariably relies on the contractor for lines and levels.

(iii) The sub-contractor who is not adequately controlled will tend to carry out that work which is more profitable to him in a given period, and is then often reluctant to supply suitable or sufficient labour to carry out the more complicated or detached work later in the programme.

(iv) Where a sub-contractor is not committed with regard to materials on site, and having completed the more repetitive or lucrative parts of the work, he will often move his labour force off the site and show a reluctance to bring it back. This is very much the case where a small company is involved or where a new contract opens up elsewhere in the area that offers attractive alternative employment.

To offset these it should be said that a number of the larger formwork sub-contractors are extremely reliable, having worked for years in close co-operation with the main contractors and having learnt many methods and techniques.

Supervision. Certain supervisors employed by sub-contractors have particular skills, possibly the ability to work profitably, or provide continuity of work for all trades under difficult conditions such as can occur in deep basements, in bad ground or on repetitious work for high rise construction. The cladding of structural steelwork requires careful co-ordination and demands a close co-operation with the steel erector and steelfixer.

Many supervisors have their own 'following' of labour, and the main contractor will be assured of a supply of capable operatives and craftsmen throughout the duration of the work. It is essential, for the satisfactory progress of the work, that the sub-contractor's supervisor co-operate with the site managers and supervisors, but he must obviously respect the interests of his own employer and those responsible to him. He must therefore be receptive to suggestions and be co-operative in his supervision such that it benefits all the main parties on the contract.

The best arrangements develop where there is mutual respect, and the most productive where formwork has been erected in line with the programme, or where facilities for handling are uniformly shared. Obviously on any successful job a number of sub-contractors are involved but it is essential to all their efforts that the prime operations of formwork handling, erection and striking should progress unimpeded. In this way all the tradesmen can come on to the scene early in the cycle of operations. For example, delay in the unloading of a load of pipe may hamper a fitter during the period of the delay, whereas the late unloading or delivery of form props may cause delays throughout the whole site for several days.

Of course, as with any work of this nature, there will be upsets and delays, but attention given to a number of the more critical points of co-operation can do a great deal to alleviate problems.

From the beginning of a job there must be an agreed programme based on an agreed method. Many sub-contractors are reluctant to make innovations or to adopt methods other than those which are traditional to formwork construction. One reason for this is that by using traditional methods, the sub-contractor can still keep himself and his men employed in moving materials and manually striking formwork when, say, a crane breaks down, or some discussion is necessary over access regarding the use of the crane. If the sub-contractor has to commit himself on more up-to-date methods, such as table forms or even mechanized systems, then he must know that he will have guaranteed access to a crane during the critical times of striking and handling forms from one situation to another. Contractors are often irritated by a sub-contractor's reluctance to prefabricate formwork. This is often prompted by the sub-contractor's system of payment, when bonus arrangements are such that the payment includes items for fabricating, erecting and striking, the money being paid on completion of the stripping operation.

Formwork quantity. Where the more traditional techniques are being employed there will often be discussions regarding the quantity of formwork to be supplied or the 'cut in'. Obviously the sub-contractor wishes, as does the main contractor if the roles become reversed, to obtain the maximum amount of re-use from a given quantity of formwork. The problem here is to establish an optimum quantity of equipment that will provide a suitable re-use factor, but which will give working space for all the associated tradesmen. For instance the quantity of decking provided is critical when considering the number of trades that will require access, and the amount of formwork which may be engaged on concreting operations and various stipulations relating to the striking times. It would thus be in everybody's interest to ensure that a greater quantity of decking is provided than originally required so that continuity of work is maintained. In this way more trades will be kept in useful work. The quantity will of course be greater in bad weather or where striking is delayed.

Obviously arguments regarding the quality of formwork, surface finish requirements, suitability of the formwork and the need for replacement of sheathing after a given number of uses often occur, but the methods employed by a sub-contractor are generally adequate providing they are carefully controlled, especially where stripping and reshoring and similar techniques are concerned. Some sub-contractors have perpetuated the bad practice of crash striking, although specifications usually stipulate that formwork should be methodically and gently eased from the concrete.

A major advantage to the contractor who employs sub-contract formwork labour is that the cost of the work is evident, much of the risk is reduced in what is otherwise a nebulous area of costing and cost control. Few contractors know or calculate exactly what their formwork costs are and indeed this is difficult where hidden costs of handling and allied work are concerned. A carefully regulated contract placed with a reputable sub-contractor can be both successful and desirable, although it cannot be over-emphasized that careful control must be maintained on all questions of the engineering and quality aspects of the work.

Co-operative working. In recent years, a number of contractors have realized that recruitment of labour can be effected via the sub-contractor who for the reasons mentioned earlier, can provide continuous employment for the formwork tradesmen. These contractors have adopted a policy of subletting their formwork operations on all suitable contracts and have designed a procedure to ensure that the utmost benefit is obtained by the engagement of sub-contractors.

The sub-contractor's representative is invited to attend the pre-contract discussions and becomes, in effect, one of the 'team' who will prepare the pre-contract estimates. Although this means that the contractor is virtually 'nominating' the sub-contractor, in the event of his securing the work, it is obvious that the element of risk in pricing the work is reduced, the sub-contractor having an increased interest in the contract and subsequent involvement when it is eventually secured. The contractor will be selective and will not invite sub-contractors who have not performed successfully in the past. A great number of substantial contracts have been completed on this basis which augers well for the experienced contractors concerned. While it is inadvisable that 'labour only' sub-contractors should be employed on formwork – other perhaps than those extremely well known to the contractor, there is a case for some degree of financial assistance being given to assist with the purchase of special equipment.

A criticism often levelled at the sub-contractor is that he is not prepared to innovate, but stands by traditional and simple arrangements, such as commercial props or a combination of ply and timber which may constitute his major stock-in-trade. This reluctance is understandable, some of the equipment required on specialized construction offers little residual value or opportunity for re-use on other contracts. By subsidizing the sub-contractor either by direct purchase of equipment or by some other financial means the contractor ensures the co-operation on the part of his sub-contractor in adopting new techniques based on new ideas incorporating the latest plant and equipment. It may be that the more affluent contractor may purchase new equipment and then supply it on hire to the sub-contractor, thus resolving problems of supply and innovation.

There is little doubt that co-operative efforts of this kind will improve relationships and result in increased outputs and improved standards, which will amply reward the contractor.

Figure 4.1 Managers must always keep abreast of technical development and innovation.

4. The management of formwork activities

General

The construction manager is responsible for all matters which relate to organization and control and the successful outcome of the whole construction project. Formwork is just one area which comes within the broad spectrum of his activities.

As discussed earlier, the formwork arrangements, the suitability of a method and choice of materials have a marked bearing on the economics of a contract and the speed at which the work is executed. A manager is judged by the results achieved, these together with the accuracy and finish acquired, help to determine his managerial abilities.

Management decisions

The early decisions on method, choice of material and labour input are as much managerial ones as technological decisions. They demand a knowledge of company policy and the resources available, and an appreciation of the conditions under which the work is to be carried out.

Their impact will have considerable effect on the performance of all those involved in the allied and associated aspects of construction. The manager must ensure that the decisions made by the formwork engineer or designer and the planning engineer, together with various tradesmen and specialists, meet his requirements with respect to:

1 Phasing and continuity of all work subject of the contract

2 Achievement of standards of finish and accuracy in accordance with company policy or specification

3 The financial plan that governs the economics of the job.

These decisions must be based on the manager's appreciation of the various methods put forward by his staff, each of whom, to some extent, will be biased during the presentation of the details.

Economic considerations

The manager, whatever background and training he may have had, be it engineering or trades oriented, must visualize the projected methods in terms of overall performance and profitability within the framework imposed by the conditions of the contract. It is essential that he has a sound knowledge of the conditions that prevail locally especially on availability of labour and the general tone of industrial relationships. He must be clear as to the standards which are to be achieved so that he will be able to make decisions based on these factors. He must also balance the factors which influence cost of materials and cost of labour, and he must decide on the degree of rationalization and standardization which can be introduced into the method by the adoption, for example, of proprietary formwork and equipment. The initial cost of a form traveller, the provision of system equipment or a purpose-made form, may on first sight appear excessive. The manager, estimator and designer will have to assess these costs against the long-term benefits of reduced labour content, taking into consideration factors such as the possible reduction in upsets through contingency or the higher predictable outputs to be achieved. This consideration can be complicated by the fact that the reduction in labour force employed renders the motivation and control of those concerned more critical.

Method considerations

The manager must ensure that not only the selected method is safe and meets the requirements of the specification, but also that the techniques can be carried out using available labour. Where special formwork is selected a great deal will depend on the manufacturing standards achieved by the supplier, as well as the allotted manufacturing period. The manager must be certain that a supplier has the production capability that can meet the standards defined in the specification.

Sound decisions on methods and materials, taken at an early stage, can by projection affect the structural design. Where the manager has a firm idea of the techniques to be employed, it is possible to facilitate construction by adjusting the design details. This particularly applies on the longer term contracts where a designer can simplify or make slight modifications which enable the re-use of items such as table forms or wall panels purchased or constructed for early phases of the work.

The manager, at the earliest possible opportunity, will want to become involved in the contract with a view to injecting practicalities of construction and matters of form design into the structural design process.

Parallel working and the letting of contracts on a

management-fee basis allow him the best opportunity of affecting the design, and provided that the points are raised early enough it will be possible to introduce minor revisions to the concrete outline and details which speed the formwork process.

Innovation

In order to obtain the best results from his team, the manager must always be ready to accept suggestions regarding new methods. Once he has adopted a method he must then continually back it and be prepared to be responsible for the outcome of a suggestion at all stages of the job to its completion. Once a suggestion has been incorporated into a working method, constant supervision is necessary to ensure that the new system or new piece of equipment is, in fact, used as intended. Failure to capitalize on money invested in new items will otherwise result in waste.

Material economy

In the formwork field waste occurs as a result of the wrong use of materials, the haphazard ordering of materials and, in the case of proprietary equipment, delays in the return of equipment to the supplier. Thus one of the most positive steps that a manager can take to control costs is to make a planned attack on all forms of waste.

The manager who finds in taking time to check that goods have been correctly ordered and properly used can effect considerable savings. These checks need not be time consuming because careful monitoring of existing information will provide useful pointers to wastage and this combined with observations made during the manager's normal daily activities on site can reduce hire periods.

Financial control

A good flow of financial information is essential to the control of all labour intensive activities and the manager needs to be advised of the following key information:

1 Number of men employed. Number of hours worked by formwork tradesmen

2 Outputs in appropriate measure of walling, columns, decking, beams and stairs (including time spent making and striking)

3 Any delays incurred by other tradesmen

4 Payments arising from bonus or incentive schemes

5 Weekly hire charges for equipment and plant

6 Payments on certificate to sub-contractors

7 Any claims or defective work.

The figures required should be available weekly, although for action a brief note of the key items could be produced daily from figures which are available, on site, taken from time sheets, clock cards and from reports produced by section foremen. Section foremen should report all equipment which is to go off hire on their daily report sheets so that considerable savings result in a positive effort to control equipment usage.

Developments

The manager needs to keep abreast of developments in equipment and techniques, and this requires a conscientious effort on his part. The pressures on the key people in the construction industry are such that there is often little time or thought given to training or the investigation of new developments. Sources of information can vary, from the trade journal to technical bulletins issued by proprietary suppliers, quite apart from the contact made with representatives of the suppliers who call on site. Organizations such as the Cement and Concrete Association and The Concrete Society run special events which are aimed at providing up-to-date information for all those involved in the construction industry. These events generally take the form of one-day meetings in which specialists either describe new approaches to problems, or outline case studies of current projects. Data sheets are published on special aspects of construction which offer an opportunity to derive information quickly and effectively. Evening meetings organized by the Institution of Structural Engineers, the Institution of Civil Engineers and the Institute of Builders are all helpful in the dissemination of information on new techniques. Full details of programmes for such meetings can be obtained from the Secretaries of the various Institutions.

The manager is well advised to interest himself with 'fringe areas' of the industry such as the pumping of concrete, or even techniques such as the thermic boring of concrete. The introduction of these methods onto a site may well influence the placing of concrete or through hole formation. Obviously new questions will arise as a result of the techniques, e.g. whether formwork can be designed to sustain the new pattern of pressures and forces imposed by pump equipment, whether it is cheaper to form openings at the time of erecting the formwork, or, possibly, whether the openings would be more economically produced by boring at a later stage.

Standards

A manager should associate himself with the pre-planning and organization of work, and he should allow sufficient time to concentrate on these aspects. Failure to plan and co-ordinate the efforts of those who work with formwork may well result in the production of sub-standard concrete. Sub-standard work imposes demands on the manager's time, usually post-mortems, through tedious meetings and discussion. Time well spent in the preparation and agreement of samples and mock-ups will help in establishing an acceptable basic standard of workmanship. Once standards have been established, and a well considered method devised and installed, the manager will then be able to channel his efforts into the basic management tasks of co-ordinating and controlling his team.

Figure 5.1 Simple models are valuable aids for planning operations. This model represents a sugar-beet platform.

5. Drawings, details and models

General

Where any degree of pre-planning is applied to the formwork arrangement appropriate forms of communication are necessary between the planner, temporary works engineer or formwork designer and those who are to manufacture and use the formwork system. The designer may be in close contact with the manufacturer – even to the extent of sharing an office in the works or on site. On the other hand he may be at main office which can be some way from the site or, possibly, even in another country. Whatever his location all drawings must be prepared such that they clearly indicate the proposed method and arrangement, and must be supported by details of erection technique, modifications during use and a schedule of all the necessary parts and ancillary equipment.

Sources of information

While the designer during his preparatory work may use the architect's drawings in addition to those supplied by specialist sub-contractors, all dimensions, particularly the critical ones that concern the profile of structural concrete and the accurate position of reinforcing bars and prestressing tendons must be based on the structural drawings alone.

The designer is often the first person to critically examine these details and undoubtedly there will be quite a number of points regarding inaccuracies, or discrepancies between both small scale structural drawings and structural details, and between these drawings and those prepared by the architect and sub-contractor. Obviously every discrepancy needs to be resolved as soon as possible.

Formwork details

The details produced by the formwork designer fall into the following categories:

1 Drawings showing proposed methods for discussion

2 General assembly drawings of agreed methods

3 Details of individual panels for the purpose of manufacture construction.

The drawings in the first two categories are used in discussion with the contractor's personnel and the authorities involved on a particular project. With the growing emphasis on quality of surface appearance, it is not unusual for the architect to request copies of the formwork arrangement drawings. The engineer may be empowered by the contract to inspect drawings for all formwork and temporary works. Where work is to be carried out over, or adjacent, to public rights of way the local authority will need to have copies of all drawings, form and falsework designs with full supporting calculations.

Drawings in the third category are mainly concerned with form construction and may be used on site, in a sub-contractor's yard, or in a supplier's workshop, and are the ones used for the actual form construction. These drawings should be accompanied by schedules of fittings and ironmongery.

Obviously those who are concerned with formwork on site must receive copies of all drawings and schedules. It is helpful if the schedules are so arranged that copies supplied to the materials clerk or storeman can be used as a check on delivery of the items. This is particularly necessary where system components are supplied by a number of different firms, e.g. ties from one manufacturer, fabricated forms from another, and so on.

Wherever possible the formwork drawings should be complete and self-contained without the need for reference to any other drawing. The size of the drawing sheets should be such that they can be handled under most site conditions without difficulty. Ideally, the drawing and schedule should be combined in such a way that, for example, a man on a scaffold high above ground should be in complete possession of sufficient information to enable him to erect a form or forms into position, incorporate fixings and fastenings for succeeding work, modify a panel or unit to its exact shape for particular use and prepare it ready for casting, without reference to any other paperwork.

This is not impossible to achieve provided that the drawings produced by the formwork consultants and by the major proprietary suppliers meet the requirements. To ensure that they do it is necessary:

1 To establish standard details and, where possible, standard techniques

2 To ensure that the drawings prepared are monitored in conjunction with a checklist.

It is well known that a checklist is an essential part of an airline crew's equipment during pre-take off drill, and, in much the same way, a checklist prepared and checked by a senior designer has a great deal to offer to those who

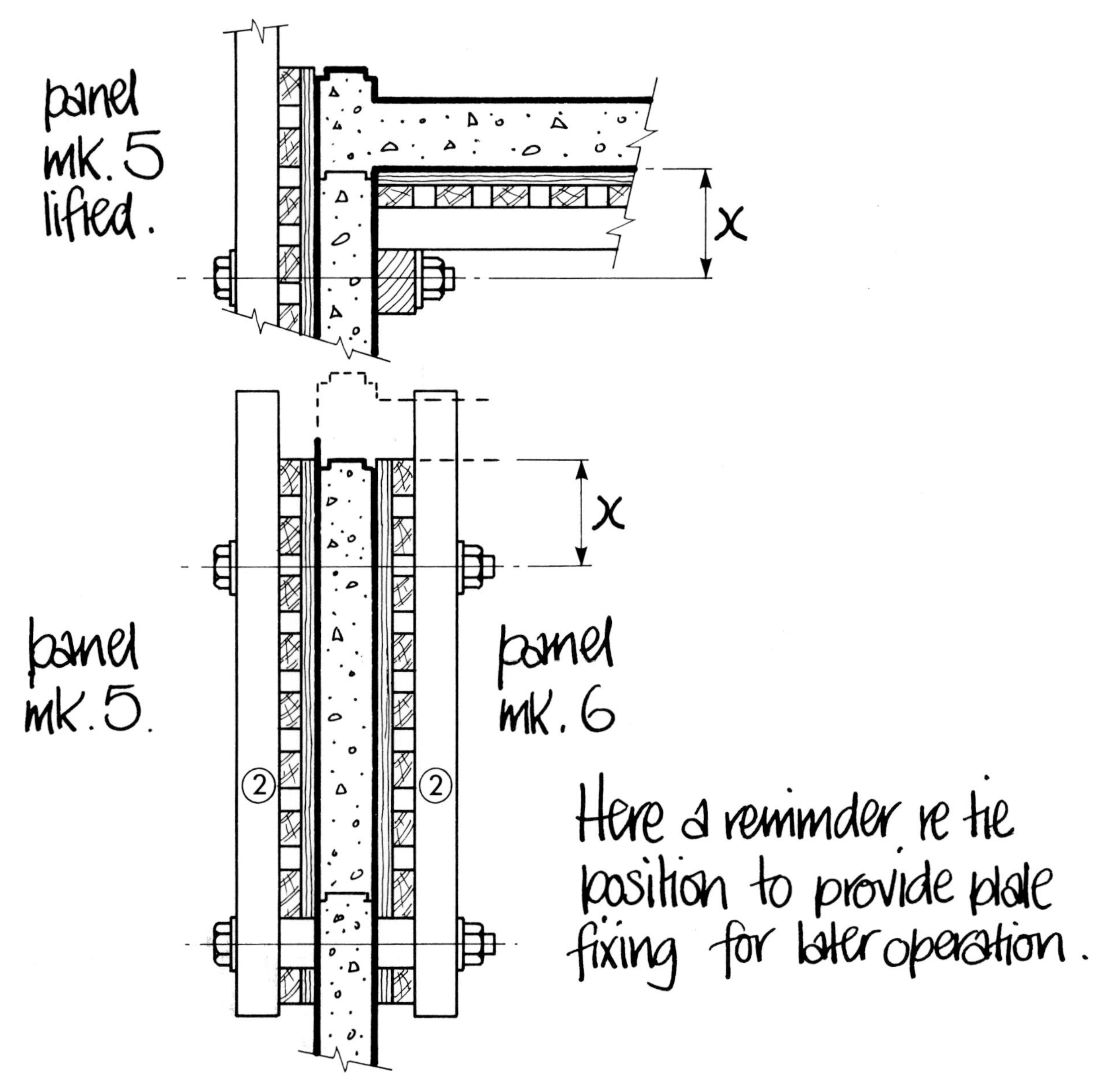

Diagram 1 Simple sketches provided by the formwork designer help to communicate method.

are concerned with communicating information or the satisfactory manufacture or erection of a formwork system.

Even a simple pre-concreting checklist for a ganger or section foreman can effect savings in time which would otherwise be wasted while omissions were rectified, or some access organized, vibrators assembled, skips prepared and such like. The basis of such a checklist is indicated further on in this chapter. Some of the items may appear to be rather specified, or even too sophisticated, however if the person preparing the communication takes the time to go through the list item by item he can be reasonably sure in his own mind that he has transmitted the essential information.

Apart from formal line drawings, 'scrap' drawings and sketches can be of immense value, especially where semi-skilled site personnel have to interpret the information. This also applies particularly where there is a great degree of geometry involved on work such as required for hoppers, tanks, silos and containers.

The use of models

Models are extremely valuable from the tender stage onwards, since they enable critical relationships between different parts of a structure to be visualized. They can help with the assimilation of the scale of some aspect of construction, or structural components in a particular frame. The sequence of operations is easier to establish 'in the round' particularly when individual lifts and bays have been established and interpreted into the block form of a model, which when completed then becomes a set of building bricks that can be assembled and dismantled when required to represent various stages of the work. Models need not be expensive to make and can be produced in the drawing office from folded paper, card or from carpenter's block scrap.

Formwork drawings are based on those produced by the structural engineer and the largest scaled, and most up-to-date, structural drawings must be used in their preparation.

Figure 5.2 A model of the column featured in the exercise on form design, given in Appendix 2.

The control of drawings

Where drawings are issued, either in the works or on site, it is essential that their 'receipt and issue' be recorded and suitably controlled. For example, assume that there is an issue of 10 drawings to various site supervisors and key personnel numbering, say 8 or 10, then a maximum of 80 drawings are out on issue and these relate only to a small part of the whole contract. As soon as there is a revision there are immediately 10 outdated drawings in circulation, some in offices, others with supervisors and some within the actual working area. If any of these were to be used errors could then be introduced into the work, possibly with dire results. The same principle can also apply to schedules or, indeed, to any written instruction. It is therefore essential that drawings which have become out-of-date should be recovered and destroyed, sufficient copies only being retained for reference purposes and where costs and methods need to be evaluated.

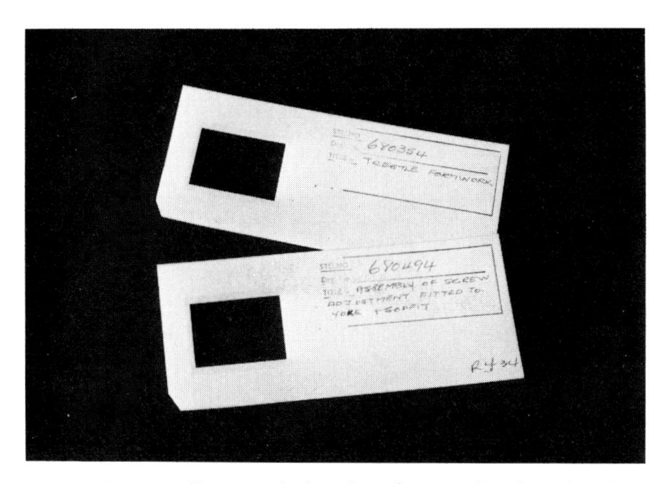

Figure 5.3 Microfilm is an ideal medium for recording form details.

Revisions to detail

There is one trap which should be avoided, and this can be caused by draughtsmen when they alter and revise drawings. Traditionally, revisions are noted, lettered and dated at the side of a drawing in some prepared space. At the outset, the notes clearly indicate what has been altered and so that anyone using the drawing can readily identify the actual change made on the drawing. But as the number of revisions increases, it either becomes impossible to note the details, or the notes become abbreviated. The result is that when a drawing is being used in which the 'revisions' column has only a note such as 'dimensions altered', the reader is confronted with a page of close detail any of the dimensions of which may differ from those employed up until that time. Other than physically checking each dimension the reader will not be sure which of the established dimensions are not correct.

Scales

With the introduction of SI units into the construction industry, draughtsmen have had to adopt not only new scales for drawings but also the way in which dimensions are indicated. The original scales of 1 in., $\frac{3}{4}$ in. and $\frac{1}{2}$ in. to 1 ft, used for general assemblies, and the $\frac{3}{4}$ in. scale allowed sufficient working information to be included on drawings for showing individual panels, while full-size details were used to indicate special points which related to fixing and fastening. The preferred SI scales now used are:

Detail drawings 1 : 1, 1 : 5, 1 : 10 and 1 : 20.

Construction drawings 1 : 50, 1 : 100 and 1 : 200.

Layout and site plans 1 : 500, 1 : 1250 and 1 : 2500.

Obviously selection of a scale is a matter for the draughtsman, although he may need to experiment to establish the scales that are best suited for the presentation of different details and construction for various materials.

Diagram 2 indicates the size and details of a typical timber form panel as it would appear when drawn to various scales. From this it can be seen that for typical timber, or timber and ply fabricated panels, scales 1 : 20 and 1 : 50 allow sufficient information to be conveyed on the general assembly drawings and layout. For constructional details of works fabrication 1 : 5 and 1 : 10 are useful scales, while 1 : 1 or full size are essential where critical fixing details or sections of machined material are to be shown.

Under some circumstances a scale which results in a

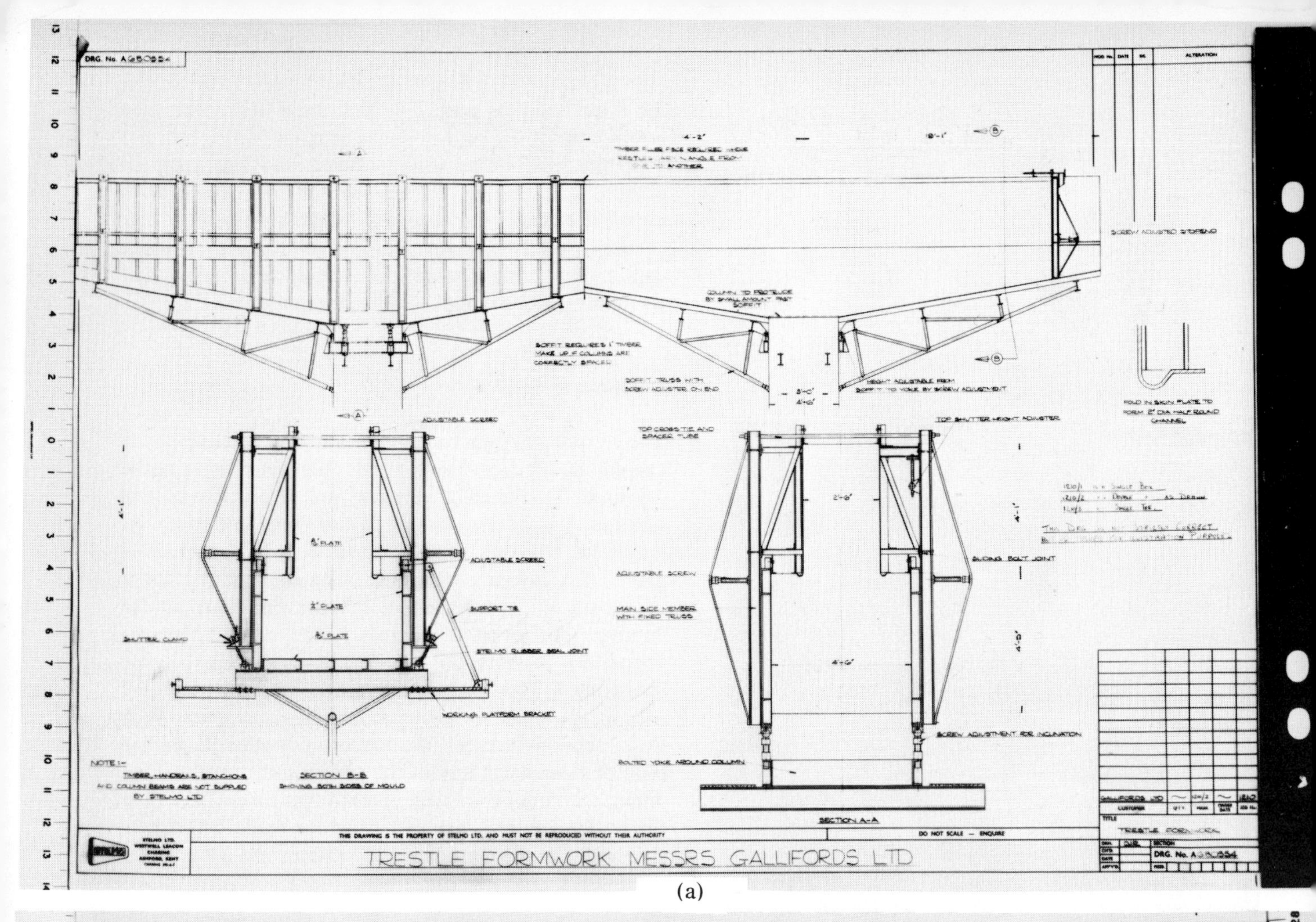

(a)

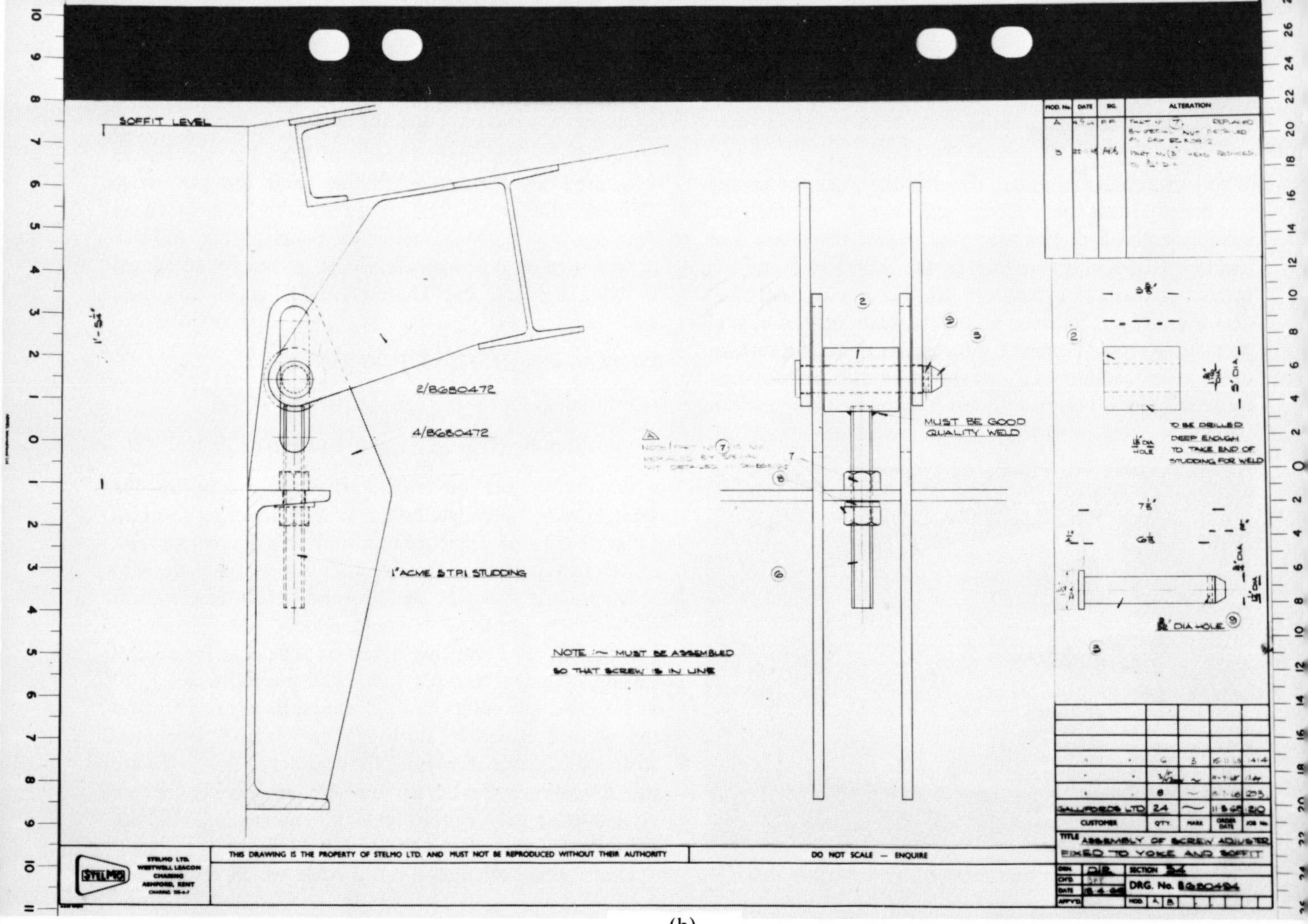

(b)

Figure 5.4 These prints have been reproduced from microfilm and demonstrate how much information can be stored: (a) crosshead formwork; (b) hinge location.

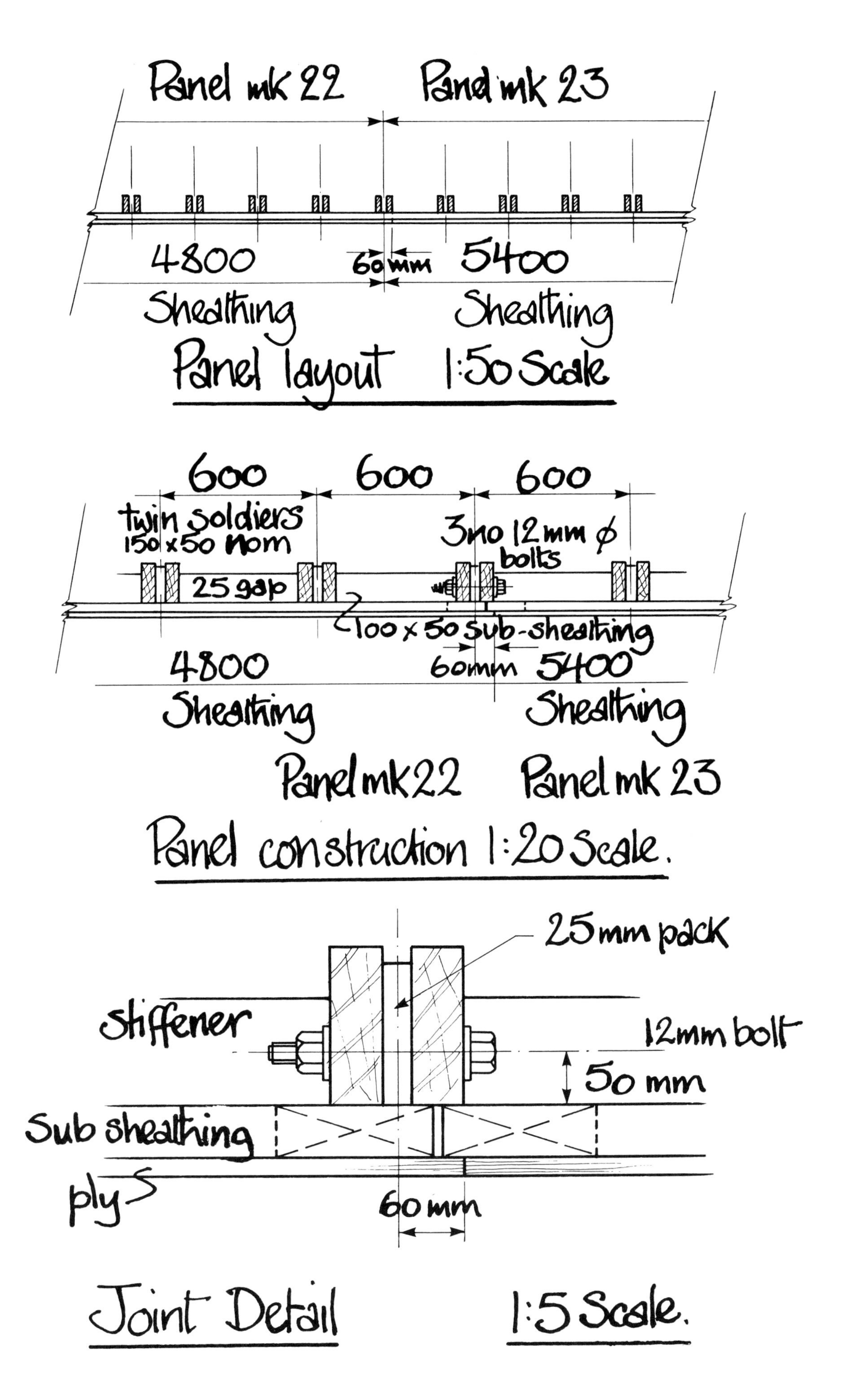

Diagram 2 Sizes and details of a typical timber form panel.

half-size reproduction is dangerous because a detailer or designer might be misled on proportion of sections, sheathing thickness and similar details. The scale selected is often related to the size of the drawing sheet, although any attempt to convey information by means of details, the scales of which are governed by the space available is, again, dangerous and it is far better to use two sheets to indicate all relevant information clearly. This will help to ease the task of interpretation under difficult conditions.

Checklist

Formwork details

Does general assembly drawing indicate position of every panel?

Are scaffold, supporting systems and similar restrictions indicated on general assembly drawing?

Does drawing indicate crane radius and lifting capacity?

Do schedules include all panels and associated ironmongery?

Do schedules indicate source of supply or place of manufacture?

Are all ties and washer plates included in schedule?

Are construction materials indicated on schedules?

Do drawings note parting agents or retarders to be used?

Do drawings indicate intended lift heights, bay sizes and construction joints?

Structural, architectural drawings and associated concrete details

Are drawings up-to-date?

Have drawings been checked?

What are special requirements regarding finishes?

Is there a special degree of accuracy?

Are reinforcement drawings and details available?

Are details of structural steelwork available?

Are details of inserts and connections available?

Are specialists' drawings of lifts and plant available?

Do drawings indicate position of adjacent structures?

Is site access indicated?

6. Timber formwork design

by H R Harold-Barry, MICE, MASCE

The formwork designer should aim to produce a satisfactory method of containing and forming the concrete within specified tolerances with maximum economy of both effort and materials. This chapter considers the forces involved and the ability of some different timbers to resist them with the appropriate degree of safety.

Loads

It is not yet possible to make full use of limit state principles in the design of formwork because we have inadequate knowledge of the statistical variation of the loads applied to formwork. There is information in Forests Products Research Bulletin No. 50, *The strength properties of timbers*[12] on the variation in certain properties of small timber specimens. CP3: Chapter V: Part 2: 1972[13] gives a statistical approach to the prediction of wind loads.

A lot has been written about the pressure of concrete on formwork[1–10] the most useful of which is *CERA Report No. 1*[8]. This has been condensed into card form as *CIRIA Formwork loading design sheet*[11] and further simplified in The Concrete Society information sheet, *Concrete pressure graphs for formwork design,* which is reproduced in Appendix 1.

This final simplification ignores the relatively small increase in pressure resulting from concrete being discharged freely from a height into the form. Forms must, in any case, be designed with an arbitrary allowance for abuse, such as being hit by a swinging skip, so there is no point in trying to be too precise in forecasting loads on formwork. The variables considered in arriving at a design pressure (kN/m^2) are:

- The height of the pour (m)
- The thickness (mm) of the wall (where it is not more than 500 mm)
- The rate at which the concrete surface rises (m/h)
- The concrete slump (mm) and placing temperature (°C)

A concrete density of 2400 kg/m^3 is assumed.

For lightweight or very dense concrete with a density significantly different from 2400 kg/m^3 it would be safe to assume a proportional difference in pressure. Alternatively see CIRIA, *Formwork loading design sheet*[11].

No firm information is yet available on the effect of admixtures but where retarders are used the effect will be similar to using concrete at a lower temperature.

The coefficient of variation of the difference between actual pressures and the pressures given by the chart is in the region of 15%. The factors of safety discussed later are sufficient to allow for this variation as well as the arbitrary allowance for accidental loading mentioned earlier.

Wind loads can be obtained from Figure 6.1 and have to be taken through the external props or ties, which are also used to align the formwork. Care must be taken to resolve all the forces correctly and to allow for their components.

In addition to the weight of the concrete on soffit forms a live load of 2 kN/m^2 should be allowed where the concrete being placed is evenly distributed, such as when hand barrows or pumps are being used. If skips or dumpers are used there is a danger of heaps of greater weight than the compacted concrete being placed on a

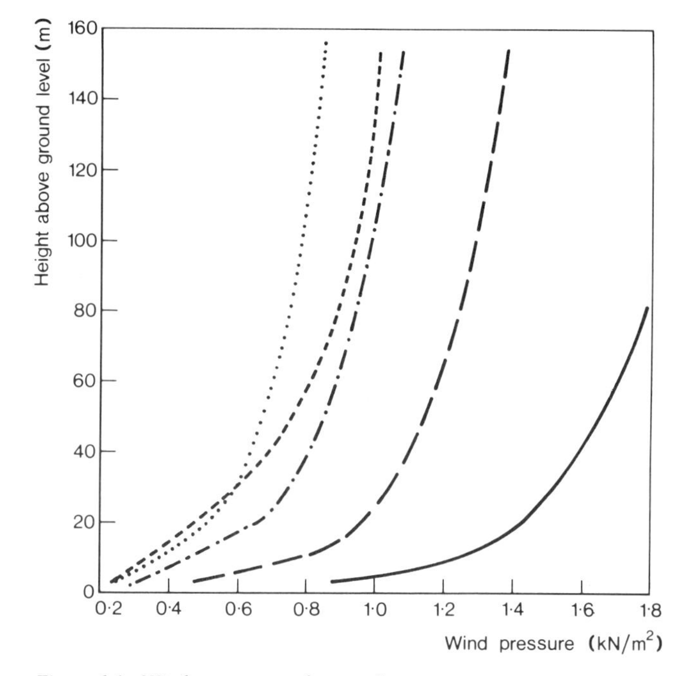

Figure 6.1 Wind pressures on formwork.

given area and a live load of 3 kN/m^2 should be taken. In allowing for horizontal surge one-third of the vertical load due to vehicles should be taken. Bracing needed to avoid buckling of a prop must be designed to take at least 3% of the prop load resolved at right angles to the prop.

Sloping formwork

The research behind *CERA Report No. 1,* on which the concrete pressure graph in Appendix 1 is based, was on vertical formwork only. However, since similar benefits will also apply to sloping formwork this graph may still be used provided the pressures on the lower face obtained for arching and stiffening are increased by the weight of the concrete resolved at right angles to the face. It will give a conservative pressure for the upper face.

Safe working stresses

Safe working stresses for many different timbers are given in CP 112: Part 2: 1971, *The structural use of timber*(14). These have been derived from the results of tests on clear specimens to BS 373:1957(15). The tests and results are briefly reported in Forest Products Research Bulletin No. 50, *The strength properties of timbers*(12). The most interesting results for formwork designers are given in Table 6.1.

Timber formwork usually has a moisture content in excess of 18% which means that the 'wet' or 'green' results are used.

For convenience softwoods are divided into three groups. Species group 'S1' is expensive and includes Douglas Fir, a high quality joinery timber. 'S2' is generally used for formwork and includes the popular Whitewood, Redwood and Commercial Western Hemlock. 'S3' includes the weaker timbers like Sitka spruce. The safest working stress in each group has, of course, to be that for the weakest in the group. If, therefore, a specific timber can be used it would be more economic to use the stresses for that timber.

The various stresses are derived in different ways. The permissible stress in bending is obtained from the stress below which only 1% of the specimens will fail, which is the average less 2.33 standard deviations. The difference between the upper and lower values in Table 6.1 is four standard deviations.

On this failure stress is put a factor of safety of 2.25 to obtain the long term 'basic stress' in bending for 'clear' timber, which means timber free of any strength reducing defects such as knots, splits or wane (the absence of timber where the bark grew). This safety factor, as well as allowing for the difference between the short term loading in the tests and the long term strength of the timber, also allows for variations from the predicted loads and accidental loading.

The same failure rate and factor of safety is also used for shear stress. However, many formwork designers consider CP 112 permissible shear stresses far too low. This is based on the observation that shear failures never seem to occur in formwork, and the stresses given are based on a shear test over an area of 20 mm × 20 mm parallel to the grain with no confining stress.

In practice the shear is perpendicular to the grain and in consequence there is an equal confining stress. In addition the larger areas usually involved reduce the chance of failure.

Table 6.1 shows that shear strength parallel to the grain is always greater than compressive strength perpendicular to the grain, and it might therefore seem unreasonable to reverse this situation in deciding upon safe working stresses. However, few people will want to ignore the code and would prefer to wait until the code is changed before adopting higher shear stresses.

Table 6.1 Range covering 95% of Tests Results on clear wet specimens of timber to BS 373 : 1957(15), from information in DoE Forests Products Research Bulletin No. 50, *The strength properties of timbers*(12).

	Compression perpendicular to the grain* N/mm^2	Shear parallel to the grain N/mm^2	Flexural strength in bending N/mm^2	Modulus of elasticity N/mm^2
Whitewood (European spruce)	1.48–2.24	3.5–6.3	28–50	5100–9700
Redwood (Baltic	1.76–3.20	3.8–8.0	28–60	4500–10 900
Hemlock (Commercial Western)	1.34–3.86	5.2–8.0	29–69	5800–11 600
Douglas Fir (Home grown)	1.82–4.34	4.2–9.4	35–71	4600–12 000
Keruing (Malayan)	4.48–6.32	c. 5.3–10.1 (interpolated)	73–93	14 400–17 600

Moisture contents at test were above the values where a change affects the strength properties.

*Obtained from the formula: $y = 0.00137x - 0.207$

where $y =$ Compressive strength perpendicular to the grain (N/mm^2)

and $x =$ Force (N) to push a 11.3 mm diameter ball 5.56 mm into the timber at a speed of 0.11 mm/s.

Having said that, the joint Concrete Society/Institution of Structural Engineers *Formwork Report*[16] does give a much higher figure for compression perpendicular to the grain than suggested in CP 112[14], which takes the $2\frac{1}{2}$% failure figure and applies a factor of safety of only 1.2. The 'basic stress' so obtained may be used in situations where there is no wane. However, this stress is so far below the traditional figure for formwork of 2.75 N/mm^2 (a direct conversion of the long-standing 400 lbf/in^2) that the *Formwork Report*[16] ignores it and suggests a blanket figure of 2.75 N/mm^2.

At this point it is interesting to study the compressive stresses which may develop under the washers supplied with bolts by proprietary formwork manufacturers. At the stated safe working load of the bolts these vary from 2.3 to 10.2 N/mm^2, the majority being between 4 and 7 N/mm^2. There are few warnings to users to reduce the safe working load in their bolts when used in conjunction with softwoods. The consequence is often very high compressive stresses, causing considerable bedding in, well beyond the 'failure' point in compression. This may be of little consequence except from the point of view of deflection.

Deflection

Deflection calculations are usually concerned with deflection due to bending and sometimes due to shear as well, but rarely is deflection due to compression perpendicular to the grain taken into account. Using the traditional stress of 2.75 N/mm^2, deflection due to compression can be as much as deflection due to bending, and more for the first use when bedding in takes place. An allowance of 1 or 2 mm per joint is often made for this so that the form ends up within the permitted tolerance after the concrete has been placed.

CP 112[14] suggests in general a limiting deflection of 0.003 of span based on the average Modulus of Elasticity where the load is distributed between members, and the Minimum Modulus of Elasticity where it is not. The 'minimum' is the average value less 2.33 standard deviations, i.e. 1% of the values will be below the 'minimum'.

Since twin soldiers or walings are almost invariably used in timber formwork the mean value is used in the graphs.

Grading timber

The stresses discussed so far are basic stresses for 'clear' timber, but since all timber to a greater or lesser degree contains defects, a reduction on the 'basic' stress has to be used to allow for the amount of defects.

Timber, being a long established building material, has been subject to numerous different methods of selection. Until the advent of Mechanical stress gradings, these were based on visual inspection of the timber. European softwood is usually classified as 'unsorted', 'fifths' and 'sixths'. The proportions in each group will vary considerably from country to country and port to port, but an approximate guide to the proportions is illustrated in Figure 6.2. The best 60% is called 'unsorted', the second best 30% is called 'fifths' and the poorest 10% is called 'sixths'. Figure 6.2 also shows the yield of various grades from each classification (R indicates the proportion rejected), and it is immediately obvious that there is not much agreement between the various ways of grading and classifying timber. Detailed information about grading can be obtained in CP 112, AMD 1265[14] and BS 4978[17].

It is generally agreed that deflection is the most important factor in formwork design, and therefore a knowledge of the elastic properties of timber would be a great advantage. This is the basis of mechanical stress grading. The timber passes through a machine which measures the deflection at 150 mm intervals under a load, and grades the timber accordingly. As can be seen in Figure 6.2 the vast majority of timber passes the highest grade (M75) and there is a strong body of opinion which suggests that there should be a higher grade which would select the best 50% or so of the timber, and allow higher grade stresses.

It can be seen from Table 6.2 that M75 costs little more than GS grade timber but has considerably higher allowable stresses. It is therefore obviously more economic to use M75, the highest grade yet available.

Modification factors

The grade stress is the long term stress which the timber with permitted defects can safely stand. Formwork is stressed for a relatively short period of time and CP 112 allows a 50% increase in bending stress for short term loading. The Formwork Report allows increases in all

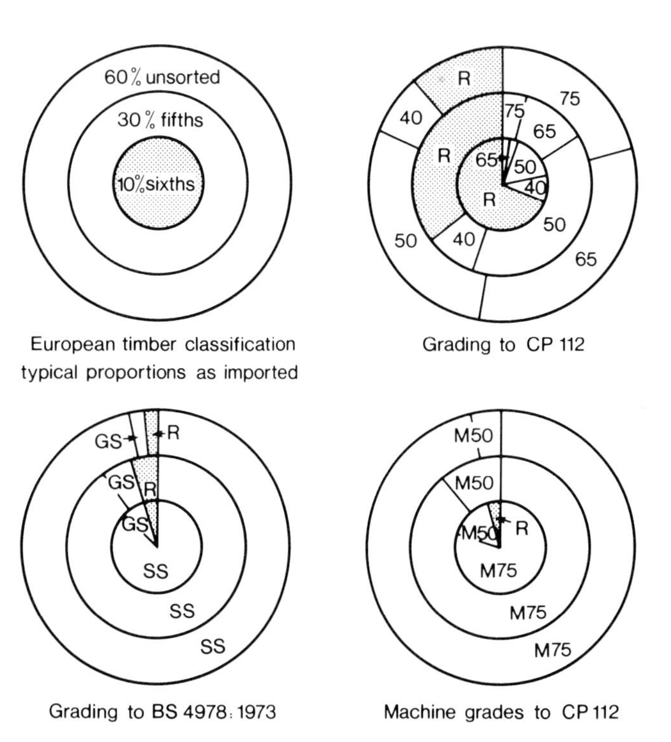

Figure 6.2 Typical proportions of imported European redwood and whitewood falling into various grades. (Area is proportional to quantity in each case.)

Table 6.2 CP112 safe short term working stresses (× 1.33) and elastic modulus for wet timber (N/mm²)

	Figure 6.4 Species S2	Figure 6.5 Whitewood (European spruce)	Figure 6.6 Hemlock (Commercial Western)	Figure 6.7 Species S1	Figure 6.8 Gurjun/ Keruing
Timber grade	GS	SS	M75	SS	75
Minimum modulus of elasticity	3900	4700	7100	4400	8300
Mean modulus of elasticity	7000	8200	10 600	9200	12 400
Bending	5.05	7.85	10.77	9.18	17.02
Shear	0.93	0.93	1.37	0.93	2.29
Compression perpendicular to the grain	1.22	1.37	1.60	1.72	3.67
Ditto, where there is no wane	1.84	1.84	1.84	2.29	4.12
Average ditto (see Table 6.1)	1.86	1.86	2.60	3.08	5.40
Traditional ditto	2.75	2.75	2.75	2.75	—
Density with 18% moisture (kg(m³)	500	510	530	570	720
Relative costs per unit volume	1.00	1.04	1.06	1.43	1.40

grade stresses depending on the duration of load of 1.5 for one day, 1.4 for one week, 1.3 for one month and 1.2 for one year. The stresses in Table 6.2 have been derived using a modification of 1.33, which was the factor chosen in the *draft* Formwork Report. Pro rata adjustments may be made if considered necessary. CP 112 also allows a 10% increase in stress where the load is shared by four or more members, but this is rarely the case with formwork and so is not included in Table 6.2.

Bearing stresses perpendicular to the grain can be increased for short bearing lengths by the factors given in Table 6.3 where the beam overhangs the bearing by at least 75 mm. This Table comes from CP 112. In most cases these do not bring the permissible stress up to the traditional 2.75 N/mm² and the designer who feels that CP 112 stresses are too low but who also wants to benefit from using a superior timber may consider using the average bearing capacities given in Table 6.2. This would be logical in the light of the fact that average elasticities are accepted and the fact that the average bearing capacity is below 2.75 N/mm² for the weaker timbers but above it for the stronger timbers.

As can be seen from Table 6.1, timber is a very variable material and the designer should be warned against assuming that stresses quoted to three significant figures imply 1% accuracy – they do not!

Plywood

CIRIA Report 37, *Factors influencing the deflection of plywood sheeting in formwork*[18] makes very interesting reading. It gives a more valid comparison between different plywoods than CP 112 which simply reproduces the manufacturers' quoted figures. Deflection is the controlling factor when plywood is used for formwork and there does not appear to be an outstanding difference between the various types of ply.

Table 6.3

Length of bearing (mm)	10	15	25	40	50	75	100	150
Modification factor	1.74	1.67	1.53	1.33	1.20	1.14	1.10	1.00

The major difference is between plywoods of different moisture contents. A safe value of modulus of elasticity for saturated plywood is 5000 N/mm² and this can be increased by 30% to 6500 N/mm² where the moisture content can be kept below 10% by sealing the faces and edges of the plywood. Some increase in load can be obtained by spanning in two directions but the extra expense of framing and the greater rigidity in the direction of the surface ply usually makes spanning in one direction more economic.

Figure 6.3 is therefore based on a span in the direction of the surface grain, fixed at one end and pin jointed at the other, with a modulus of elasticity for the section as a whole of 5000 N/mm². No account is taken of the cumulative effect of repeated loading of plywood under hot wet conditions such as is found in soffit formwork designed for several re-uses under a thick slab. The author has seen deflections of 6 mm in 19 mm ply spanning 225 mm under a 1 m thick concrete slab after 13 uses. According to Figure 6.3 ten times this load is required to give this deflection, which means the elastic deflection was only 0.6 mm, the rest was creep under the hot wet conditions. It may be preferable to use a material like steel in these conditions. In any case plywood must be suitably bonded with 'weather and boilproof' adhesive.

The safe load graphs are for standard metric timber sizes (BS 4471: 1969, dimensions for softwood[19]) generally available; 10% of the timber may be up to 1 mm below nominal size before being regularized by having the depth of all pieces reduced to 1 mm below nominal size. The nominal size is used in calculating the geometrical

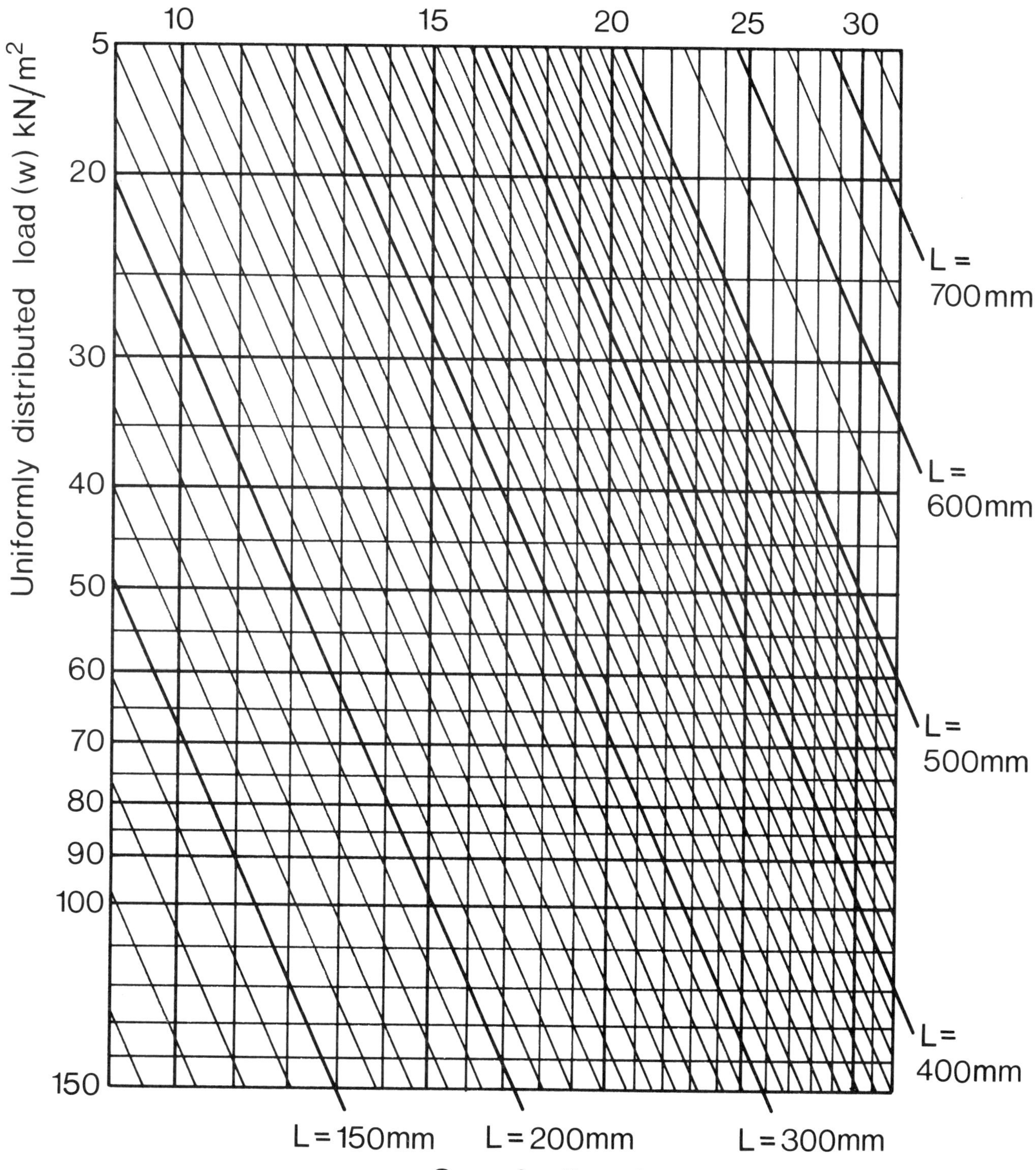

$$\text{Deflection} = \frac{3 \times L}{1000} = \frac{12 \times w \times L^4}{185 \times 1000 \times Ed^3}$$

$$\text{If } E = 5000 \text{ N/mm}^2,\ w = 231\,250 \frac{d^3}{L^3} \text{ kN/m}^2$$

A 30% increase in load can be taken by 'dry' (10% moisture) plywood ($E = 6500$ N/mm²).

Figure 6.3 Safe load graphs for saturated plywoods.

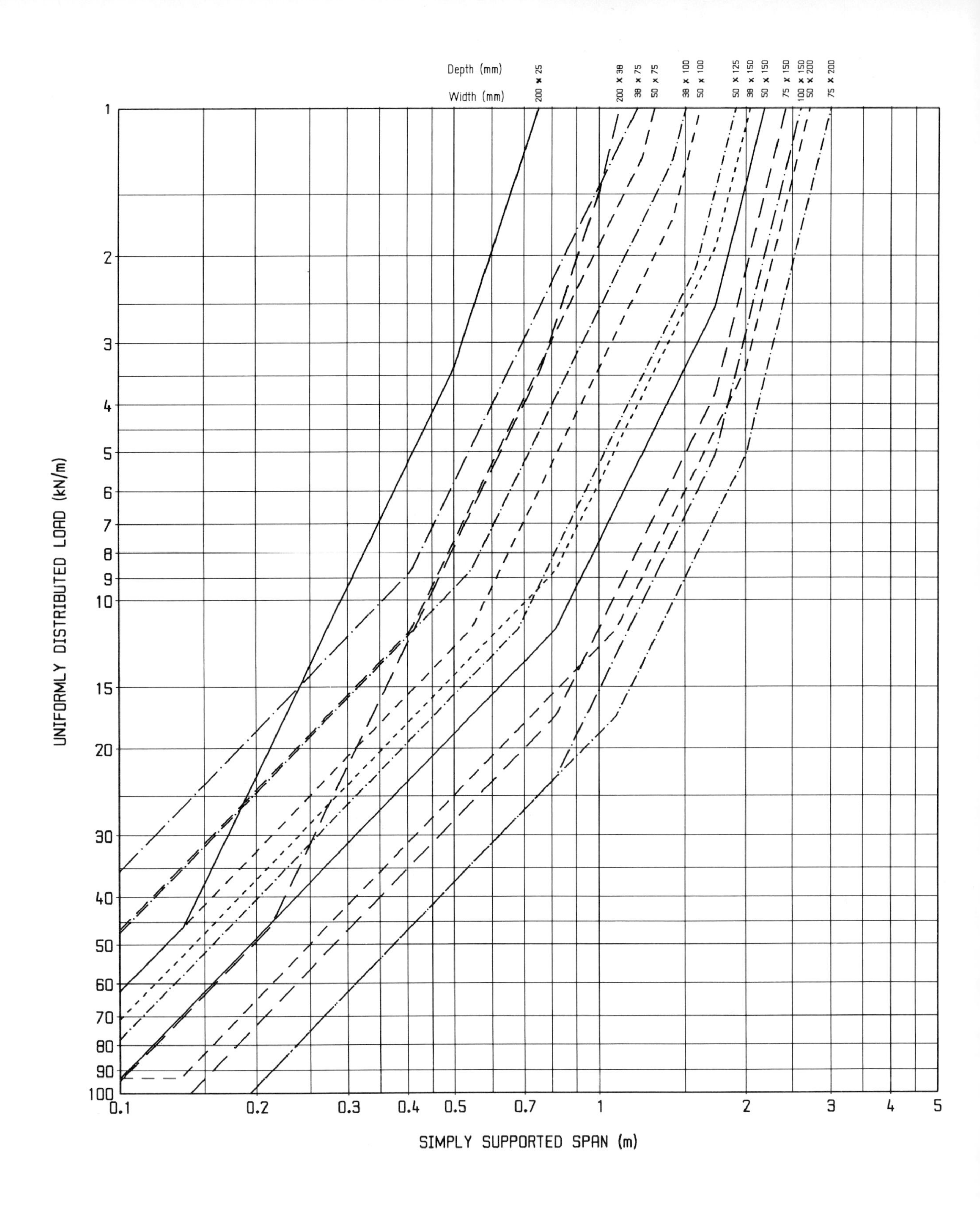

Figure 6.4 Safe load graphs for rectangular beams of wet grade GS species S2 timber.

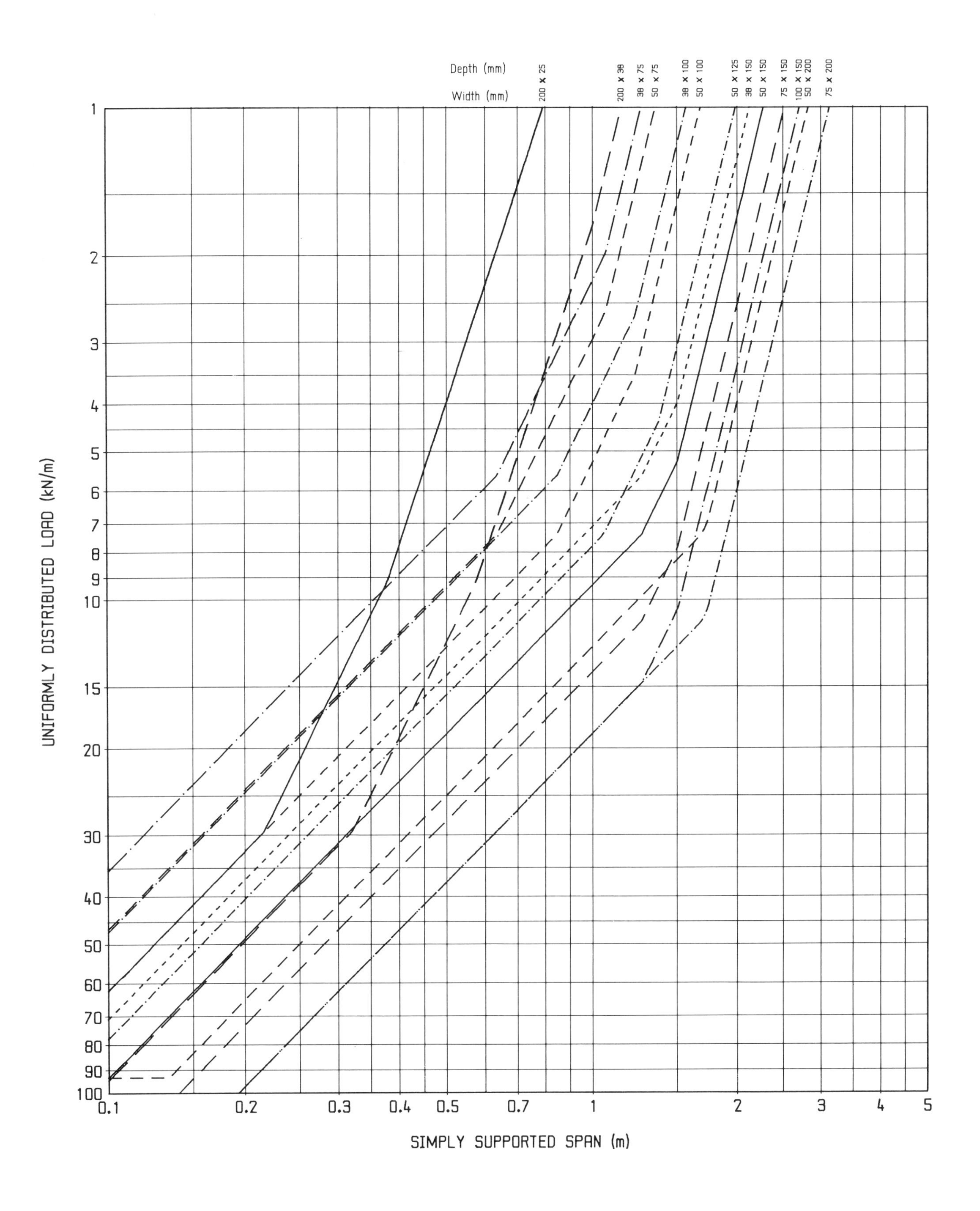

Figure 6.5 Safe load graphs for rectangular beams of wet grade SS whitewood (European Spruce).

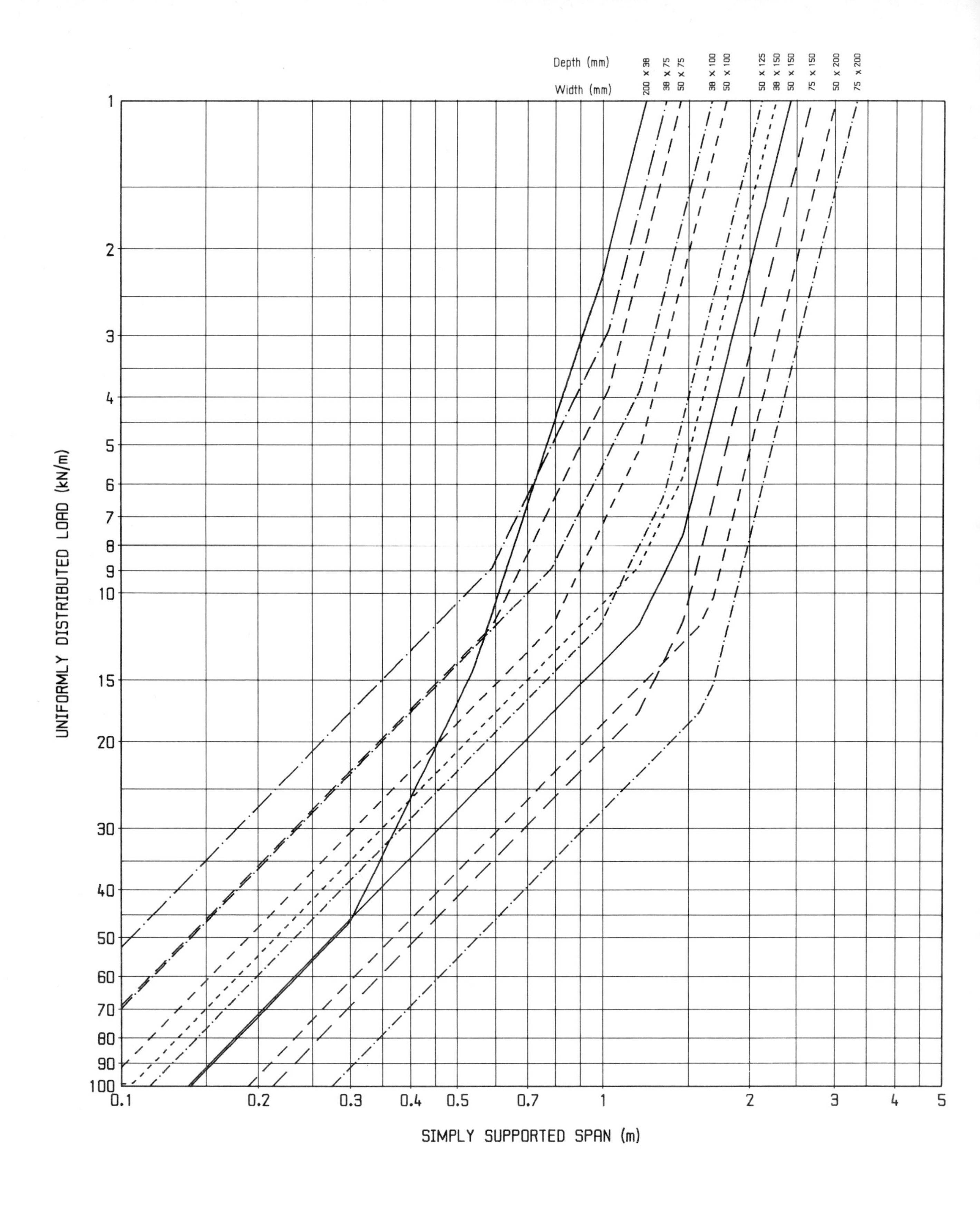

Figure 6.6 Safe load graphs for rectangular beams of wet grade M75 hemlock (Commercial Western).

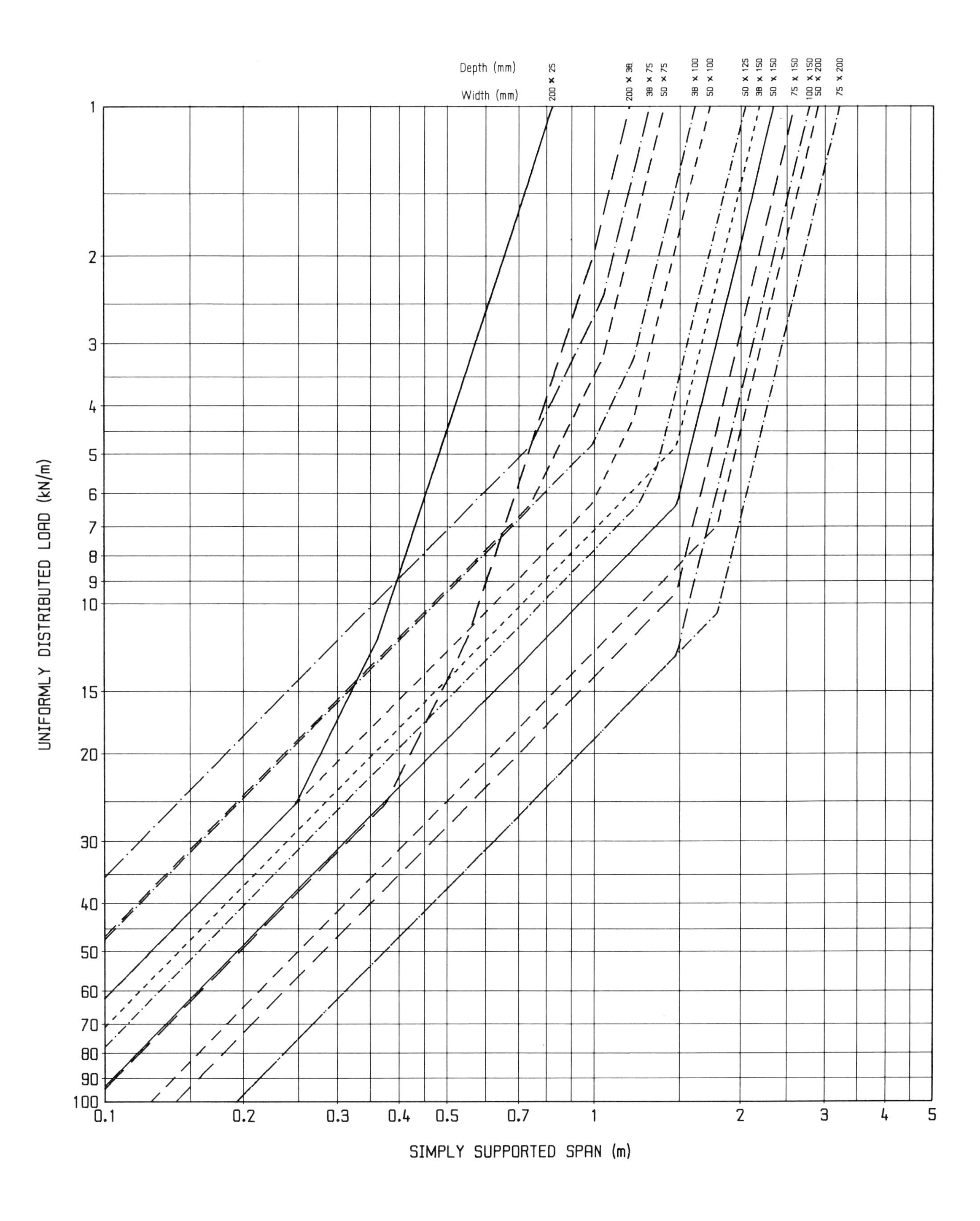

Figure 6.7 Safe load graphs for rectangular beams of wet grade SS species S1 timber.

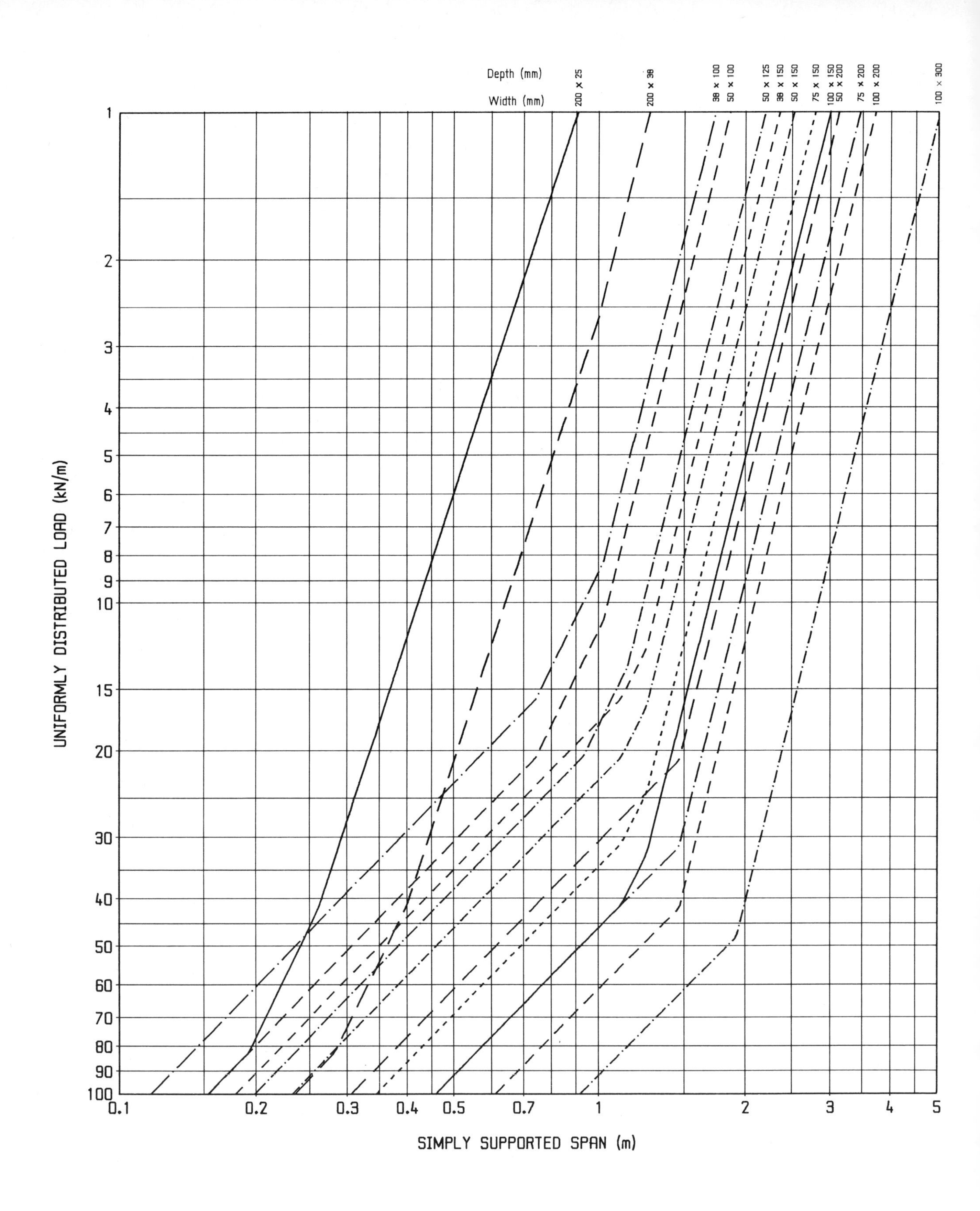

Figure 6.8 Safe load graphs for rectangular beams of wet grade 75 Gurgun/Keruing.

properties. The graphs assume simply supported rectangular beams of wet timber spanning from centre of bearing to centre of bearing. The stresses used are those given in Table 6.2 for mean modulus of elasticity, bending, shear, and average compressive strength perpendicular to the grain. Deflection is limited to 3 mm for spans greater than 1 m and 0.003 times the span for smaller spans.

The slope of the graph indicates the limiting criteria: 4 in 1 for deflection = 3 mm; 3 in 1 for deflection = 0.003 × span; 2 in 1 for bending; 1 in 1 for shear and horizontal (0 in 1) for bearing perpendicular to the grain under the load. The bearing stress at the supports should be checked separately.

Use of the graphs

1. Find the pressure on the formwork from the concrete pressure graph given in Appendix 1 and Figure 6.1, make any additions necessary for sloping formwork.
2. Decide on a plywood thickness from experience; or start with 18 mm. Using Figure 6.3 find the span; this is the stud spacing. Alternatively, boards could be used and their spans can be found using Figures 6.4, 6.5, 6.6, 6.7 or 6.8.
3. Calculate the distributed load per stud (concrete pressure (kN/m^2) × ply span (m)) and choose a span to give a convenient stud size from Figures 6.4, 6.5, 6.6, 6.7 or 6.8.
4. Calculate the load on the soldier or waling (concrete pressure (kN/m^2) × stud span (m)) and halve it to find the size when using a pair of timbers to form the member. The span chosen should generally be between 1 and $1\frac{1}{2}$ times the stud span.
5. The tie spacing will be the soldier or waling span in one direction and the stud span in the other direction. The tie load (kN) will be the concrete pressure (kN/m^2) × stud span (m) × soldier span (m).
6. Check the bearing pressures under washers and between timbers.
7. Decide upon the bracing needed to resist wind forces, to keep the formwork alignment correct and to resist striking and handling forces.
8. Repeat the above design steps using different sizes and select the most economic. The prices given are only a guide; larger sizes will obviously be cheaper per m^3 than smaller ones.

For economy in fixing and striking, it is desirable to have the minimum number of ties and therefore the largest convenient size of tie. For ties larger than 20 mm diameter the walings or soldiers will usually have to be of steel.

Example 1

Choose formwork for a vertical wall in a windless position 500 mm thick and 2 m high with a concrete density of 2400 kg/m^3, a slump of 50 mm and a placing temperature of 15°C to be filled uniformly in 1 hour.

Calculation:

From concrete pressure graph, Appendix 1, the maximum pressure is the *least* of:

(a) Hydrostatic pressure = height of pour × 24

$$= 2 \times 24$$
$$= 48\ kN/m^2$$

(b) Arching

Rate of placing = height of pour ÷ filling time

$$= 2 \div 1$$
$$= 2\ m/h$$

From the graph (Appendix 1), a rate of placing of 2 m/h crosses the thickness $d = 500$ mm line at 71 kN/m^2.

(c) Stiffening

From the graph (Appendix 1) a placing rate of 2 m/h crosses the 15°C $s = 50$ mm line at 57 kN/m^2.

The smallest of these is (a) = 48 kN/m^2.

Since this is the pressure developed by the concrete in the liquid state it increases uniformly from zero at the surface to 48 kN/m^2 at the 2 m depth. However, it is not normally economic to design for a lower pressure near the top of the form, so the design pressure is therefore 48 kN/m^2 all over.

From Figure 6.3, 18 mm ply gives a span of 300 mm for the load of 48 kN/m^2.

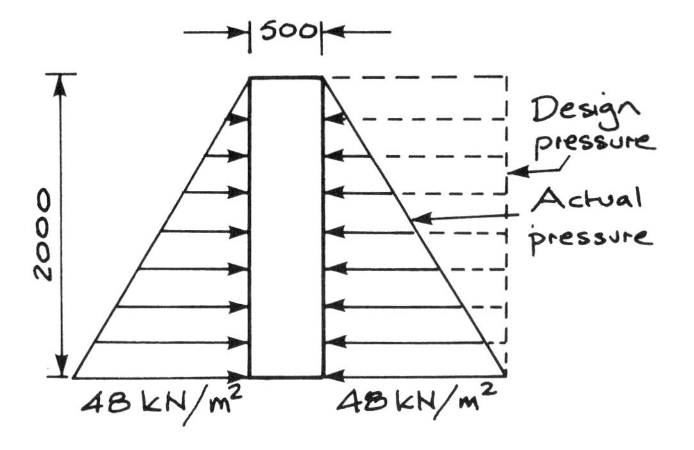

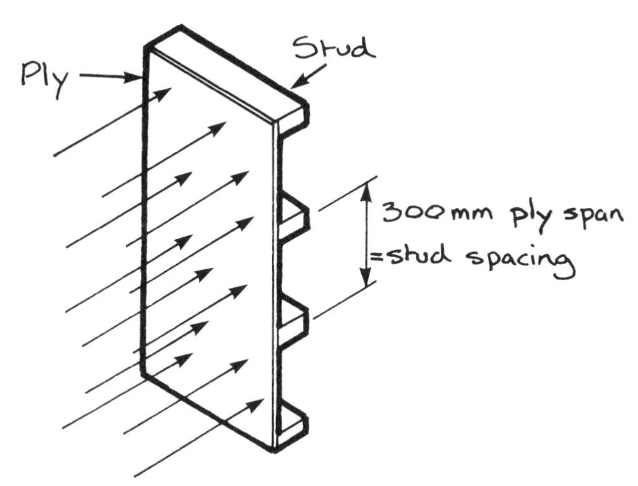

Stud load = concrete pressure × ply span
= 48 × 0.300
= 14.4 kN/m.

From Figure 6.5 choose 50 × 100 mm studs of grade SS Whitewood to span 0.430 m.

Half the soldier load = concrete pressure × stud span ÷ 2
= 48 × 0.430 ÷ 2
= 10.3 kN/m.

From Figure 6.5 choose 50 × 100 soldiers of grade SS Whitewood to span 0.600 m.

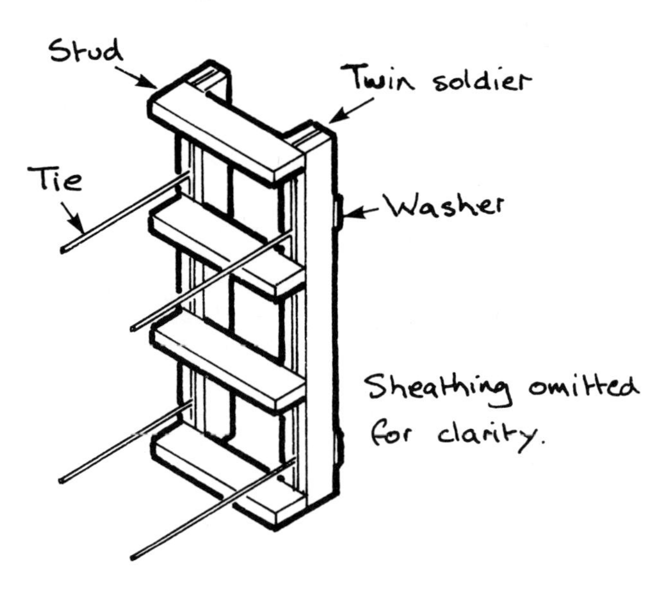

Tie load = concrete pressure × stud span × soldier span
= 48 × 0.430 × 0.600
= 12.4 kN

Bearing area required under washer

= tie load × kN to N ÷ compressive strength of timber
= 12.4 × 1000 ÷ 1.86
= 6667 mm².

Washers 100 mm square will give this area provided the soldiers are not more than 33 mm apart, and the washers are placed square to the soldiers.

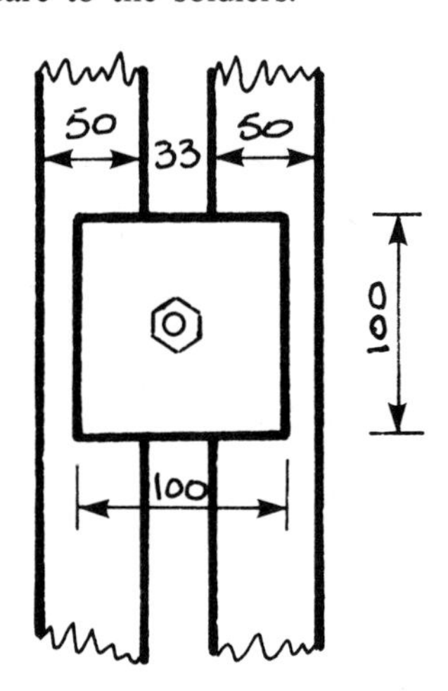

Area between washer and timber

= (100 − 33) × 100
= 67 × 100
= 6700 mm².

Compressive stress between stud and soldier

$$= \frac{\text{load on half stud span} \times \text{kN to N}}{\text{width of stud} \times \text{width of soldier}}$$

$$= \tfrac{1}{2}\,\frac{14.4 \times 0.430 \times 1000}{50 \times 50}$$

= 1.24 N/mm² which is less than the permitted 1.86 kN/m².

Example 2

Select props, runners, studs and ply sheeting for a slab 600 mm thick, 3 m high, to be placed by pump with concrete of density 2350 kg/m³.

Calculation:
From the Concrete Society Prop Selection Chart (Appendix 1), choose No. 3 props each supporting an area of 0.7 m². Assume a grid of 1 m × 0.7 m.

$$\text{Pressure} = \frac{\text{concrete density} \times \text{kg to N} \times \text{thickness}}{\text{N to kN} \times \text{mm to m}}$$

$$= \frac{2350 \times 10 \times 600}{1000 \times 1000}$$

= 14.1 kN/m²

Add 2 kN/m² live load allowance

= 16.1 kN/m².

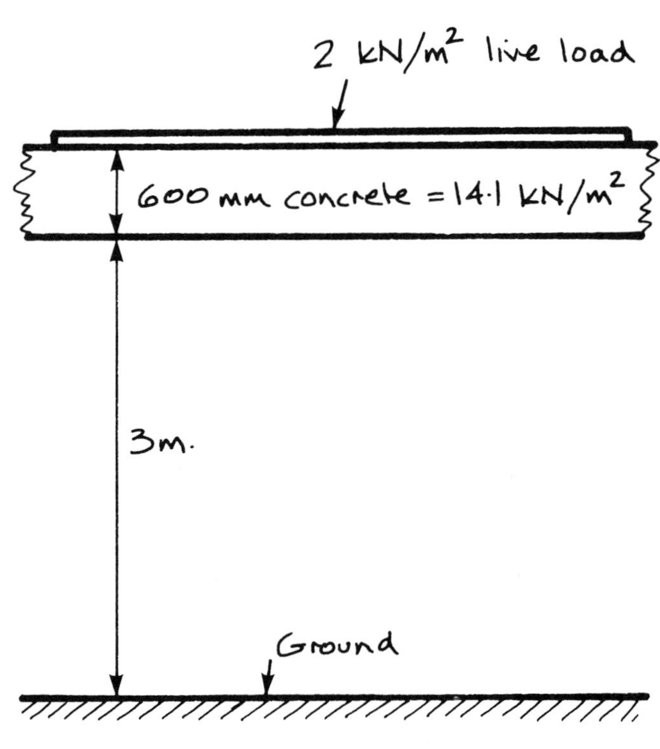

Load on main runners = pressure × runner spacing
= 16.1 × 0.7
= 11.3 kN/m

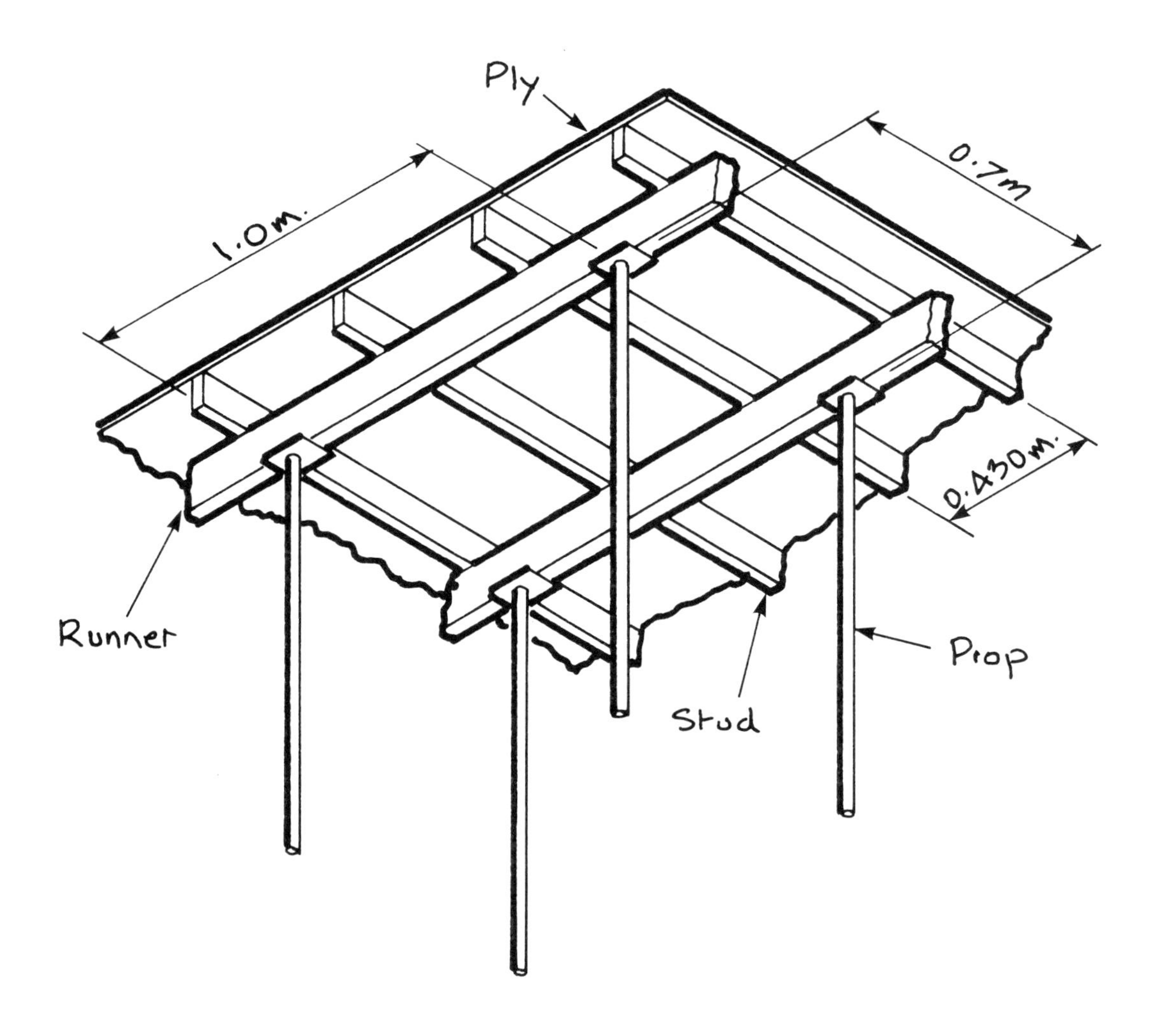

From Figure 6.8 choose 50 × 100 grade 75 Keruing to span 1.0 m.

From Figure 6.3 the load of 16.1 kN/m^2 can be taken by 18 mm ply over a span of 430 mm. (For repeated use the faces and edges should be sealed to prevent creep in the ply under wet conditions.)

Load on studs = pressure × stud spacing

$= 16.1 \times 0.430$

$= 6.9$ kN/m

From Figure 6.6 this load can be carried over the span of 0.7 m by 50 × 75 mm grade M75 Hemlock studs.

Example 3

Design formwork in timber and ply on a 2.440 × 1.220 m module for the upper and lower faces of a 7.000 m high wall, 400 mm thick, sloping 30° from the vertical at an elevation 100 m above the ground in a small Midland town. The 100 mm slump concrete has a density of 2400 kg/m^3 and a placing temperature of 20°C. The filling time is 3 hours.

Calculation:

Rate of placing = height ÷ filling time

$= 7 \div 3$

$= 2\frac{1}{3}$ m/h

From the graph (Appendix 1) the concrete pressure on the upper formwork surface is the *least* of:

(a) Hydrostatic pressure

$$= \frac{\text{concrete density} \times \text{kg to N} \times \text{height of pour}}{\text{N to kN}}$$

$$= 24 \times 7$$

$$= 168 \text{ kN/m}^2$$

(b) Arching from the graph

Pressure = 62 kN/m^2

(c) Stiffening from the graph

Pressure = 69 kN/m^2

The concrete pressure is therefore 62 kN/m^2.

Wind pressure from Figure 6.1 is 1 kN/m^2.

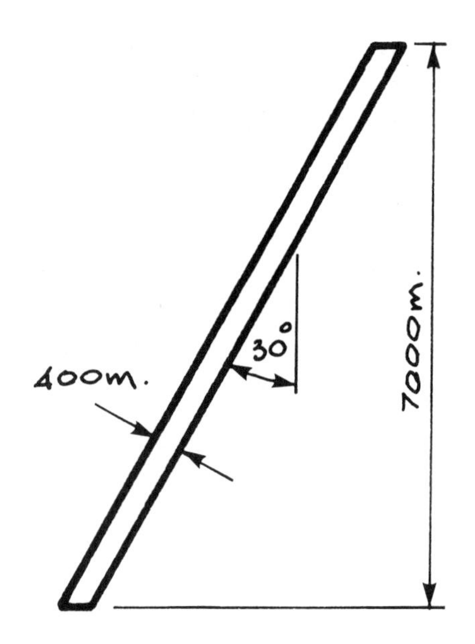

Resultant load resolved at right angles to the lower face due to the dead weight of the concrete

= Concrete density × volume on 1 m² of form × kg to N × sine angle from vertical ÷ N to kN

$$= \frac{2400 \times 1 \times 1 \times 0.400 \times 10 \times \text{sine } 30°}{1000}$$

= 4.8 say 5 kN/m²

Therefore the pressure on the lower surface

= 62 + 1 + 5

= 68 kN/m²

From Figure 6.3 this pressure is taken by 18 mm ply over a span of 271 mm, which gives 9 spans to a module of 2.440 m.

Stud load = 68 × 0.271

= 18.5 kN/m

From Figure 6.6 choose a 50 × 124 mm grade M75 Hemlock stud to span 0.610 m, giving 2 spans to the module of 1.220 m.

Half the soldier load

$= \frac{1}{2} \times 68 \times 0.610$

= 20.8 kN/m

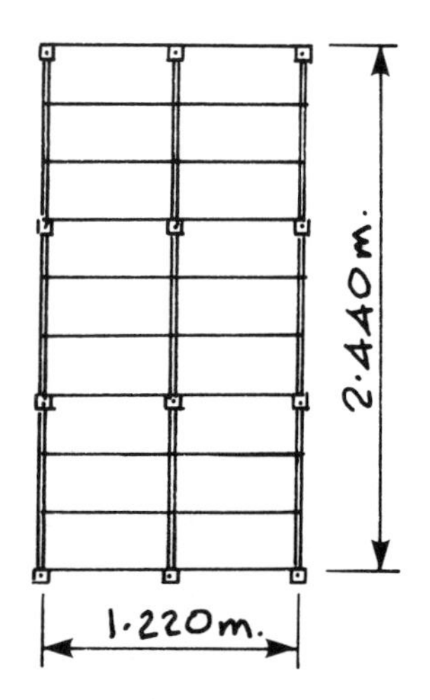

From Figure 6.8 choose two 38 × 150 mm grade 75 Keruing soldiers to span 0.850 m, say 0.813 m to give 3 spans in the 2.440 m module.

Bearing pressure between stud and soldier

$$= \frac{18.5 \times 0.610 \times 1000}{2 \times 50 \times 38}$$

= 2.97 N/mm²

The average compressive strength perpendicular to the grain from Table 6.2 is 2.60 N/mm², and this can be multiplied by a modification factor of 1.33 from Table 6.3 to give a permissible stress of 3.46 N/mm².

The tie load = concrete pressure on the upper surface × stud span × soldier span

= 62 × 0.610 × 0.813

= 30.8 kN

The bearing length under the washer to carry the tie load from the Keruing soldiers at a stress of 5.40 N/mm²

$$= \frac{30.8 \times 1000}{5.40 \times (38 + 38)}$$

= 75 mm

Therefore 100 mm square washers will be ample.

For the upper surface the span for 18 mm ply could be increased to 278 mm to carry the reduced load of 62 kN/m², but should be kept at 271 mm for simplicity of construction and to give 9 spans in the modular distance of 2.440 m.

The stud load = 62 × 0.271

= 16.8 kN/m

This has to be carried over the span of 0.610 m dictated by the tie spacing. From Figure 6.6 this could be taken by a 38 × 150 mm M75 Hemlock stud but without bracing a depth/width ratio greater than 3 is considered too unstable and the next suitable size is 50 × 125 mm which is the same as for the lower face.

Half the soldier load is $\frac{1}{2} \times 62 \times 0.610 = 18.9$ kN/m.

This still requires two 38 × 150 mm soldiers, as for the lower face.

Check self weight for ability to prevent empty forms being blown away. Weight of 1 m² of both faces

= 2 × (ply weight + stud weight + soldier weight)

$$= 2\left[(600 \times 0.018 + \frac{530 \times 0.050 \times 0.125}{0.271} + \frac{720 \times 2 \times 0.038 \times 0.150}{0.610}\right]$$

= 2(10.8 + 12.2 + 13.5)

= 73 kg/m²

The component of this weight acting at right angles to the face

$$= 73 \times \frac{10}{1000} \times \text{sine } 30°$$

$$= 0.36 \text{ kN/m}^2$$

which is not adequate to resist upward wind pressure by its own weight. The falsework supporting this formwork will therefore have to hold the form down when empty, as well as taking the dead weight of the form plus concrete plus wind in the opposite direction.

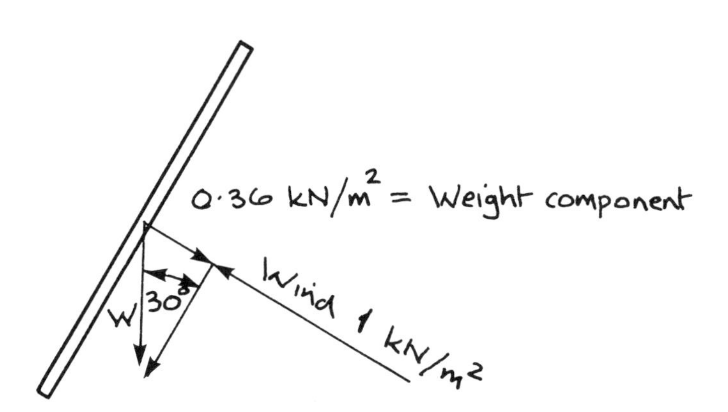

References

1. RODIN, S. Pressure of concrete on formwork. *Proceedings of the Institution of Civil Engineers.* Vol. 1. November 1952. pp. 37.
2. ACI COMMITTEE 622. Pressures on formwork. *Journal of the American Concrete Institute.* Vol. 30, No. 2. August 1958. pp. 173–190.
3. RITCHIE, A. G. B. Pressure developed by concrete on formwork. Part 1. *Civil Engineering and Public Works Review,* London. Vol. 57, July 1962, No. 672. pp. 885–888.
4. RITCHIE, A. G. B. Pressure developed by concrete on formwork. Part 2. *Civil Engineering and Public Works Review,* London. Vol. 57, August 1962, No. 673. pp. 1027–1030.
5. RITCHIE, A. G. B. Research on pressures developed by concrete on formwork. *Structural Concrete.* Vol. 1, No. 10. July–August 1963. pp 454–463.
6. RODIN, S. Pressures of concrete on formwork. *Structural Concrete.* Vol. 1, No. 10. July–August 1963. pp. 445–453.
7. PEURIFOY, R. L. *Formwork for concrete structures.* McGraw-Hill Book Co. Inc., New York, N.Y. 1964.
8. KINNEAR, R. G. et al. *The pressure of concrete on formwork.* CERA Report No. 1. April 1965. pp. 44.
9. LEVITSKY, M. Analytical determination of pressure on formwork. *Journal of the Engineering Mechanics Division.* ASCE, Vol. 100, No. EM3. Proc. Paper 9418. June 1973. pp. 551–564.
10. LEVITSKY, M. Form pressure and relaxation in formwork. *Journal of the Engineering Mechanics Division.* ASCE. Vol. 101, No. EM3. Proc. Paper 11379. June 1975. pp. 267–277.
11. CERIA Formwork loading design sheet. Construction Industry Research and Information Association.
12. LAVERS, GWENDOLINE M. Department of the Environment, Forest Products Research Bulletin. No. 50 (Second edition, Metric units). *The strength properties of timbers.* London; HMSO 1969. pp. 62.
13. BRITISH STANDARDS INSTITUTION. CP3: Chapter V : Part 2 : 1972. *Wind loads.* CP3. pp. 49.
14. BRITISH STANDARDS INSTITUTION. CP112: Part 2: 1971. *The structural use of timber,* ADM 1265 (Amendment Slip No. 1) London. September 1973. pp. 123.
15. BRITISH STANDARDS INSTITUTION. BS 373: 1957. *Testing small clear specimens of timber.* London. pp. 32.
16. THE CONCRETE SOCIETY/INSTITUTION OF STRUCTURAL ENGINEERS. Joint Report on *Formwork.* April 1977. pp. 77.
17. BRITISH STANDARDS INSTITUTION. BS 4978 : 1973. Timber grades for structural use. London. pp. 18.
18. MAYNARD, D. F. and MACLEOD, G. *Factors influencing the deflection of plywood sheeting in formwork.* CIRIA Report 37. London. Construction Industry Research and Information Association. October 1971. pp. 33.
19. BRITISH STANDARDS INSTITUTION. BS 4471 : 1969. Dimensions for softwood. London. pp. 12.

Figure 7.1 The capability of system formwork, as demonstrated by this experimental casting. Concrete has been pumped into a 600 mm × 600 mm × 10 m high standard formwork arrangement. The fill was completed in 12 min under near freezing conditions. The resulting column was satisfactory both structurally and visually.

7. Formwork equipment

General

Early in the history of formwork operations, what are now known as the 'traditional' methods of formwork construction, proved to be wasteful particularly where little repetitious use of the materials was being achieved. In the United Kingdom and the United States, companies were soon created to market patented equipment for formwork use. Arrangements were designed to fulfil a number of construction requirements and were known and sold as 'systems of formwork'. Designs now have been improved while the companies that supply the equipment have expanded such that there are more than one hundred operating in the UK alone, each with its own 'system' component or arrangement. Considerable impetus has been added to the movement towards 'systems' by European producers of sophisticated and heavily engineered arrangements of formwork, many of whom have incorporated some form of mechanical operation for erecting or striking the equipment.

Some system formwork arrangements are intended to be completely universal, the component parts being designed to be clipped and bolted together to form column forms, wall forms and slab formwork. The non-mechanical systems are constructed so that large areas of wall form and tables or tunnels can be used to form slabs or wall and slab combinations. The more sophisticated systems incorporate handling equipment, castors, jacks and slinging arrangements which allow the modular units to be crane-handled from one use to another with the least labour being expended on the dismantling and re-erection processes. Many of the companies that offer the systems also provide a design service, quite apart from selling the system or hiring it to contractors on site. So specialized has the proprietary system industry become that certain firms exist purely to sell tie arrangements or special form components.

So that the formwork designer and planning engineer can operate efficiently and be able to make intelligent selections from the range of systems and components available, they must make a careful study of formwork problems and the advantages to be gained by any particular system.

The formwork supplier

It is first necessary to understand something of the proprietary formwork supplier and the basis of his business. When this is known the formwork designer can examine the claims made for the equipment and assure himself of the quality and design of the components offered. The following should be remembered when a company and its products are being assessed:

1 The formwork supplier is a specialist and, since formwork is his business, he will naturally want his system to be successful as it is unlikely that he will have any other form of income

2 He will have invested time and money in developing the system, and knows that only the more useful systems will survive. However he must constantly design new equipment to keep abreast of the times in a rapidly changing specialized market

3 Until recently the proprietary supplier will have been probably the only person to carry out a formal research and development programme on any real scale

4 He has to maintain a staff of civil, structural and mechanical design engineers. The first two categories to deal with design of formwork for tender purposes, the latter to deal with design detail and manufacture of the equipment offered. Formwork manufacturers often employ a greater number of qualified engineers than the contractors working on a particular construction

5 He invests a considerable amount of cash in equipment which offers in terms of financial risk a relatively slow return with slim margins that pertain in a competitive market

6 He operates in a market with little or no guarantee of continued demand

7 He can only survive and prosper by maintaining a flow of work which has been won on a competitive basis as a result of considerable speculative work in design and what amounts to the supply of a 'free consultancy' to contractors.

A brief look at the operations of suppliers soon confirms these points and serves to underline the impact of scale of production on the market. There are very few small suppliers mainly because the market is such that only a limited number of firms can operate on a large scale. The smaller supplier usually concentrates on a particular component or accessory. The resources and margins required make it virtually impossible for any small firm to enter the large formwork system market. Contractors' demands are such that large quantities of equipment must be made available, at extremely short notice, almost anywhere within the country The

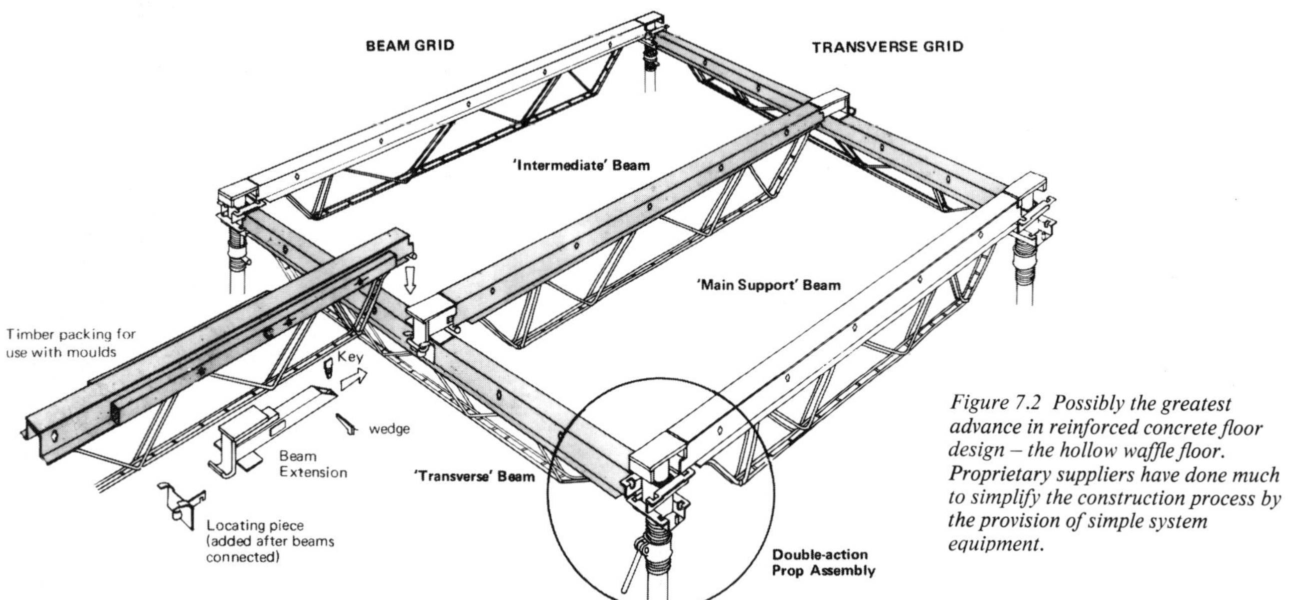

Figure 7.2 Possibly the greatest advance in reinforced concrete floor design – the hollow waffle floor. Proprietary suppliers have done much to simplify the construction process by the provision of simple system equipment.

equipment must be new or nearly new and complete, while the sale or hire price must be such as to bring the operation into the area where savings can be made over available traditional methods.

The object of this chapter is to examine how well the systems meet these requirements. In all fairness it can be stated that however well the system meets the performance specification, it is the way in which it will be applied by the contractor's site staff that ultimately governs whether the resultant concrete satisfies the contractural requirements with regard to economy and construction standards.

One major consideration that has slowed the introduction of systems into general construction has been the lack of standardization in construction. Recently with the introduction of modular philosophies and the opportunities presented by the change to metric and SI units, standard setting-out grids, modular planning and co-ordination of sizes have tended to favour the systems, and certainly in certain specialized areas, such as those concerned with waffle floors, major steps have been made in connection with the unification of sizes by close co-operation between structural designers and the proprietary supplier of formwork equipment. Previous writers have taken great pains to explain the potential economies to be achieved through standardization of sizes of items such as dimensions of fillets, incremental increases in beam depths and column faces. These recommendations, combined with the British Standard recommendations and the influence of the manufacturers of building boards, form sheathing, ducts and service trays, have slowly introduced a discipline which tends towards the economic rationalization of design sizes for structural components. Remarkably, it has always been the architect, the client's authority, who, in his efforts to achieve the design that best suits his customers, has complicated the situation and done most to upset the programme towards standardization.

The supplier, then, has always been presented with the problem of assessing the modular arrangement which will be most useful in the majority of cases, rather than being able to select and manufacture components ideally suited to a carefully calculated preferred dimension system. During the early days the proprietary market saw the offer of adjustable components which relied on overlapping plates or adjustable screws, in an attempt to achieve a total solution to variations in the design sizes of columns and beams. These components, for example, often varied by increments of inches, and even half inches on succeeding floors. While some of these arrangements are used today, they are generally precluded by the functional expression of the visual concrete and the specifications with regard to the surface finish.

The early systems depended entirely on steel sheet as a facing material, although at present while steel panel systems are available for civil engineering works, the greater number of panels incorporate plastics laminates on ply sheathing due to the simplicity of fixing sub-assemblies – timber formers, blocks and fixings for connections and such like. It may well be, however, that the cost of materials may once more preclude the use of timber-derived sheathing and the industry may see a return to the steel-faced panel. The rise in the cost of raw materials may also prevent the use of temporary formwork, and suppliers may well turn to the manufacture of permanent formwork arrangements using prestressed concrete plank and panels which incorporate fibre reinforced cements.

At the time of writing such developments as fibre reinforced products only offer marginal savings and it is likely that the formwork designer and planner will for some years to come be concerned with proprietary systems and their applications.

Factors determining the use of system formwork

Having considered the supplier and his interests, the formwork designer will focus his attention on the advantages that can be achieved through the adoption of a system, these are:

1. Assistance in the design of a method of construction
2. Availability of components designed to meet special requirements of construction, particularly heavy construction
3. Immediate access to quantities of materials of a good standard
4. Known re-use factors
5. Disciplines on ancillary materials usage, with a calculable waste factor for make-ups and infill
6. Predictable output for a given circumstance that simplifies programming
7. Reduction in the number of skilled operatives required for a given structure
8. Some assurance regarding safety, inbuilt access and such like
9. A known degree of attainable accuracy and surface finish which can be gauged against specification
10. The possibility of replacement when required standards fall off.

A rider can be added to these factors in that certain considerations will arise from policy decisions as to whether the equipment will be hired or purchased. These decisions determine whether the equipment is to be stored, cleaned and maintained between uses on succeeding contracts. Further considerations will apply with regard to cash flow and investment where contractors who purchase equipment must suffer considerable delays before returns are achieved on the capital involved.

While it is not unusual for formwork purchases to be costed against a particular contract, factors of finance, maintenance and storage usually require that a contractor be able to achieve 40 to 50 uses on a particular contract, with a forecast of a considerable number of uses on succeeding contracts to justify the outright purchase of proprietary equipment. This applies especially where there are increases in maintenance costs and rents for storage. Hired equipment can be ordered on to site for a precise date and put off hire immediately it has been released from the structure. Indeed one of the skills in using hired equipment is the controlled time of hire – extended hire, which is often the result of poor control of construction techniques, drastically affects the margins achieved by the contractor. There are many cases where equipment hired for a specific part of a contract remains on hire for many weeks after the completion of the work as a result of some failure to appreciate the urgency of releasing it for some other job, and often because those responsible for materials control do not put it 'off hire' as soon as it has served its purpose.

The formwork designer needs to constantly remind himself, and others concerned with planning and method engineering, that no system offers the best or complete answer to all problems in every situation. The designer of commercial equipment is constantly concerned with achieving a sound balance between function and adaptability, and the more responsible formwork suppliers will advise against the use of their equipment in certain situations where a competitor can achieve better results. The designer is, however, warned against the problems that are likely to arise from the random use of *parts* of systems. Often the components of a system are complementary and any one part may be less effective when used in isolation, or in combination with parts from some other supplier's system.

The diversity of the available range of components and equipment precludes the mention of any specific manufacturer in this chapter, so that the author has been confined to enumerating the factors to be considered when the use of systems is reviewed. The following detailed considerations arise with regard to specific components and it is intended to remind the designer and planner of the key points that must be observed when suitable equipment for various types of work are being selected.

Figure 7.3 Proprietary equipment. Shoring installed as permanent support for a heavy reinforced concrete slab.

The formwork panel

Generally, the panels used in sheathing applications are modular and can be either steel-faced or faced with a ply sheet. Ply sheets are often faced with a plastic laminate that gives a smooth impermeable face to the concrete. They are designed to be manhandled between uses, while attachment to adjacent panels and such soldier or waling members as are necessary are made by various types of clip. The sizes of the panel normally allow sheathing arrangements to be fabricated to within 50 mm of the required configuration. Some manufacturers describe their panels as 'universal' since they are suitable for sheathing to columns, walls or slabs. In some systems slabbing, or horizontal uses, require additional beam members to be incorporated for use in the sheathing plane.

Proprietary panels provide excellent means of forming plane surfaces, while certain systems also cater for curved walling. The supplier will, as part of his service, provide a detail which shows the best arrangement of sheathing and, where applicable, the drawing will indicate how the make up or infill is to be fabricated from available angle members and auxiliary sheathing material. The supplier must be considered an expert on his own product and his draughtsmen are trained to indicate the simplest method of sheathing an area using their particular equipment.

It is essential, where any degree of visual concrete is to be achieved, that the contractor advise the supplier of the intended specification. The lines and marks that become imprinted from panel joints onto the surface of the concrete are evident once the panels are struck and it is important that these marks be sensibly related to the wall, column or slab areas formed. The contractor must discuss with the architect or engineer these aspects of finish quite early on in the contract, so that samples can be prepared and agreed.

The imprints need not be too unsightly, particularly where foam strip or tape sealers have been applied, although it must be emphasized that it is impossible to eliminate the joint lines other than by a complete lining of some semi-rigid material. Joint lines and prints are particularly noticeable where form stiffeners, or the faces of primary beams, pass between adjacent sheathing panels and comprise a part of the overall sheathing surface to the concrete. Similarly, where 'quickstrip' arrangements are used, the head of the standing support or the support beam becomes a part of the sheathing face, and it is almost impossible to mask the differences in plane and line of the sheathing face locally at intersections, or where the head passes through the general plane of the sheathing.

Many authorities accept panel marking on concrete in public areas such as car parks and subways, especially if they are invited to advise on the layout of the panel joints related to beams, columns and returns in walls. What they are reluctant to accept, however, are black stains or honeycombing where grout has been allowed to leak from poorly constructed joint positions.

The establishment of standards has been discussed elsewhere in this book, but the importance of early establishment and confirmation of known standards cannot be over-stressed, as failure to achieve them and to provide approved samples will result in problems occurring

n formwork

products - from 70 local depots

B Metriform
e leading metric formwork system for wall and fits. Gives a first class finish.

SGB Marketing Division,
Willow Lane, Mitcham, Surrey CR4 4TQ.
Telephone 01-648 3400

FixScaf -in support work

Bovis Civil Engineering Ltd., main contractors for the Stormy Down-Groes section of the M4 Motorway for the West Glamorgan County Council have certainly met and overcome many problems concerning the construction of the Kenfig Viaduct. Indeed, it was decided to construct the Viaduct mainly because of the ground conditions at this point. Bovis were confronted with soft clay, gravel and poor alluvium below the route of the section. It would have been necessary to have stripped up to 6M of unsuitable material in order to have placed an embankment, so the Viaduct was the logical choice.

Time as always was of utmost importance, and when C. Evans and Sons Ltd., (part of the Thos. W. Ward Group) approached Bovis stating that they could save considerable time and labour in the construction of the pier crosshead beams, Bovis were naturally interested. Evans' design section duly prepared a scheme for falswork to the crosshead beams employing the Fixscaf Shoring and Scaffolding system, and immediately the first potential benefits became apparent when it was seen that, compared to the existing faslework design, there was a reduction in component parts of 50%. The design and calculations were, of course, carefully vetted by West Glamorgan's Engineers and those of Bovis. On approval of the scheme by both parties, the possible savings in time and labour were then carefully evaluated.

Halving the number of components to be transported, craned, erected and dismantled was obviously a good thing. (In practice labour savings in erection and dismantling alone have proved to be in the region of 22%). But how had this been achieved?

First of all, Fixscaf Heavy Duty vertical members were used to support the pier crossheads which allowed the propping grid size to be opened up from ·615M x ·615M to a maximum of 1·538M x ·923M. Normal duty components were employed in providing an integral access scaffold. The heavy duty components (60kN) are manufactured from 60 mm dia. steel tubing,

FixScaf in its supporting role on the Kenfig Viaduct

whilst the normal duty (40kN) are in 48 mm tube. All horizontal members are standard Fixscaf components, and because of the unique captive fixing method (no loose fittings), the expected "creep" of 12 mm when connecting two different diameters of tube does not exist. The much larger grid size also provided better access for Bovis' personnel working within the falsework structure, a factor much appreciated by them.

One of Bovis' particular requirements concerning the falsework design was that the thousands of feet of scaffold tube and the accompanying fittings should be eliminated when considering the diagonal bracing. When it is considered that the average height to the undersides of the crossheads is 14·3M, that the crossheads are generally 2·3M deep and 1·8M wide in section and that horizontal wind loadings of 105 m.p.h. had to be absorbed by the falsework structure within the confines of the pile-caps, then some idea of the diagonal bracing requirements can be imagined. Due to the inherent rigidity of the Fixscaf vertical members, it was not necessary to transmit all the horizontal forces through the node points, so accordingly all the diagonal bracing was achieved by using standard Fixscaf system braces. The only tube and fittings used were on the inevitable tying-in around the pier columns.

Several factors also contributed to more speedy levelling both at fork-head and base levels. At the base, quick-action adjustable jacks were used both on normal and heavy duty members, where it is usually necessary to employ solid threaded jacks. The need to wind up and down up to 500 mm of solid thread was thus eliminated. The telescopic puncheon unit at the top of the structure provided up to 1M of "pin and hole" adjustment in 250 mm increments. This meant that the adjustable fork heads were never extended beyond the 300 mm maximum recommended in the Bragg Report before tying together of forkheads becomes necessary. All tube and fittings were thus eliminated at the top of the structure as well as in bracing.

For West Glamorgan County Council:

W. J. Ward, F.I.C.E., F.I.Mun.E., F.I.Struct.E., F.I.H.E., F.R.T.P.I.
(County Engineer & Surveyor)

W. A. J. Sketch, B.Sc., M.I.C.E., M.I.Mun.E., F.I.H.E., F.G.S.
(Assistant County Engineer & Surveyor)

F. W. Williams, B.Sc., C.Eng., M.I.C.E.
(Chief Resident Engineer)

For Bovis Civil Engineering Limited:

P. G. M. Bigby, C.Eng., M.I.C.E.
(Project Manager)

K. Paddock
(Structures Agent, East)

throughout the whole contract. Each concrete panel would then become the subject of contention from the moment of stripping until the completion of the job. As mentioned earlier it is almost impossible to mask such discrepancies; at the best there will be colour differences between the original concrete and any remedial paste or concrete used for repair purposes, these differences becoming much more apparent with time.

Provided care is taken with the panel faces and edges, and provided the joints are well made using appropriate clips or wedging arrangements, there is no reason why a perfectly adequate visual face should not be achieved. Where the concrete is to be clad or sheathed, the speed at which the areas can be formed, using proprietary plates, will far exceed outputs achieved by any other means, apart from where special forms such as angle form or table forms, have been employed.

One advantage of the proprietary panels is that they can be used individually, as manhandled elements, or they may be combined using soldier or waling members to provide large areas of sheathing which are suitable for crane handling between uses.

Panels can be used with any of a variety of tie arrangements. While steel panels dictate that ties should be inserted at joints, ties can where necessary be passed through the sheathing of the ply-faced panels at positions that are determined by the configuration of the concrete being cast.

It is in a contractor's own interest to minimize the amount of drilling and fixing carried out on the panel face. While studs and fillers can be used to make-good through holes a small imprint will be left on the face. When panels are hired, charges are levied according to the size of holes or openings cut in the face of the panel during the time of hire.

To achieve the greatest number of uses from panel forms, care must be taken between uses to clean the faces and apply the appropriate release agent. It is essential that the corners of panels should not be allowed to score the face sheathing of other panels. Bad stacking and handling can result in the faces becoming so badly damaged that they cannot be used again. Where large frames of panels are assembled for crane handling they should be stored flat or in racks to avoid the joints becoming wracked or panels distorted. In framing up such panels it is usual for the waling members to be applied to the head of the panel to provide stiffness and maintain line, the lower edge being sufficiently flexible to conform to the line of the previous lift or kicker. The fabricated panel should thus be stored without undue strain being applied to this lower edge.

Individual form panels framed onto proprietary soldier members offer support such that the panels can stand on a flat surface between uses, although care should be taken to ensure the panels do not blow over – particularly from the upper floors or a multi-storey structure.

Flexible panels are available and consist of simple steel sheets framed onto vertical ribs which are perforated for connecting clips. These panels are used in conjunction with shaped scaffold tube to allow the formation of special profiles. Tubular walings are used for the casting of walling which is circular or geometrical on plan, the use of such flexible forms offering what is probably the most economic method of circular wall casting. They can also be used where there are variations in radius, or changes in the line of a wall within its length.

The results that can be obtained from sheathing panels depend to a large extent on the conditions of the panels, and to ensure that standards are maintained, the materials as delivered on site should be carefully inspected. Damaged faces, distorted edges or angle framing can cause discrepancies, and the supplier should be advised about special requirements with regard to surface finish at the time of ordering in equipment.

Column clamps

Column clamps are useful for tying column forms. The normal configuration is that of a set of individual steel bars or tees, so slotted and hooked, as to allow for adjustments for a variety of column sizes. They can be obtained in sets which allow them to be used on columns up to 1200 mm square, while their simplicity makes them invaluable in all small frame construction, and they are readily understood by site staff.

For the larger sizes, clamps must be used carefully to avoid failure caused by the buckling of the flat metal members. When in use the clamps are usually lodged either onto cleats, or on temporary supports formed by nails driven into a form. The wedges are inserted and driven into the slots until the whole assembly is sufficiently tight enough to resist grout leakage.

Sometimes formwork carpenters so orientate the individual sets of clamps as to ensure that the arrangement of the form sides is not forced out of square, as can happen when each set being assembled with the projecting ends point the same way.

Occasionally column clamps are used to support beam sides, but this is not to be recommended other than where smaller beams or casings to rolled steel joists are involved because, locally, this sets up excessive bending stresses within the assembly.

The adjustable steel prop

Perhaps the most versatile component available from the proprietary supplier of formwork is the adjustable steel prop. It can be used to support and brace formwork in a number of ways. Originally the prop was developed from the practice and use of square timber puncheons. Clips or clamps were introduced to allow the lapping of standard timbers to achieve variations in height and thus overcome what had up until then been the problem of cutting-waste between uses at various heights. The lapping arrangement being superceded by concentric tubes arranged to slide and adjust.

Steel adjustable props (telescopic props being a deprecated term according to BS 4340 *A Glossary of Formwork Terms*) are now available in standard sizes calculated to cater for storey heights of between 2 and 6.5 m. The range of sizes and capacities overlap, and the designer thus has the option of ordering one of two sizes which are capable of coping with various storey heights within the accepted range. This needs to be studied to ensure that the prop selected is capable of being used through the heights for which it is intended on a particular contract. The selection of props is made simple for the designer and planning engineer because the relevant information has been issued in a chart (see Appendix 1) which sets down the safe loads for different props at various heights.

Research has provided some startling facts which have caused considerable consternation among designers and those responsible for formwork construction. It has been proved that proprietary adjustable props where loaded eccentrically by some 40 mm, or placed 1° out of plumb (a situation which has occurred on many contracts) are capable of sustaining only 60% of the load that can safely be applied axially to an upright prop. It should be remembered that eccentric and out-of-plumb propping often occurs on many sites, and situations can arise where the props have to be specially set on an incline to allow for height adjustment having to be made adjacent to a cross wall. Also, raking props are often used beneath balconies, under stair soffits or where gradients and ramps are being formed. As a result of recent research and the publication of the Concrete Society Prop Chart mentioned earlier designers are now looking more closely at details, while construction staff are gradually becoming aware of the importance of well designed and carefully executed support arrangements. In spite of the emphasis placed on these matters, risks are still being taken, particularly where props are poorly founded or where packs of materials are used to overcome minor discrepancies of support heights which are achieved by props being located in various situations. Props are still used where, as a result of loading, bending forces are applied to the top or bottom plates. This means that the form arrangements are not concentrically loaded upon the prop tube. To overcome this, fork-head reveal pins or system props can transfer the load from the runner or joist. Wedges inserted between the cheek of the fork and the member being supported rotate the fork until the timber member is forced into the axial position above the tube. Special wide forks are used where the supported members need to be lapped to avoid cutting waste.

All prop systems require lacing while large areas have to be diagonally braced. Lacing at the centre or head of the prop can be effected by using normal scaffold couplers. Diagonal bracing requires the introduction of special couplers that are capable of being applied both to the standard tube diameter of the inner member of the prop, and the oversize tube used as the outer member. Where props are laced or diagonally braced, projecting scaffold should be in contact with floors or previously cast concrete walling, although all bracing of this kind can only be tightened when the final adjustment of the prop to the correct level has been carried out.

It is essential that the pre-concreting check should cover the plumb and line of props, it being sufficient to plumb just one prop while the others can be aligned by eye. The check should include a provision for an inspection to ensure that there are not blocks or packers at individual prop feet. Continuous plates are of course desirable when there is any doubt about the bearing capacity of the ground.

For multi-storey work, props should be laid to a pre-determined pattern based on the load. The systematic arrangement should help with the regulation of plumb and also provide ease in checking the spacing. The pattern should be repeated on succeeding floors as this ensures that continuity of support is achieved in such standing supports as are incorporated in the design.

Much has been written with regard to the desirability or otherwise of re-propping after striking, and the major consensus of opinion is against the complete removal of the supporting system and the re-introduction of supports below a relatively, green concrete slab. This arrangement must not be confused with the necessary re-introduction of supports where some exceptional load is to be applied during subsequent construction processes on succeeding levels.

The urgency of such measures cannot be over-emphasized and indeed where, for example, some special equipment is being used for the location or support of precast cladding components, the design of the floor slab should be checked to assess whether such propping is necessary, particularly where a fast cycle of floor construction is to be achieved.

It is generally satisfactory to insert an additional row, or rows, of props under some part of the sheathing that remains securely in place while the general removal of props and sheathing is carried out. These standing supports sustain the floor during the curing period and cater both for construction loads and the dead load of the floor, as well as controlling deflection during the hardening process. Unfortunately, again as a result of fast construction cycles, these loads can be excessive. Their transmission through previous floors, combined with the dead weight of previously constructed floors which, because of the standing support system do not contribute to the load-bearing system, can result in excessive stresses being set up in the supports to the lower floors. This is an engineering consideration and the advice of a competent engineer should be sought on some system of 'easing of supports' to allow the previously constructed elements to make a contribution towards the load-bearing system.

The use of 'quick strip' arrangements built into the prop effect considerable savings in the amount of form materials used in soffit sheathing. Quick strip systems provide for the continuous support of a floor slab during the striking operation and prevent sudden deflections of the green concrete. The higher of the two 'heads', in the simplest arrangement, remains in contact with the concrete while the lower support can be dropped to free the remaining sheathing which it has supported during concreting and the initial stages of curing. The slab is isolated and while at say, three days, 2/3 of the 28-day strength has been realized the slab may only be required to span 1/5 or 1/6 of the distance for which it is designed. This idea is quite convincing and provided that the loads from the construction process above are considered, formwork can often be removed from the largest and most expensive areas, such as slab soffits, within three days of casting. The standing supports must be formally set out and should coincide on succeeding floors to ensure suitable distribution of loading. As with the previously mentioned standing support arrangements, progressive easing of supports needs to be considered where fast cycles are being achieved.

Over the years various methods have been employed to make the pinned joint at the point of adjustment, so that whatever the design only the correct pins supplied with the prop should be used. If a check is not made that the correct pins have been inserted then failure will occur. It has been known for site operatives to use the first piece of steel that comes to hand, perhaps a piece of reinforcement or even heavy gauge nails, and this has had dire consequences.

Props should always be spiked into position. This prevents the props from falling sideways when a progressive cast is carried out, or where bearers tend to lift in one bay as they deflect in the previous bay under load.

The prop most likely to cause trouble is the one that has become bent or distorted due to some mis-use on a previous operation. Props can be distorted by the misapplication of jacks, by being overloaded when used as puncheons between kickers, when used for push-pull arrangements in bracing formwork, or where any lateral load has been applied while the prop is already under an axial load. Apart from the initial difficulty of adjusting a bent or distorted prop, it may well go unnoticed, which means that a substandard component could become incorporated into the system. On one particular site some strained props were inserted into a system and, because of distortion, they remained extended without the insertion of a pin at the point of adjustment. Fortunately in this particular case the absence of the pin was noted during the pre-concreting check, and a disaster was thus avoided. The use of inner members of incorrect lengths can contribute towards bending and distortion. With care and regular maintenance, however, the prop provides unlimited re-use, and gives a thoroughly predictable and versatile piece of form equipment.

The reader is advised to consult the relevant sections of the Concrete Society/Institution of Structural Engineers Joint Report, *Falsework* (Published in 1972), which contains a lot of useful information on the use of props and scaffolding in support systems both for falsework and formwork.

Beam clamps

Some manufacturers market a steel beam clamp, and in some cases it is so designed as to provide a bearer member which can take direct support from the props of the supporting systems. It is particularly useful when concreting protective casings onto steel frames, as no through tie is required. With the increasing costs of alternative materials, the adjustable steel beam clamp can also prove to be economical for frame concrete construction, since it eliminates unnecessary cutting of timber. It can be used satisfactorily with either steel or timber form panels.

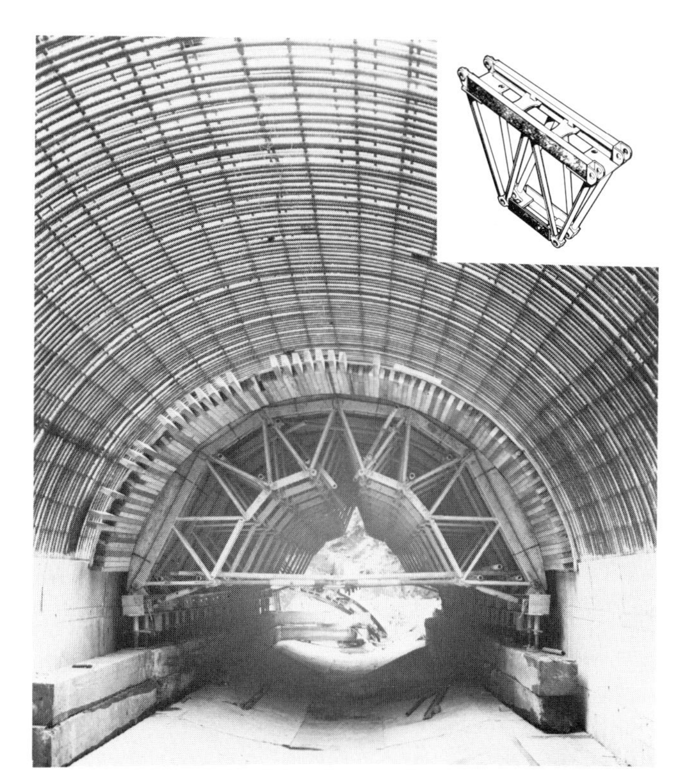

Figure 7.4 Centring equipment which can cater for heavy loads and a range of soffit configurations.

The centring girder

The telescopic centre, which was originally used in the casting of simple slabs supported on brick-bearing walls, now constitutes a major component of falsework and temporary supports. The early versions were of box construction, but lately a combined box and lattice-type construction has been adopted in general, as this facilitates the establishment of a system of modular components which are capable of being combined into both straight and arched girders of varying lengths and geometrical form. The modular arrangement of the larger component system caters for end-bearing arrangements where RSJs are used as supports, where concrete walls are pocketed at the abutment to provide support, or where brick-bearing walls support a girder.

Large lattice-type girders usually incorporate a turnbuckle arrangement which allows the profile to be adjusted to impart camber into a concrete slab, or assembly to profile of arched girders.

The heavy systems allow double-banking on the 'Bailey Bridge' system, the resulting double depth girder providing for the formation of supports to spans of 30 m in normal practice. For special cases or where greater spans are contemplated the suppliers can always be consulted.

Where exceptionally long spans are to be formed, manufacturers introduce additional tie frames which, of necessity, encroach on the headroom of the opening, a matter of concern in many bridging applications. As the heavier systems have been introduced, the manuals which describe their assembly and use are extremely informative, well illustrated and thus provide a complete guide not only for the user but also for the engineering designer who is concerned with formwork or falsework applications.

The systems can be simply assembled and erected onto

the bearing arrangement, while levelling jacks are used to ensure accurate installation. Considerable care must be exercised in planning the striking and removal of the girders once the concrete has matured. The suppliers of centres make available a bearing bracket, a hanger bracket and shoes which help with the establishment of supports that are level and capable of transmitting the reactions from the centres into the structure.

The simplest means of removal is that in which the girders, either individually or in framed sets, are lowered to the ground by tackle that passes through openings in the newly formed deck or slab. Alternatively, anchors can be cast into the soffit of the newly constructed slab and the form lowered on bars which are screwed into these sockets. Where the weight of the formwork is to be imposed on the slab, in this way, the structural engineer must check to ensure that the relatively 'green' concrete is not overstressed. It is essential in all applications that no 'temporary' intermediate support is applied to the lower chord of the support as this can completely upset the load-bearing characteristics of the system and so cause failure.

Where girders are to be removed from beneath a newly cast deck, a simple system of runners extending laterally beyond the edge of the deck can be used to facilitate crane handling. The girders are lowered 50 mm or so from their casting situation, the sheathing arrangements stripped out and the girders then slid laterally or winched along these rails until they clear the slab edge and can be removed by crane using long 'brother' slings.

At the lower end of the span range, system suppliers provide adjustable floor centres which range in span from a little over 1 m to the mid-range of 8 to 10 m. These units are fully adjustable over the intermediate range by simply sliding one component part within the other and checking against a template, rod or tape. They are used in conjunction with supporting systems to provide a variety of soffit and beam form solutions. At this end of the scale it is quite usual to support the centre members from plates or steel sections bolted to sockets cast into previously cast concrete walls or beams. Where the end lugs bear on previously cast concrete – facilitated by the formation of pockets in the concrete – the pockets should be packed with expanded polystyrene or paper for easy removal once the concreting operations have been completed.

The spacing of all centres must be determined from the manufacturer's tables. A carefully designed sheathing system which incorporates joist supports can ensure the most economic use of the centres for given spans and thicknesses of concrete slab. The use of the larger girders and telescopic centres are particularly advantageous where clear spans have to be formed because of some structural design or location, where access must be maintained below formwork for concurrent operations and where only isolated support towers are to be established.

Ancillary items

There are some small ancillary items of hardware which can help the formwork tradesman to deal with the problems that arise from construction and form re-use. Small plates can be hammered into the faces of ply forms to cover redundant tie holes while simple plastic studs similar to batchelor buttons can be used for the same purpose.

Sealing tape which will adhere to form faces even in the event of residual oil or parting agent from previous uses can help to control surface defects. Similar tape which carries a layer of closed cell foam can be used for joint sealing between panels, or even for individual boards of high quality construction.

Some of the hardware, which at first sight might appear to meet an invented need, can prove valuable in solving what are quite real problems on site. As an example, the small bracket or cranked threaded rod allows steel or timber waling members to be attached to the soldier members of a formwork panel, normally a clumsy arrangement to make. But the bracket or rod saves considerable juggling with quite substantial formwork members. The fitting offers a cheap, speedy and efficient connection which can save hours of site handling.

Ever since formwork was created, people have produced fittings with some sort of gimmicky attachment. The formwork designer needs to evaluate the various devices, and even if he rejects an idea, he should keep a record of its availability so that on some future project when conditions or requirements may be different, he can

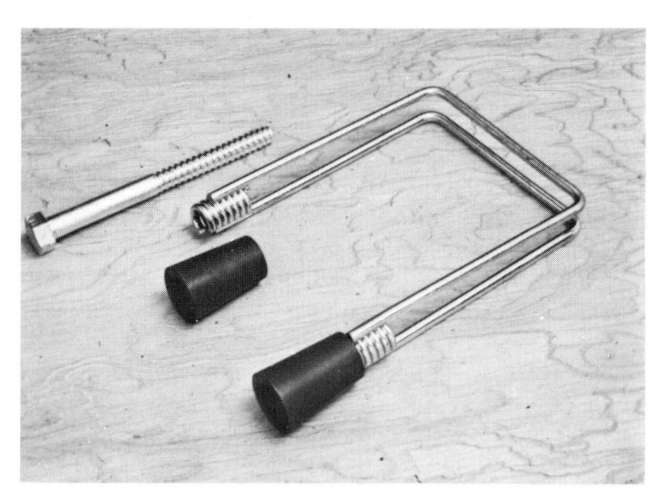

Figure 7.5 The hanger – a useful item for fixing forms which have to encase structural steelwork.

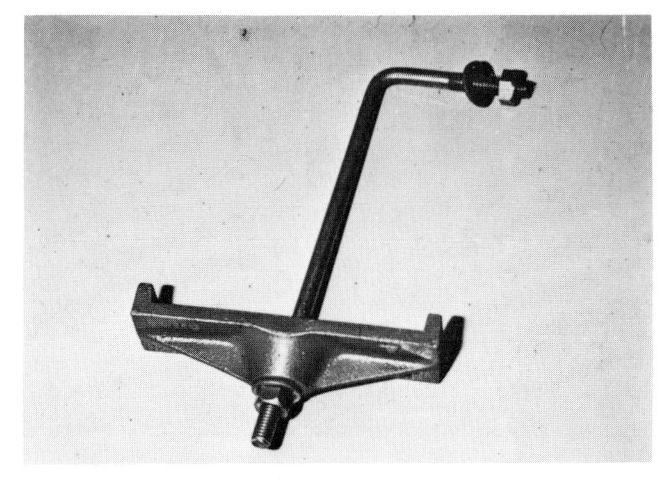

Figure 7.6 This simple but useful component allows timber walings or steel strongbacks to be attached to a form panel.

re-assess and perhaps introduce the component into his scheme.

Soldiers and strongbacks

These components fall into two main categories:

1 The channel or fabricated member, normally used as a component within the sheathing system.

2 A fabricated soldier or strongback member of considerable substance used to provide external support for sheathing arrangements.

The simple channel member is normally slotted to conform to the fastening arrangements of the system with which it is to be used. Generally it forms a part of the sheathing face which is clipped or coupled in the same plane as the standard modular sheathing panels.

A channel serves three purposes. First, it allows the overall length of the formwork to be adjusted. Channels are provided in a range of widths, and this is helpful when the designer sets out his sheathing to the concrete profile. Corner members, internal and external angles facilitate the construction of complete forms, including returns, and in some cases splayed corners. Secondly, it can be used to provide continuity of the vertical face line. The channel member, which can be obtained in lengths up to 3 m is continuous and thus stiffens the fabric of the assembled form panels as a whole.

Figure 7.7 Traditional materials used with proprietary soldiers for deep lift construction. The simple construction provides a speedy casting cycle.

Thirdly, it fulfills a further function in providing a facility for the use of standard panels in climbing formwork. Here the channels extend over more than one lift and the intermediate panels are moved upwards from one lift to another between the long channel members which remain bolted to the concrete.

To maintain vertical line the joints in the channel members at each side of the wall are staggered vertically, thus avoiding changes of line which would result from coincident joints in the channel or opposing faces. The channel members are within the same plane as the standard sheathing arrangement and thus the normal horizontal waling tube can be used to connect panels and maintain the line of the form.

Soldier and strongback members, while often forming part of the sheathing in contact with the concrete, are quite deep, and so provide a stiffening effect over the depth of one or a series of lifts. The heavy duty range of soldiers are generally long enough to allow a storey-high lift to be achieved using the minimum of through ties. The configuration of a member will vary from supplier to supplier, e.g. some reproduce the form of the bowstring girder, while others are simple lattice girders, but all will be fabricated from steel, many of them utilizing angle or channel sections in conjunction with mild steel plate and bar. Some systems utilize the face of the heavy duty soldier in sheathing the concrete and provide through holes for tie arrangements.

Strongbacks can be used with scaffold tube, rolled steel sections, or sometimes with timber waling members, and the most appropriate use is where girders, walings and sheathing panels are fabricated into large forms for crane handling between uses.

Heavy duty soldiers are extremely useful where a vertical cantilever member is used to support formwork, the tie being achieved from previously cast, lower lifts of concrete. The stresses which develop in soldiers render timber members uneconomical for such applications. Massive deflections can also occur, these having to be eliminated by the use of the appropriate steel member.

Modular super soldier units are available and comprise three basic units which are capable of combining such that 3582 mm height of pour can be achieved with only two ties in the height, the plumb being maintained by a plumbing foot. These units can be used on cantilevers which have to support lifts of some 1219 mm. For higher lifts and especially for 9 m lifts it is recommended that push-pull props are used for plumbing.

These sort of systems are ideal for the construction of cross-walls where through ties have to be kept to a minimum. The upper tie, required for normal storey heights, can pass above the concrete and can thus be recovered. Connections with either timber framed panels, proprietary panels or fabricated steel forms can be made by using clamps and bolts which are obtainable as standard components from the supplier of the soldier members. In many cases, soldiers can be used to provide a working platform for placement purposes and thus eliminate costly scaffold arrangements.

A further range includes rolled steel or fabricated steel channels fixed back to back in such a way as to allow the tie bolts that pass between them at almost any point along their common length. As with the lattice soldier members, components for plumbing scaffold walkway support and connections are available, as are clamps for attaching the soldier members to various materials that can be used as walings.

Channel-section soldiers have the advantage of reduced depths with less projection from the back of the form than heavy duty, lattice supports. This means that working space, restrictive in shafts and individual bays of cellular construction, can be maintained and the fabricated forms can be lifted within a space of some 450 mm even when used in conjunction with the heaviest timber waling member.

A recent development of this type of formwork component has undoubtedly been prompted by the dramatic increase in the cost of timber, particularly in sizes of 150 × 75 mm and the general formwork sections. This component comprises the steel joist member which is fabricated from pressed steel sections. The suppliers suggest these members be used for normal application such as soldiers for supporting timber walings. Indeed the joist members can be used horizontally in the waling situation while capping plates at the ends of the sections allow 90° connections to be made, the flying ends being staggered as with the normal column clamp. Another application is where runners or beams support the joist system in floor soffit applications. Here stripping forkhead supports are used in conjunction with a scaffold, a system support or adjustable props.

Table forms

The term 'table' originated with respect to framed arrangements of formwork that were used to form the soffits to floor and ramp slabs. The early versions were assembled from scaffold tube and fittings used in conjunction with castors and levelling arrangements. Table sizes were related to the size of the bays in cross-wall construction, to reservoir roof slab construction and other situations where considerable re-use of modular arrangements could be achieved. Basically, tables were slab forms with combined travellers. On civil engineering works cases have arisen where complete areas of formwork for reservoir roofs have been mounted on a lorry chassis for transportation between uses, although jacking arrangements were necessary to immobilize the forms while the steelfixing and concreting operations were carried out.

Table form systems were given great impetus by the increasing use of tower cranes for multi-storey, multi-cell construction although for some years tables were only used for flat slab construction where they could be simply extracted at the perimeter of the formed slab. Commercial suppliers have now introduced options, however, and it is possible with standard equipment to extract a table form from underneath a slab which has both downstand beams and a spandril upstand on the floor below which may limit the opening to about 1 m. Tables can be raised and lowered manually, although for large areas of flooring, as a means for reducing the labour required to carry out table striking, hydraulic rams are used for this operation. The rams are mobile and can be coupled to jacking points on the support system.

Considerable ingenuity has been introduced in handling table forms, the simplest arrangement being that of the 'c' hook which is so arranged as to support the table from a hanger frame in the bay under a newly cast slab while the shackle and crane hook are connected in line with the centre of gravity of the table. This arrangement allows a table form to be clearly withdrawn without the need for any intermediate re-slinging operation.

The re-slinging system, generally used for very large tables, requires careful control especially with regard to safety. It involves the movement of the table out of a structure, temporary resting or landing while being supported by the crane, and attachment or transfer of brother chains or slings to allow the continued extraction from beneath a newly cast concrete slab.

During the temporary 'landing' stage particular care is required to ensure that forces are not imposed upwards under the newly cast concrete, and that operatives do not move about on an inclined table surface in order to move slings.

In both cases the designer has to consider the strength of the attachment points for hooks or slings, and must provide the necessary fixing or bearing plates. Apart from being shown on the drawing they should be clearly marked on the table equipment itself. Tables can be badly wracked by bad lifting arrangements which includes the wrong attachment of lifting hooks. During the past few years much imported equipment which ranges from individual system form components to specialized tables, and parts of form systems has become available.

The simplest systems are those of wall forms which are used to cast cross-walls and incorporate such a system of supporting points into the concrete that roller suspended tables can be installed direct, the levels being achieved by the accurate positioning of the wall form arrangement. The tables can be lowered and rolled out from beneath the bay of freshly cast concrete by crane, the advantage being that the floor below the table panel can be completely free of all obstructions. When the cross-walls are being cast, kicker blocks are also cast into position which serve to position steel members that form the kickers and control the height of wall and thickness of the floor slab. These blocks have through holes into which are passed the tie rods for the wall panels, some small adjustment being made within the clearance hole allowed.

Tunnel and angle forms

As crane capacities in normal constructional work have increased, a tendency has developed towards the adoption of large panel angles and tunnel formwork. Crane loads should be optimized such that they are consistent with manoeuvrability. Large panels which include bracing and levelling arrangements and which are sufficiently stable to stand on the structural slab immediately on release from the crane, provide extremely economical construction methods. Naturally flexibility of the more traditional formwork arrangements is lost so that the form arrangement becomes much more a consideration of the planning engineer and even the design engineer. It may become necessary for the structural designer to consider particular aspects of the way in which the formwork loads are to be transmitted into the structure. Certainly in some forms of construction it may be necessary to incorporate additional reinforcement to facilitate the early removal of large plate-type tables where it is difficult to ensure continuity of support during the striking operation.

Angle forms and tunnel forms which are stripped in 'L' shaped sections can still be arranged for either continuous support or carefully controlled re-shoring.

Heavy trestle supports

At the end of the Second World War a large amount of bridging equipment, and numerous trestles became readily available. Among the equipment could be found Bailey panel structures, transomes and so-called military trestles. Despite the heavy cost of development over the years, they have now been modified and improved into what must be the most dependable units of heavy supports for civil engineering works. These units mainly were devised as parts of self-erecting or manually-erected systems since these ensure that when cranes and normal constructional equipment are used high rates of erection can be achieved. Because of their military purpose, the units of modular design could be used in compound or gang arrangements to the stage where quite massive loads could be supported, and where large structures and components were concerned, the reactions at the support positions became extremely massive.

Bailey units can be used as span members and for support towers and while trestles are usually adopted only for their original designed purpose of tower support, they can be utilized for shoring and bracing.

For formwork applications panel and trestle systems are mainly used but applications sometimes arise in building construction where substantial loads have to be supported over openings for access, i.e. where space is limited and plant and equipment have to be supported on gantries. On such occasions the supplier should be consulted regarding applications and erection, although for the latter, the activities mainly resolve themselves into matters of material handling and the insertion of simple bolts or pins at connections.

Steel sections used in formwork

Steel sections are sometimes used at the head of props or support towers to assist with the distribution of concrete and construction loads applied to formwork. Often the materials used are secondhand, i.e. they have been used on some previous contract or have been brought onto site from a contractor's plant yard where they may have been stored for some time. Their use particularly in situations where they have to bridge working areas, access ways or

Figure 7.8 This 29 m high L-trestle supports formwork for the in situ *casting of hollow box section bridge components in motorway construction.* (Mabey & Johnson Ltd.).

public throughfares can be extremely hazardous. It is likely that the steel members will vary in thickness, or be sub-standard. Where there is doubt, the designer will do well to inspect the materials or in the case of detached working have a survey carried out by some suitably qualified person.

Tie arrangements

Apart from the sheathing selection with regard to surface finish, the maintenance of line of face and accuracy of the form, few considerations can be more important than those of the selection of the tie arrangement.

The pressures and forces which result from the concreting operations are transmitted via the sheathing, into the carcassing or substructure and then to the tie or support system. A tie system, as well as sustaining the resulting forces, can be designed such that it helps with the location of the formwork and also provides support for ancillary equipment which may be required for subsequent operations.

Simple ties

The simplest tie arrangements incorporate bolted cleats which connect the soldier members above the concrete; wedged systems, so arranged use the soffit bearers to sustain the forces and windlassed wire ties.

Various arrangements of trussed members have been used to form clamping arrangements which clip the form panels back to the top of previous lifts of concrete. Incidentally, these avoid the need for through holes.

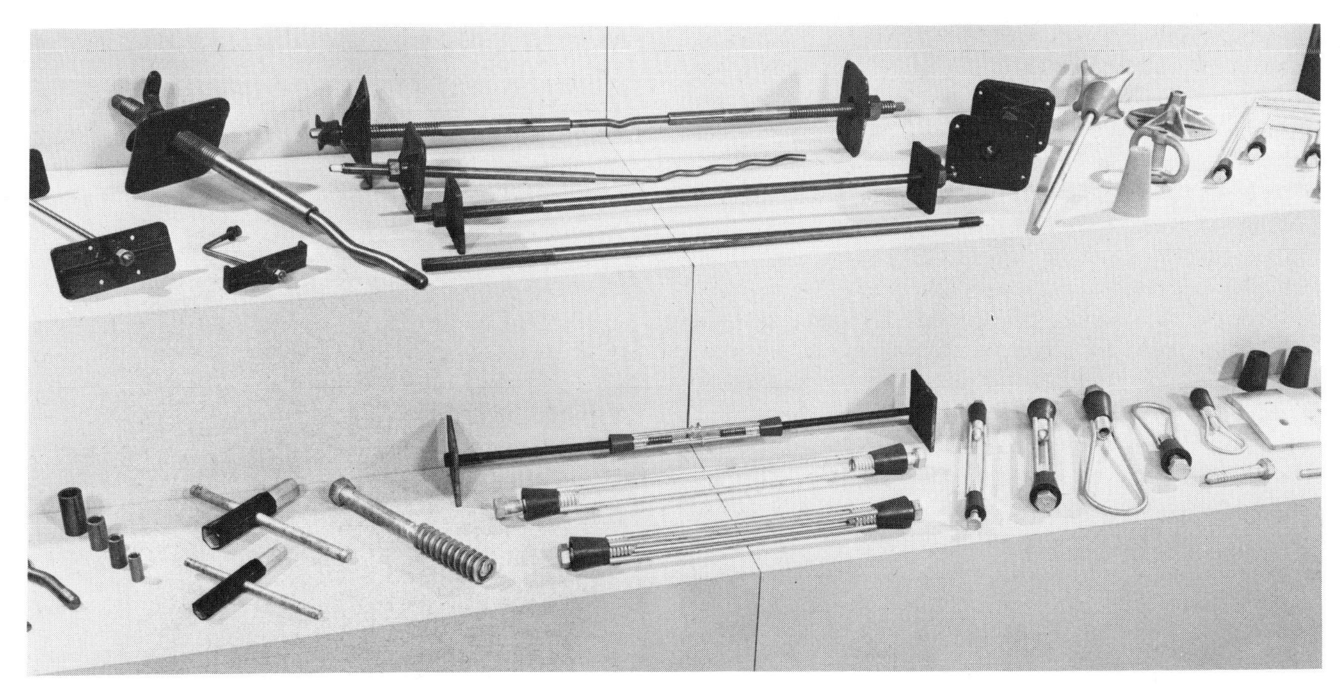

Figure 7.9 Some of the many items available for tying and fixing forms.

Many designers fight shy of through ties because of difficulties in tie removal, or as in the case of wiring of the problems associated with loss of cover at the tie position. Care taken in the selection of suitable sleeving, or some appropriate means of providing recessed tie ends, can prevent these difficulties. Removable through ties form probably the cheapest, simplest means of tying opposing form panels. Wire ties are certainly still appropriate where ground beams and general foundation concrete is concerned.

The through tie that comprises a steel rod with some form of thread and a suitable washer plate and nut provides an excellent tie for most building and civil engineering work. Square, rolled or similar threads which can be readily cleaned should be used especially as they are simple to install, economical to use and easy to maintain.

Tapered ties

Where firms specialize in water-retaining structures and with certain proprietary tie systems, the tie rod is tapered from one end to the other. This helps with the removal of the tie from freshly cast concrete and avoids the need for any form of sleeve.

The taper profile of the through hole will achieve a good fill when the holes are made good, the plugs being driven from the larger diameter into the hole before the sealing material is compacted either by rodding or mechanical means. The taper is oriented such that water pressure on filling tends to drive the plug material further into the tapering hole.

Ideally tapered bolts should be of substantial section and it is imperative that they be withdrawn either by the use of a purpose-made extractor or by a jacking action applied to the large diameter thread. Attempts to drive the tie from the end with the smaller diameter will result in failure due to bending of the bar or damage to the thread.

Threaded bar ties

Where simply threaded bars are used, they should be sized to the job to avoid using excessive amounts of packing between the carcass of the form and the washer plate. Generally about 75 mm of thread are sufficient for most adjustment purposes. Threaded bars of less than 12 mm diameter are subject to bending in normal construction and can cause difficulty and danger should they become embedded in the concrete.

Extreme care should be taken to ensure that any kind of tie is used in such a way that it is not subject to an excessive bending force induced by splay, batter or even the mis-alignment of the substructure members through which it passes. Commercially available ball joint type washers and plates can be used to avoid such problems.

Useful commercial through tie equipment includes those which incorporate crudely cast nuts suitable for the configuration of indented bars as adopted for prestressing.

These systems often are fitted with wing nuts which, because one nut is larger than the other, avoid any tendency for them to unwind as a result of vibration. Where simple tie rods are employed a wing attached to one face of a normal square nut can overcome this problem.

Where a simple tie rod becomes embedded because of a jammed or collapsed sleeve, a simple jacking arrangement, such as that used to extract nails from crates, can save considerable time and effort.

Snap ties

Some expendable ties are so fabricated that by twisting or bending, the part of the tie which projects beyond the depth of steel cover can be broken off. Certain systems incorporate a cone arrangement which forms a neat recess in the concrete face for easy plugging or dry packing.

Snap ties can be used either with individual proprietary manually handled panels of formwork or with fabricated assemblies such as cross-wall forms, that are handled by crane. For the latter case an extended tie is used to facilitate snapping prior to form removal. They are expendable and may thus be directly costed to the area cast. Their low cost makes them particularly applicable for commercial concrete in situations where surfaces are other than to the highest visual standards.

Snap ties are inserted at close centres and form arrangements for normal storey heights and thus require only a waling member at the top of the lift to maintain the required degree of accuracy. The floating bottom edge can nest closely against the preformed kicker and can adjust to any local inaccuracy.

Possibly from the designer's viewpoint the only detraction is that snap ties having performed their function of spacing and tying forms, offer no further contribution to the support of succeeding equipment or forms.

Coil ties

The coil tie is highly recommended both for spacing and tying and for general construction processes. The coil tie and anchor, as their names imply, comprise coiled steel sockets joined by steel bar tie members. The configuration of the coil matches the rolled thread on the complementary bolt. This rolled thread provides an easily maintained threaded connection which, when used in conjunction with spacer cones and large washers, comprises a complete and economic tie and support system.

Various ties are available, including those that incorporate a water baffle. Some coil ties are designed in such a way that the cone can be threaded onto, and thus made captive, to the coil nut thus facilitating insertion in deep lifts and awkward locations. Where increased cover to the coil is required to match that applied to reinforcement within the structure, then either longer cones or plug-in extensions may be used together with standard cones.

There are several points of design which may promote the adoption of one or other of the coil arrangements available. The captive cone threaded to the coil, which thus is released from the concrete by turning, is a simple and direct piece of equipment and one which will leave neat recessed holes with a minimum of spalling which could detract from the final surface finish.

She-bolts and pigtail anchors

In heavy construction work where special formwork arrangements impose substantial loads on a relatively small number of ties, she-bolts have now become standard items of proprietary equipment. They are available in small diameters and are used as ties with traditionally constructed forms in addition to commercial systems. The she-bolt, a tapered bolt threaded at one end for the nut which transmits the tie load through washer plates into the system and tapped for a stud or bar at the reduced diameter end, forms a neat hole at the concrete face which can be readily and simply filled by plugging on dry packing with a cement mortar.

The stud which is indented to prevent rotation within the form can be obtained in various lengths and configurations. Connectors can be used to allow various wall thicknesses to be formed and to accommodate taper or batten in the wall profile.

The she-bolt member is passed completely through the sheathing and carcassing and the tie force is then transmitted via a matching washer plate (sometimes a combined washer plate and nut) back to the waling or soldier member part of the form substructure. This fact of passing a completed assembly through the sheathing of both forms is valuable where there would otherwise be difficulty of access for inserting and threading the components due to the position of the tie within the form. The she-bolt can still be used both as a tie and spacer by adopting a circlip or special combination nut and washer which provides a positive connection to the form carcass.

The she-bolt offers a facility where massive tie loads are involved, although considerable care must be employed to avoid loss of what are in effect very expensive re-usable components. It can also be used to anchor back succeeding lifts of formwork or to secure ancillary equipment and temporary support for later operations.

One problem associated with the use of internally threaded pieces of equipment, is that of ensuring full engagement of the threaded portion into the she-portion. The capacity of a she-bolt tie arrangement may easily exceed the safe load which can be applied to the timber in compression below the washer plate with the result that crushing could occur. In general, however, this does not cause a problem as the larger ties are mainly used where the stresses in the formwork are such that they require the use of steel walings or framing members.

A recent innovation is a system which depends on special adapters that allow deformed bars to be used for both the she-bolt body and the buried stub. The claims made for this system, which appear to meet a number of formwork design needs, are that the 'Acme'-type threaded bar allows speed of threading, has very high strength and can be obtained in lengths up to 6 m which are advantageous in certain civil engineering applications. The system allows the use of either a buried stud technique or a completely recoverable tie arrangement when a suitable sleeve is used through the concrete.

Patent tie-systems

There are a number of components, clips, clamps and arrangements which can be used for tying concrete – many of them depend on the use of offcuts of reinforcement or bars of special configuration which act as the through tie members. Space prohibits detailed examination of these components, although the following points should be observed when the formwork designer makes his selection:

Are the components sufficiently simple to be used in bad weather by unskilled operatives?

If a spring or part of a mechanism fails, is the load still restrained?

Are there any loose parts which may become separated or lost?

Does the mechanism depend on an expensive tool for tightening or removal?

Provided that the equipment is favourable in view of these points it may well prove to be suitable for formwork practice.

Strapping

Considerable use has been made of patent strapping particularly by building contractors. The strapping systems were originally developed for packaging and palletization, but have now been adopted for tying forms such as those used in column and beam casting. They are also used in the fabrication of formers for hollow components, usually in conjunction with timber batten and expanded or foamed plastics materials.

It is important that strapping be utilized in ways which exploit its high tensile strength characteristic which, together with 'instant' fixing, can be achieved by the crimping or clamping tool. Where possible, tie arrangements must be such that the loading in the form of 'ring tension' is applied axially to the strapping, the backing members on the form being so disposed and shaped to promote this situation.

For fast site erection of small column forms of commercial quality, panels can be pre-assembled into a complete box which is then dropped over the column reinforcement in one piece. Wheel spacers on the links adjacent to the main reinforcing steel ensure that the correct cover is maintained. For larger columns, say 600 × 600 mm or even greater, or where columns are over 2.4 m in height it is more practical to assemble the panels around the kickers, starters and steel reinforcing cage. A few nails at the corners of the forms will hold them correctly positioned while the first straps are being applied. Strapping can be carried out by one operative although obviously time is saved if men work in pairs.

In all cases the backing members should be disposed in such a way as to avoid sharp changes in the direction of the strapping, thus allowing equal distribution of the stresses. The normal tightening action, using the appropriate tool, is sufficient for backings with rounded edges to ensure this uniformity, and thus allows the strap to work to the best advantage. For larger columns, where horizontal cleats are used, backing members should be designed to span between column corners and cleats shaped to provide a smooth profile around which the strapping will bear in service.

Where vertical soldiers are used the size of the members should be selected such that when oriented the centre of the panel projects furthest from the back of the sheathing, thus establishing the rounded plan profile which ensure ease of strapping and the development of a 'ring tension system'.

Economically, little is gained by using the smallest available strap and indeed the existence on site of more than one size of strap can be dangerous. The largest proportion of strapping used in the construction industry appears to be of the order of '¾ in. × 031 grade'. While it can be used more than once, this is not desirable other than where those employed work under the highest degree of supervision. Failure to ensure proper supervision may place an operative in the position where he must decide whether a rusted or crimped strap is suitable for re-use, with the result that some sub-standard lengths may be inadvertently used in critical positions.

A number of factors need to be considered when comparing the cost of using a strapping system with those of other, more traditional methods of tying forms. Strapping avoids the need to store and maintain items such as column clamps which need transportation from yard to site. In the case of such material hire these are frequently retained for longer periods than those for which they are actually needed on site and consequently either double handled or subject of loss or damage during the storage period. Column clamps are particularly cumbersome to use an projecting ends have often caused personal accidents.

The clips that are used with strapping are, of course, subject to wastage. They may be lost in quantity, while bands or straps can be wasted simply by bad storage or handling. Generally, however such equipment is easy to handle between operations, and as the tools are not bulky and provided they are kept clean, they can be used countless times. Companies which specialize in the supply of strapping systems often lend this equipment against some deposit, and will maintain the tools as required.

Operatives who use straps must be instructed in the normal safety aspects especially with respect to handling in order to avoid cut hands or eye accidents. The problems associated with sudden release of force when clipping the bands at time of release also need to be understood.

Strapping can be used for tying other types of forms, such as column and beam members, particularly where some kind of fire protection is being cast around structural steel sections. When banding is used for this it is essential that the beam side members and beam soffits be supported either by adjustable props or a proprietary supportine system, rather than suspended by straps from the structural member. This avoids movement of the forms should the banding crush the timber during concreting.

Bands or straps can, due to their sectional proportions, be passed through sheathing joints, and this is particularly useful with the location of formers or ducts where flotation otherwise would occur. As this only applies to other than visual concrete, the formers are spaced from the steel reinforcement, and this in turn is carefully supported from the soffit or side member. Straps are then passed round the hollow former and through the sheathing, and around some substantial carcassing member where the clip or seal is attached. In this application and indeed in any other where semi-rigid formers are being constructed or secured, the edges and corners against which the straps bear, should be reinforced with timber batten or offcuts of ply or hardboard, as these resist any tendency for crushing. If this is not done a misformed aperture or some displaced opening will result where slackness due to crushing has allowed some movement of the former.

While it should be noted that straps which pass through the sheathing are useful when the concrete is not of a visual or exposed nature, there is no reason why, provided some suitable recess is formed at the point of emergence from the face, the straps should not be cut back below the surface to conform with the normal stipulations with regard to cover. The recess should be subsequently filled as is the case with coil ties or other buried tie systems.

Perhaps more in keeping with its packaging role, strapping can be used in conjunction with laths or battens to fasten together individual pieces of expanded plastic or card forming hollows or ducts in concrete.

The selection of smaller centres is one of achieving a sound economical balance between numbers of supports and sensible spacing of centres. Construction systems which particularly tend towards the adoption of centres include supporting of floor forms from structural steelwork, support of slab forms from concrete or steel cross-walls, and corridors where access, unobstructed by supports, can be maintained below the formwork.

Figure 8.1 The designer of temporary works constantly looks for alternatives to established methods and materials. Here brick piers form the support for long term application.

8. Formwork materials

General

It is essential that those who are associated with all aspects of formwork construction, erection and dismantling should achieve a sound understanding of the basic properties of the available materials. A knowledge of mechanical attributes and properties is necessary for the appropriate design of the formwork arrangement, and the reader is advised to make reference to the bibliography given in Appendix 4 for further information. This chapter deals with the properties of materials, as these are so often critical for formwork use. Generally they become apparent from their practical application, but there are properties which govern the selection of a material. The formwork designer when selecting a material has to consider several factors, quite apart from the physical characteristics. Each must be identified and suitably weighed and rated.

The factors not covered by Codes, recommendations and specification are those which involve preference, adaptability or cost. Naturally for a given set of circumstances and as a result of certain requirements each material may have some particular attribute that will resolve a particular constructional problem.

Although moulds constructed of balsa wood sandwiched between glass reinforced plastics (grp) have recently been used for the repetitious casting of very heavy precast components, it is unlikely that balsa wood would have much to commend it in a heavy casting situation.

On first sight foamed plastics does not appear to possess any real structural value and would not apparently be considered as the formwork for an 80 ft diameter dome. But given the appropriate techniques of construction and a suitable method of support, the material can be used. The shaped plastic segments act as permanent formwork to the outer skin of ferrocement while providing the insulated inner sandwich layer which receives subsequent plastered finishes.

The formwork designer is fortunate in that he can 'borrow' materials, and indeed technology, from industries that employ resin, steel and plastics. For example he can use methods normally adopted for light and heavy engineering, ship and aircraft building. For many designers this provides an interest in the work, while keeping abreast of developments in materials technology will provide a considerable and continuing challenge.

The materials discussed in this chapter have all been used for forming or moulding concrete. Some are widely used and well understood and provide the basis for various trades and skills. Occasionally other materials are used where the various demands of the formwork or mouldwork situation warrant their use.

Obviously in this book space is not available for considerable detail, so that each material is only briefly described together with some suggested uses which may determine where the particular material can be applied. Virtually any material can be used to form and mould concrete and this is a commentary on the inventive skills of the formwork designers, supervisors and tradesmen involved. One factor, however, is paramount – that of safety – and in any decision as to method, the formwork designer must seek precedents and design criteria which validate his final selection of material.

Recently because of some accident or failure the performances of traditionally acceptable materials have been questioned. The designer must always remember that a material which is apparently suitable under normal service conditions of average temperature and humidity, may present quite different characteristics in some unusual condition, to the extent where movements, deflections and failures can occur.

Innovation should always be encouraged and indeed it is often forced by economic considerations. Where a designer uses some new found material or where he uses an existing material for the first time, he must, obviously, ensure its suitability. He should carry out tests, even to the extent of testing materials to destruction, under conditions similar to those in which he intends to use them in practice. By doing this, guidelines can be established and published for the information of all concerned.

Timber

The traditional material for form and mould construction, timber, can be readily worked and understood. Because of the rapid rise in costs of basic raw materials and the effects of inflation, timber is no longer a cheap material. The designer is now faced with economic implications in using timber. Timber-derived materials and plywood, as sheathing materials, have provided the best uses and have had a considerable impact on proprietary formwork systems; these are discussed elsewhere in the book.

Economic formwork arrangements can still be created and timber used in bulk and in stock sizes, and for framing, carcassing and similar situations can, with care, provide hundreds of uses. Although at first sight they may appear to be relatively shortlived it would be interesting to follow runners or bearers through their uses on site after site. The estimator's allowance is usually based on one job or one contract alone but there are frequent instances where stock timbers are used on a succession of contracts.

While a designer is interested in the mechanical properties of timber he should not however overlook practical considerations and site preferences. He must attend to certain details which may otherwise be disregarded by the purchasing officer in his search for the 'best buy'.

Timber should be purchased in the common sizes that can be achieved by sawing or machining. Many applications depend on uniformity of carcassing material thickness to generate a plane form face. Cases have arisen where considerable time and effort has been expended in packing and shimming members to overcome thickness discrepancies.

Ideally, material used for formwork should be stress graded, and with the advent of mechanical stress grading techniques, this has now in fact become usual practice.

The mechanical properties of timber vary considerably with temperature and humidity, and while calculations may validate quite slender sections for a particular application, the designer is well advised to, in effect, 'over design' for situations where, for example, accelerated curing or steam techniques are employed. While it may be assumed for design purposes that loading on formwork may be considered to be of short duration, the designer must appreciate that timber is often subjected to considerable stress for long periods particularly when used as falsework. Where considerable forces are encountered the bearing characteristics can be supplemented by the inclusion of large thick washer plates or hardwood, since these are capable of spreading the forces over a greater area.

Drillings or checkouts soon degrade timber. The intelligent designer will so dispose his structural members as to avoid unnecessary boring. Twin members avoid the need to bore and allow greater scope for variation in through-fixing positions.

Proprietary plates help to distribute heavy stresses from bolts and ties, while a carcassing member can be used advantageously where shear plates are included at bolted connections. Length cutting should be avoided, timbers being arranged to pass each other to allow for variations in bay width. Careful purchasing to schedule will of course minimize end cutting waste. Form panels should be designed to correspond with commercially available sections and lengths. The designer can benefit by taking advantage of the developments which have resulted from the advances made in resin and adhesive technology. 'One-part' adhesives enhance the strength of even simple nailed joints and, when used in the workshops in conjunction with prepared timbers, can produce almost indestructible joints. These jointing techniques are more applicable to mouldwork than formwork, although glued joints may prove advantageous for work involving critical visual aspects.

With regard to construction, the carcassing members for a formwork panel will provide better re-use values if properly jointed, while even the apparently expensive techniques of housed and shouldered joints will bring about re-use economies when employed in shop-made forms or panels.

Timber can be readily machined on site by the use of power hand tools or a sawbench although with the increasing cost of timber, cutting needs to be carefully controlled to avoid any wastage. Timber can be simply fixed into panel form and is ideal for small numbers off for non-standard column and beam forwork. Infills around system panels and make-ups in system arrangements necessitate large quantities of timber. It can be simply framed into supports for plywood and hardboard and used as a sheathing material. When employed as the supporting medium for decking and walling, the framing members rarely come into contact with concrete and thus are easily preserved for re-use. Where used as a sheathing material – either for small numbers of uses or where the timber is required to impart special board finishes, mould oils or parting agents must be applied to ensure clean striking from the face of the concrete.

As timber is particularly absorbent, it is advisable to coat the face with a sealer in order to reduce the absorbency. This is particularly important where sawn timbers are being used to produce board marked concrete. Differences in absorbency will produce permanent variations in the shading or colour of the concrete. The main problem with timber is the inherent mobility of the material, for example variation in section and dimension due to moisture content and stresses set up in rigidly fixed boards by differential wetting or drying. Special surfaces cast against boards require joint sealing to avoid hydration staining, dark marks and discoloration at joint positions. The old technique of tongue and grooving, or the provision of loose tongues would appear to combat this defect although the use of a foam strip gasket is probably a more economical way of retaining moisture and fines.

As timbers, particularly the heavier sections, are reduced in length by end cutting and adjustment, the offcuts can be utilized for bearers and cleats. Board offcuts can be re-cut for wedges provided care is taken to avoid unnecessary wastage. Although these techniques may be time consuming with regard to supervision, the savings thus achieved can quickly provide margins beyond the main costs. It must be remembered however that any reduction in the section of the timber can drastically affect its structural characteristics.

Fastenings in timber can be nailed, screwed, coach-bolted or bolted by use of normal bolts and washers. The formwork designer and estimator must take care to include the cost of these items. If it is omitted any calculation or comparison of the economics of particular form will be meaningless. On an important safety aspect, everyone on site should ensure that nails and spikes which project from struck timber are withdrawn or bent flat once the dismantling operations have been completed.

Timber consumption should be monitored and any excessive usage constantly checked. Usage can be balanced with earnings where for example bonuses for operatives are easily achieved at the expense of the material. It is often simpler and quicker to cut-in new material than to re-use existing concrete-soiled board.

While boards may be sprung to provide curvature, it is generally more advantageous to arrange the individual boards to run parallel to the radii or axis of the curve. For smaller radii and certainly for double curvature it may be necessary to lay two thicknesses of board, in order to fit or fill the face in contact with the concrete.

Where substantial sections of timber are used to support sheathing kerbs, achieved by sawing the back face, will reduce any tendency for the boards to cup or bow and so disturb the plane of the sheathing.

When assembly boards are adopted as sheathing or as supports for ply skin, the end joints between the boards should be staggered over a number of cleats, as this will help to maintain line and provide more substantial panels.

Polythene sheeting or film

A number of construction techniques have evolved as a result of the manufacture of polythene sheeting. Polythene is being used more and more as a vapour barrier, and even as a water-retaining membrane. It can be used in place of blinding concrete in good ground because it excludes clay and dust which would otherwise contaminate the reinforcing steel, and it may be utilized as a liner, being wrapped around a form panel. Blemishes have been caused where the sheeting has become wrinkled and trapped due to the movement of the concrete during compaction, but an interesting effect can be achieved on precast work where polythene sheet is laid on damp sand or plaster, the whole surface then being moulded and/or imprinted with patterns. The resulting concrete surface has a glass-smooth texture and the overall appearance is somewhat similar to the quilted texture of upholstered furnishings.

Polythene sheeting has the effect of reducing suction between the form face and the fresh concrete and can be used to good effect where early striking is carried out in some attempt to achieve special finishes.

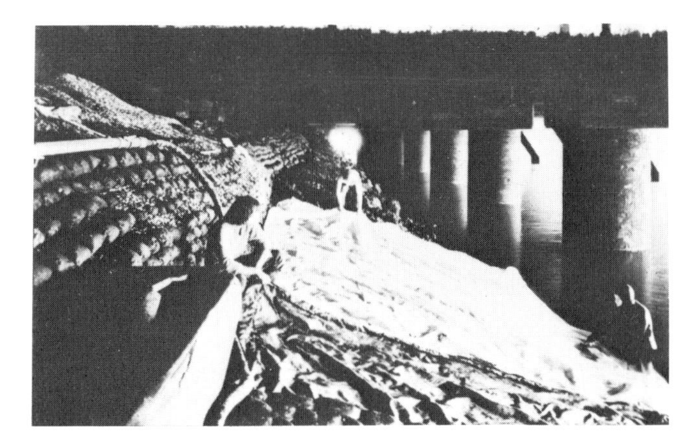

Figure 8.2 An interesting application of plastics in formwork. The envelope can be simply positioned and filled with concrete between tides. (Intrusion Prepakt (UK) Ltd.).

Rubber

Specially formulated casting rubber can be melted and re-melted, and in its molten state can be poured onto masters or plugs made from a number of different materials. Rubber moulds or forms are usually employed when intricate shapes are to be cast, and when sculpture and classical details have to be reproduced. Because of the high initial cost of the material, it is usual to use it in combination with plaster or some suitable secondary mould material in the form of a container.

In techniques that employ plaster as a secondary mould material, the master or model is coated with a layer of clay, and then the secondary mould or plaster container is cast over this. The thickness of clay depends on the depth of detail and the qualities of the rubber to be used to cast the mould. Thus the plaster container forms only a crude envelope around the master, leaving space for the poured rubber. The rubber compound completely fills the space between the secondary mould and the model and because of the support offered by the secondary mould is capable of withstanding the pressure of concrete during the actual casting process. Where there are re-entrants, and where even the extremely flexible rubber mould may be over-stressed by stretching in the demoulding process, a mould which comprises several parts can be cast. By carefully removing parts of the clay lining in a pre-determined sequence, and thus dividing the mould into a number of independent pieces that are still capable of re-insertion into the secondary container, a composite mould can be prepared, the parts of which can be removed easily from the freshly cast component or feature.

There is one commercial system that depends on the use of a very thin film of rubber which is vacuum stretched over a mould or form face, which after casting, the mould or form can be stripped by introducing compressed air. The resultant surface texture is elegant and the technique eliminates many of the defects which occur where complicated shapes are cast using more conventional techniques. Thin rubber films are also used in conjunction with vacuum or compressed air to form featured concrete and to allow the embedment of brick tile and cast-in textures. Heavy rubber sheeting in pre-textured form is available for imparting decorative finishes onto concrete surfaces. The textures include crepe-type patterning and decorative motifs which because of their stretch/shrink characteristics can be readily removed from concrete. Heavily ribbed rubber provides a key for plaster or for applying mortar on tendering, tiling or mosaic work.

Plaster

Plaster, like concrete, can be simply prepared by using the equipment and skills that are readily available to a contractor or precaster. Large moulds can be constructed over an armature of mesh or lath, while high quality of finish can be achieved by skilled labour. For architectural features and details such as niches, small domes and such like, plaster can be struck or 'horsed' to profile using zinc templates which are reinforced by steel or timber frames. The surface of a plaster mould can be hardened by coating with shellac. The application of a chemical parting agent ensures clean stripping. As with concrete moulds, care must be taken over re-entrant forms and the moulds must be suitably split or jointed for release.

Expanded plastics

Foamed plastics such as expanded polystyrene or foamed polyethylene are useful materials for form applications. They are not cheap, especially if used only to form openings in walls and slabs. In such cases blocks of foam are often crudely cut and hastily jammed into position, and located by the steel reinforcing cage so that the only way to release them is to break them out or dissolve them from within the concrete.

There are some materials which can be used to coat foamed plastics to build up a dense, smooth outer skin that allows withdrawal from the concrete. Commercial foamed plastics formers incorporate a plastic skin which renders the face suitable for the production of good visual concrete and allows re-use.

Expanded plastics come ready formed for use in forming holding-down bolt sockets, apertures for switch boxes and lighting fixtures and a variety of other shapes. The precast concrete industry employs disposable foamed plastics moulds for the production of ornamental garden products and street furniture. Bulk supplies of foamed plastics material often proves wasteful and schedules of cut sizes or moulded shapes are much more economical.

Large sections of expanded plastics are used to form cores and apertures in structural concrete components such as bridge beams, being extremely satisfactory and fast to install. It is necessary however when using large sections to guard against lateral displacement and uplift. When securing cores of expanded plastics sheet material or timber lagging should be employed to distribute the forces which will otherwise be localized at the tie position, as this avoids tie-cutting through the foam and allows displacement of the former.

Several of the most recent innovations in precast concrete production have become developed as a result of foamed plastics. One system of core production where cored precast slabs are produced relies on the foaming action induced by a mixture of resins to fill a plastics container within a mould. The moulded former can be produced continuously in lengths suitable for insertion into the precast floor unit moulds on a long-line bed. Another flooring system includes pieces of foamed plastics which are bonded to prestressed planks to provide a prepared, located former for producing a cavity in the in situ concrete laid as topping.

Because suitable formulations of resins can foam, and thus provide a closed cell structure which is many times the volume of the unmixed components and which rapidly hardens and sets, they can be used to advantage in the production of 'instant moulds'. A master or plug is set into a suitable container and the resins are mixed and introduced around the plug; within a few minutes a rigid mould is available for filling with concrete. At present the cost of the material limits the use of this technique for other than the smallest or most urgent products, although the rising cost of alternatives is tending to make this technique economically viable.

Plywood

Plywood provides many advantages for the formwork designer. Standard board sizes reduce the joints in sheathing faces, while standard weather- and boil-proof bonding provides sufficient heat and moisture resistance to render plywood suitable for all but the most arduous formwork applications. Provided the relevant BS Codes are observed, plywood is a material which possesses better mechanical properties than the timber from which it was manufactured. By skilful design, and bearing in mind the mechanical properties, the utilization of the principles of box and folded plate construction can provide economical solutions for most formwork problems.

The advantages of using plywood can be enhanced by the adoption of suitable glued or bonded construction techniques that merit the use of composite construction. This becomes apparent where, for example, a ply sheathing is used in conjunction with machined edging members that possess rebates. When glued or bonded the strength attained equals that of a channel section, and thus considerable mechanical gain is achieved.

The use of plywood provides joint and warp-free surfaces and when suitably coated or treated yields little grain marking or impression on the final concrete surface. A considerable re-use value can be achieved particularly from framed panels, while an unframed sheet of ply can be conveniently handled by one man. It can be used more if the treatment includes edge sealing because it is the end grain where moisture is most readily absorbed. Edge damage should, where possible, be minimized although ply sheets can be trimmed on site. Any holes and defects can be readily filled with epoxy fillers, although for most commercial applications a simple plug driven into a hole should provide a satisfactory seal.

The use of plywood sheets should always be planned even if this only means that a simple line diagram which sets out the location of the various sheets is prepared. This ensures that random cutting with subsequent waste is avoided. After plywood has been used several times to form visual concrete, it can then be used further for commercial concrete finishes which are to be subsequently covered. After this the sheets may be used again for work in the ground, and eventually for making disposable boxings and 'lost' formwork.

Where possible all fixings into coated or laminated plyboards should be kept to a minimum and cut faces or borings should be re-coated prior to use to prevent moisture ingress, fines penetration and local degradation of faces.

An advantage of plywood is that it can be bent and secured onto framing for casting, flowing and curved surfaces. Heavier plies need to be considerably restrained to maintain the curvature and consideration should be given to either using two layers of thinner plyboard or one layer on an open jointed framework of timber.

Recently, manufacturers have introduced plyboards with applied textured faces and faces which will impart board marking and indents onto the face of the concrete. While these provide excellent surfaces the designer must stipulate that care be taken in the cleaning process and during the application of oils or parting agents since small concrete particles may collect in the features and mar the finished appearance of the final surface. Every precaution should be taken with constructional details to ensure that the end grain of plywood is masked either by housing joints or by hardwood or softwood fillets and features. This will prevent grout ingress and adhesion to the concrete face.

As plywood is a sheet material, it can be used to span a timber or steel framework. The ply should be oriented to obtain the maximum advantage of its mechanical properties. Potential deflections must also be carefully considered. Situations may arise where, due to the mix design and workability, the flexibility of the material is such that it causes flutter or local vibration thus spoiling the concrete face even though the plywood has met the requirements of specification regarding deflections.

Similar affects, usually indicated by a high incidence of fines or even water runs, can result from poorly framed joints which cause differential vibration.

Designers should always arrange any carcassing and framing to plywood panels in such a way that the ply is securely fixed throughout the face and at panel edges so that local deflections do not allow the escape of fines or moisture, or the ingress of fines between the carcassing and the sheathing.

Plywood is currently used in Europe in mould construction for situations where massive vibratory effort is applied to the mould. Moulds are constructed using box-ply techniques, typical of structural timberwork, and multi-ply moulds comprising ten or more veneers of 18–20 mm ply are laminated up by pressing, and subsequently machined to provide the more exotic forms which at present are popular in precast concrete construction.

Fastenings and fixings

A few years ago nails were commonly regarded as cheap forms of fixing for formwork purposes, but today the price of the simple wire nail is considerable, so that careful materials control is necessary to keep costs to an economical level. In the past the construction industry has been particularly wasteful and unfortunately on many sites this has been perpetuated. However a handful of nails which was once so wastefully discarded or dropped into muck on site may now equate to an operative's wages for half-an-hour's work. Time should be taken on materials control to ensure careful issue of appropriate quantities and ironmongery items in order to effect economical savings.

Often the fastenings that are used are far too heavy and, technically, incorrect. Because nails have been used traditionally to fasten materials, there have been times when no other fixings have been considered and while nails may well be the correct fixing solution they may need to be of different gauge or configuration. Small gauge nails, ring shanked or cement coated, provide good resistance to potential pull-out forces, while slender nails with smaller heads provide neater fixings with less noticeable blemishes at the concrete face.

Stapling machines or pneumatic staplers with their extremely high rates of fastening through ply and board, are useful where sheathing needs to be fixed. Modern stapling equipment is capable of driving 75 mm nails of varying configuration and these are used on frame construction and general fabrication. Staples up to 60 mm length can be driven at a high rate from small portable equipment, and provide excellent fastenings for sheet material, being less likely to cause splits. The small indentation which remains in the face can be filled using resin. The configuration of certain staples is such that a dovetail-type fixing can be achieved which possesses high resistance to direct pulling. Where sheet material is to be fixed as a liner, staples become an automatic choice because there is less disfiguration of the basic panel when compared with other nailing techniques. Self-tapping and thread-cutting screws are useful in that they provide instant fixings for timber to steel and ply to steel. The pilot holes required can be drilled by means of a suitable morse bit, while the self-tapping is screw-driven into place manually or mechanically.

Figure 8.3 Staple and nail guns are particularly useful for panel construction and other formwork applications.

Where vibration is a key factor in design, a screw known as a 'Taptite' produces an extremely tight cut thread. Shotfired fixings can be used in formwork operations but not generally for primary attachments. However shot fixings are used to locate blocks against which props and braces are borne and they can also be employed for locating screed boards and kickers.

Great care is needed in the selection of appropriate nails or studs to avoid end-splitting or failure of the cleat or timber that is being fixed.

In addition to normal wood screws the 'twinfast wood screw' has a twin start thread which produces increased holding power in low density materials.

For screws which are to be removed for re-use it is advisable to use flat washers under the head, and to apply some lubricant to the thread to help with removal. Ideally, for special applications, pilot holes should be drilled into the timbers that are being secured.

Bolts tend to be particularly expensive and considerable care needs to be taken in scheduling fastenings so that the most economic lengths and diameters are used. Every panel detail should be accompanied by an ironmongery schedule which notes the exact quantity of bolts and fastenings required.

There are certain rules applicable to the use of fixings in timber that are often overlooked. The formwork designer who is acquainted with recommendations given in CP 112:1967 may seldom think it necessary to detail joints with regard to nail fixing positions and numbers. A study should be made however of the way in which joints are made in practice on site, and even in inaccessible locations. The designer may then wish to be more specific in his details, if not in run-of-the-mill work, then for specially constructed panels and formwork that is detailed for outside manufacture. There are a few details which when examined from a practical point of view clearly require attention:

1 Nails, driven fixings and bolts should be placed uniformly about centre lines

2 Attention should be paid to edge and end distances

3 For each screw size there is an optimum penetration of thread

4 Any detail should be based on the smallest sized member at the joint

5 Most joints can be improved by including a toothed plate or shear ring.

Hardboard

Hardboard which is often neglected as a form material has long since had the image of a short-lived disposable material. Good quality oil tempered hardboard can, however, provide a considerable number of re-uses, 10 or 12 not being exceptional. The detailing of the form panels must be carefully carried out, particular attention being paid to the thicknessing of sub-sheathing and carcassing material where any discrepancy would be reflected in the extremely smooth form faces that result from the hardboard sheathing.

Hardboard, if not suitably wetted down before fixing, does tend to move on contact with concrete and indeed the fixings can cause problems. The ideal fixing is an oval brad or staple which is slightly driven below the face level.

It is particularly useful for forming shaped or circular surfaces where, with support and frequent fixing, it can be bent to fairly sharp curves. A surface that approximates to double curvature can be achieved by the use of hardboard in strip form. For stopend construction hardboard is simply notched around steel reinforcement, but it requires considerable support however to avoid local deflections and breakage at checkouts.

Hardboard can be used in many situations where thin slivers of virtually non-compressible materials are required, while shims for levelling forms are of course a standard construction technique.

The reverse side of hardboard possesses a textured surface which can be advantageously used for certain formwork and mould applications. The texture provides a good key for a retarder when used on vertical faces. It also gives an excellent key for the mixture of sand and adhesive when preparing boards of aggregate for transfer, where the aggregate transfer method is used to provide localized areas of exposed aggregate finished concrete; pegboard or pierced hardboard is also suitable for this application.

The cohesive nature of hardboard and lack of grain proves useful where small ribs and margins are to be formed within the mould and form. The material can be ripped and planed by means of sharp tools, and when pinned into position, can form recessed margins or ribs. Depending upon the frequency with which pins are to be inserted, a hardboard fillet can either be arranged to come away with the form or remain with the concrete during the striking operations.

Where there are stringent requirements for surface finishes, and where for example forms have been used a lot during earlier operations, hardboard can be employed to line forms and so improve the final surface. Fluted and textured hardboards can provide decorative faces on concrete, since the low cost of the hardboard liner ensures that it can be replaced when any wear occurs.

Steel

When steel is used for formwork construction it will be for some particular requirement or feature of the concrete structure. Apart from fabricated proprietary steel formwork systems steel is selected because:

1 A considerable number of uses can be obtained from the formwork

2 Particularly close tolerances can be specified for the finished concrete work

3 High stresses are involved

4 There are particular requirements with regard to conditions of use, tidal zones and such like

5 Some degree of mechanization can be introduced into the formwork system.

Each of these factors can determine some aspect of the construction and requires the formwork designer's consideration.

Steel is generally used in standard sections and plates, as purchased, and careful specification of sheet material for sheathing purposes is essential. Steel fabrication falls within the province of the engineer or plater, particularly where a considerable amount of welding is involved. The skills in maintaining straightness and correcting distortion which can be introduced by welding are specific to certain areas of the construction industry. The designer will find it uneconomical to design other than the simplest fabricated steel moulds without consulting a commercial manufacturer. The design of a form must be carried out in the light of the equipment available for manufacture, and the techniques which have been developed for its use.

Initially, the purchase of material is a specialist matter and while sheet from mills up to $\frac{1}{4}$ in. will generally give faces suitable for the production of good quality concrete surfaces, plate above that thickness can present problems with scale, pitting and local defects. The purchase of plate (known as 'cold rolled and pickled plate') while adding to the general cost does ensure a better surface quality, although the specialist manufacturer often finds it necessary to inspect plate at the stockist and thus introduce some degree of selection. Selection continues at the manufacturer's works where the sheet and plates are cut, and material which may have blemishes likely to affect the concrete surfaces being used for substructure and gussets for example.

Where formwork is to be used sporadically, perhaps due to phasing of a contract, it may be worthwhile to take the precaution of having the mild steel form zinc-sprayed to prevent rusting between uses. While this is not over expensive, the process involves grit blasting and the total cost must be set against the benefits on site.

Steel provides little or no insulation in the thicknesses when used in form construction, but may be an advantage where accelerated techniques for curing rely on the transfer of heat into the concrete through the form face. When concreting is being carried out in exposed positions, however, the designer will have to introduce some kind of insulation to combat heat losses due to radiation. Steel members often enhance the properties of other form and mould materials, for example, steel-faced concrete moulds provide highly durable, stable and extremely accurate methods where a high degree of accuracy is required on repeated castings. Steel is used to reinforce joints and connections between other form materials and steel plates, and ferrules set into timber provide faces which experience little wear on successive re-uses. Where extreme accuracy is required, form faces machined to limits after fabrication ensure the achievement of required quality. Attachments to steel form faces can be simply made by drilling and tapping or bolting from outside the form.

One particular problem of form construction, that of a satisfactory anchorage for the vibratory equipment, can be easily resolved by the use of welded brackets direct to the main structural members of the form. Designers should take care over the location of external vibrators, particularly with sheet-faced forms, because the vibration of the membrane of the sheathing between supports can cause problems in the surface finish – promoting bleeding and aggregate transparency or mottling.

Steel sections are often used as a plant item being transferred from site to site and stored in a plant yard. One note of warning – it is not unusual to find that contractors hold stocks of second-hand materials in their plant yards, and with the multiplicity of sections that have been manufactured over the years it is possible some substandard section may be supplied for a particular purpose. This has caused problems on site in the past and should be avoided. Selection must be carried out by a suitably skilled person before despatch, against a requisition, the sizes and sectional dimensions being noted and checked.

The added advantage of steel forms and moulds is the considerable inherent strength exhibited, more so where members are welded and form an integral part of the form. This is an advantage where fully assembled forms, complete with steel reinforcement assembled within them, can be handled into position as on heavy beam and gantry construction. Similarly completed assemblies of forms and reinforcement can be lowered into position for marine construction. Steel forms can be designed to incorporate chambers for steam curing using air, water, oil or steam, and moulds can be so designed to allow the prestressing of the cast component such as slabs for industrialized buildings, and moulds for producing prestressed products such as pipe and slender lamp standards.

Special steel forms are often extremely heavy, but this should not be too much of a problem provided a traveller is used or there are means of attaching bogies. The use of steel generally reduces the number of through ties required to retain a given mass of concrete and in fact this may even become a factor for consideration in material selection.

Aluminium

Where lifting equipment is extended and weights of components become major factors, aluminium has a great deal to offer the formwork designer. Its structural formulation is such that it has many of the attributes of mild steel, and is used extensively in the United States to construct table forms.

As a sheathing material, some problems may arise because of reaction with the fresh concrete due to the presence of moisture and alkalis.

The formwork designer should always contact the Aluminium Trade Association whose function is to promote the use of alloys. The following formulations for particular uses are recommended:

Sheathing material. Alloy NS4, in accordance with BS 1470

Carcassing or framing material. Alloy HE 30, in accordance with BS 1474

Cast components, formers and such like. Alloy LM6, in accordance with BS 1490.

Figure 8.4 Aluminium soldier members used where access is somewhat confined. Manual handling made possible due to the lightness of the material.

Figure 8.5 A large aluminium 'flying form' or table, as used in the United States, being raised to the next location.

Aluminium is used for pipe moulds, tile pallets in extrusion processes, and castings for intricate feature formers where fine detail precludes the use of timber.

Chemical parting agents can be used successfully in conjunction with aluminium, and to keep corrosion to an acceptable minimum it is advisable to clean cement paste and fines from the face prior to storage. Fixings to other materials are best effected by using heavy galvanized or sheradized screws or bolts.

Concrete

Concrete can be site mixed or it can be brought to site ready-mixed. It is relatively cheap and of course the skills necessary for its use on site are readily available. It is stable, massive and can be trowelled and cast to a fine finish. It is excellent for mould manufacture as evidenced by any pile-casting yard or tunnel segment manufacturer, and where component manufacture for heavy civil engineering works, such as those involved with oil drilling platforms are required.

When producing concrete moulds a master unit built of timber, plaster, plastic or even concrete is required to generate the first mould. Where the end product is a continuous section or of modular design, the mould may be cast in sections using cast or post-tensioned connections for subsequent assembly.

Concrete moulds can be used to overcome problems of line, geometry and texture, and provided that good quality concrete is used considerable economy in terms of uses can be achieved, hundreds of casts being obtained from each mould.

The moulds must be carefully constructed in good well-compacted concrete, and the casting surfaces dressed-in to fill any blemishes; finally the whole unit must be carefully cured. Although high strength concrete is not essential the small, extra investment in quality will undoubtedly give a more durable mould. The weight of the concrete can, in fact, prove helpful during demoulding process, while its mass certainly helps to maintain the line and accuracy without disturbance. For formwork applications, some sites by having the appropriate crane capacity have produced precast concrete beams and then used them to provide a formwork for the construction of in situ reinforced beams.

For all cases chemical parting agents are excellent for releasing freshly cast concrete from the concrete moulds. Permanent concrete formwork made up of precast slabs or precast planks is now becoming increasingly competitive, the formwork remaining in position to provide a structural or decorative support or liner. Support grillages are required to ensure that a satisfactory line is being obtained and, in slab construction, to support the load of the fresh concrete topping until the whole arrangement of form and slab or infill becomes monolithic. It is probable with the recent introduction of glass reinforced cements (grc) and fibre reinforced concrete that permanent formwork arrangements will become more popular. Sections could become thinner, but additional care and attention will be required in the design of the supporting systems. A large proportion of glass and fibre reinforced materials will be produced by a wet folding process and this will result in particularly slim sections. At present these materials are used to cast-in aggregate surfaces for decorative purposes and require careful support in order to avoid damage.

Plastics

There are few materials the technology of which has more to offer the formwork designer than plastics. Already its technology and production has made a considerable impact on methods necessary to form and mould concrete. Glass laminates and sheets of thermoplastics are constantly used to form intricate moulded surfaces. Casting resins and various formulations of plastic materials are used by contractors and precasters to produce moulds for complex concrete shapes. Plastics for mould manufacture are generally such that they can be mixed and moulded by an experienced tradesman, and provided that the manufacturer's instructions are rigidly followed, good durable plastics moulds or liners should result. Ready fabricated liners are also obtainable in various formulations that range from the highly elastic to rigid sheet materials. The maker's instructions regarding oil or treatment must be followed as must directions with regard to the thickness of the material to be used.

Plastic liners can be used to provide featured surfaces directly or as a means of transferring tile or similar casting surface materials onto the concrete member. In their simplest form cast liners are either taped into position on the form or back to the form. Where exotic finishes are cast employing deeply relieved liners, or where the matrices for transfer of tile are to be attached to the form, this can be achieved by casting into the line of reinforcing bars which are then caught back by hook bars to the form face. The casting of the liner onto a perforated backing material rigid enough to allow bolting or screwing from behind the form is a further useful method of construction. Complete moulds can be cast from resins or, alternatively, just critical parts of the mould or form which would otherwise be damaged by scour or the demoulding forces.

Indent formers and small components which are particularly vulnerable either in the compacting process or during form handling or striking, can be simply produced – on site if necessary by casting from timber moulds. Simple patterns suffice for the production of plastics moulds which can rapidly be used for multiple production of the cast components.

Thermoplastics sheets can achieve modelled textures and features on concrete. The materials are usually only available in quite restricted sheet sizes although the individual pieces can be welded or bonded into larger pieces to suit form panel sizes.

Thermoplastics sheets present a glazed surface to the concrete face and this may lead to crazing, the smooth impermeable face tends to impart some local darkening to the concrete face and, depending on the fixing or backing which is installed, the flexible nature of the sheet may cause local upset of the concrete mix. Grit blasting or mild acid etching will improve the appearance of faces moulded from thermoplastics sheets.

Moulded sheets can be manufactured by heating sheets of thermoplastics materials such as vinyl, and then vacuum-forming these over or into some suitable master cast in plaster, resins or fabricated form timber or steel. The sheets are quite brittle but have the advantage of cheapness and can be easily replaced.

Glass reinforced plastics

These offer the formwork designer considerable opportunities to meet architectural and engineering requirements. They can be moulded over, or into plugs, or washer moulds to provide intricate formers, complex textures and shapes that provide a considerable number of re-uses. They have provided the basis of many special formwork arrangements where troughs and waffles have been required that are greater than the normal trough and waffle standards.

Generally speaking, moulds made of glass reinforced plastics (grp) are only as good as the master from which they are produced. Their manufacture is labour intensive and the materials used are expensive, although quantity production by specialists can be more economical. Production requires careful control and must be carried out within a works or shop environment.

The stresses that are encountered in removing and handling the forms usually establish the design criteria, although performance specifications can result in the design of a grp form being carried out by a specialist supplier on the basis of laboratory tests and his previous experience. Deflection is critical in most situations and sufficient thickness of laminate, probably of the order of three times the required thickness of steel panels must be allowed to achieve satisfactory results. The stiffness can be improved simply and efficiently by the inclusion within the laminate of low density rib formers made of polyurethane foam and balsa wood. Apart from the considerable re-use value, other advantages include dimensional stability, low weight and a smooth blemish-free surface finish which can be moulded to specific contours and textures.

The designer will always need to make a careful assessment of the number of forms or moulds required, because there is an economical break even point between the number of moulds or forms, and the cost of the moulds and forms, this being largely a function of the cost of the model or plug which the grp manufacturer must produce for use in lamination. The designer should note the following practical aspects concerning the material when used as a form material.

The forms should be loaded uniformly to avoid excessive local loading and distortion, and care should be taken to avoid excessive local deflections which could result in the flexible forms becoming trapped by re-entrants. Sharp corners and arrises should be avoided, although the insertion of glass tape or rope can help to prevent the corners of moulds suffering excessive spalling or damage.

The best results are obtained where adjacent sides of a column are joined to make an angle form, the base and sides of a precast unit being formed in one piece to provide a box or container. By doing this the number of form or mould components to be handled is reduced,

Figure 8.6 Steel and glass reinforced plastics permanent soffit formwork being installed between bridge beams. (E. M. J. Co. Ltd.).

while accuracy and stability of the form is improved. The laminate can be strengthened by the inclusion of continuous or angled stiffeners. This is particularly applicable on circular and conical forms where the form pressures can be translated into hoop tension.

All bolting and fixing positions require reinforcement and the insertion of local stiffeners to avoid flexural loading to distortion and subsequent breakdown of the laminate through fatigue and corrosion of the fibres.

Grp can simply be repaired on site using the usual techniques with which every tradesman should be familiar. Where some wear has taken place, surfaces can be improved to some extent by the application of a resin coating, although sound decisions with regard to the quantity of moulds required should avoid this situation ever arising.

Grp is not the best material to use for forming flat or low relief surfaces although of course it can be used in conjunction with ply or timber for the construction of featured surfaces. It is ideal for printing facsimiles of various textures such as grain, weave and such like onto concrete when the natural material can be used as a master for form panel lamination.

For mould applications grp can be used successfully in conjunction with concrete, the resulting forms presenting accurate stable containers with excellent surface characteristics.

Whenever possible the form designer must increase lead and draw so that scour of the gel coat is avoided at the time of striking from between reveals. Some advantages can be obtained from the inherent flexibility of the material in striking by controlling warping or bending to free the mould from the concrete surface.

External vibrators can inflict considerable damage on a mould and it is essential that suitable backup members should be laminated in to facilitate vibrator attachment at *designed* positions. Moulds which are to be used in conjunction with vibrating tables should be suitably stiffened to resist impact stresses imparted by the table surface, and all internal vibrators should incorporate rubber hoses to avoid gel coat damage. The surface of a mould quickly matts down after a few uses because of the abrasive action of the concrete, and this will improve the cast surface of the concrete and facilitate the application of parting agents and retarders.

The formwork designer should contact the specialist mould manufacturer early in the course of the contract to ensure that due allowance is made for the various properties of the glass laminate.

Complementary materials

There is now a wide range of ancillary materials which when used with primary formwork materials can produce various finishes and forms. These include impregnated papers, coatings and oils.

Perhaps the most widely used, yet least understood material, is mould or form oil. The Cement and Concrete Association has carried out considerable research on the ways of achieving high quality surface finishes, and there is little doubt that the selection of the appropriate oil or correct treatment is of the greatest importance. The treatment must be compatible with the material from which the form is manufactured and should include a surfactant or wetting agent which is capable of achieving a uniform coating on the form. Preferably the oil should be in an emulsion which can be conveniently spread over the form by brush, roller or spray. Oil-phased emulsions are particularly desirable as they ensure an even film of oil across the form.

Chemical parting agents are useful in the production of high quality finishes, since they react with a microscopically thin film of concrete paste to allow easy form removal. The cost of a chemical agent must be a selection factor although if applied by fine spray, all but an extremely thin film is removed by squeegee sponges, the treatment becomes economical. Apart from improving the appearance of the concrete, this technique reduces any tendency for retardation to occur.

Surface treatments can be effected by the use of rubber-based paints, plastics and polyurethane-based substances, and a variety of chemical compounds. Each will have some feature to recommend its adoption for a particular situation and the designer obviously must be the judge of the best one to use. Whichever is used the material must be capable of sealing the form face and bond to it, providing a surface onto which the oil or parting agent may be painted or sprayed.

The purpose of the coating is to improve or harden the face of the form and to reduce differential absorbancy which otherwise drastically affects the appearance of the concrete surface. The place of application of both coating and oil and parting agents, must be selected carefully to avoid the transfer of dust onto, or disfigurement, of the concrete.

For some years surface retarders have been viewed with distrust by various authorities. However more recently commercial producers have upgraded their products, so that they are now reliable and can be purchased in forms which determine the degree of exposure of aggregate to be produced at the concrete face.

Release agents are available which, when applied to the form prior to the application of a retarder, ensure the removal of the retarded skin from the form and thus reduce the cleaning operations which were once required. These usually became stuck to the form face on striking with a coating of fines. Paper impregnated with a retarder provides a uniform distribution of the retarding agent, although the composition of most retarders is such that the earlier problems of runs and displacement, particularly on sloping surfaces, have now been virtually eliminated. This impregnated paper is particularly suitable for mechanized production of precast concrete.

Coatings are often used to provide a skin or wearing face on the surface of expanded plastics. Previously, expanded plastics had to be regarded for single use or as a limited re-use material. A painted coating, however, provided good lead and draw are included, allows polystyrene formers to be re-used, perhaps five or six times for simple boxings and checkouts.

Tapes which will adhere to most surfaces such as steel, timber ply, grp and such like are used to seal the joints in the face of moulds and forms, and leave only the mildest

of imprints on the face of the concrete, provided that they are not inundated with oil or parting agents.

Foamed plastic gaskets, which are similar to domestic draught excluders can seal the gaps between form panels. Here they have the advantage of not marking the concrete. These gaskets eliminate leakage and prevent dark staining which often occurs on form joints. Provided that the seals are made at the correct positions within butt joints, they can provide many uses without the need for replacement.

Foamed rubber and plastics sections, neoprene gasketing and similar materials in preformed shapes can be utilized in recesses, grooves and rebates so that by slight compression due to bolting up of the form, they ensure grout tightness.

Rigid and semi-rigid plastics extrusions are used to form features, indents drips and arrises within a formwork face, and large scale plastics extrusions can form shear keys, joggles and joints. Because of the nature of the plastic material these require a minimum amount of cleaning or oiling. They are also extremely strong and highly resistant to impact.

Steel and plastics plates and studs of various configurations can be employed to fill tie and service holes in the face of formwork. These are fixed by simply hammering into position when the teeth on the studs engage in the form material and permanently cover the hole. The filler position will of course be visible on the resulting concrete face, but this defect is nominal on commercial concrete applications which are to be subsequently rendered or decorated. The fillers help to prevent leakage and combat honeycombing and loss of durability.

On water-retaining structures tie holes at the face of the concrete can be caulked by certain types of plastic plugs and stoppers, the plug being driven into position prior to the inserting of the dry-pack concrete mix. Other types of plug are available for insertion into tie holes through visual concrete, here the plastic is a push fit and can neatly mask the hole, but which remains visible in the finished work.

While reinforcement spacers do not constitute formwork, the impact of their uses on results achieved from various items of formwork is sufficient to warrant a mention here. Spacers can be made out of plastics, concrete and asbestos. The main requirement for their use is that they should positively locate the reinforcement with reference to the concrete profile and thus the form face. The spacer material should be stable and its configuration suited to its intended location. Stools on flat surfaces and ring spacers or clips on or against vertical surfaces.

Other materials used include cast-in plugs, and plastic studs and plastic bolts which are designed to shear on striking for the transfer of cast-in components. Mastics are used for plug location, the plug being stuck to the form face by the mastic and transferred at the time of striking as a result of the bond with the concrete.

Glass reinforced cement

A recent development in concrete technology has been the introduction of glass reinforced cement. This material which is made of cement mortar and glass fibre is produced in much the same way as glass reinforced plastics. The inclusion of glass fibre provides a material which can be formed into thin sheets, of plain or moulded profile and having good impact resistance. The sheet thickness can be regulated to suit the engineering requirements of the application, and in general, a thickness of 12 to 15 mm is used where formwork is involved.

The process calls for the manufacture of a suitable former or plug as for grp and these plugs are generally produced to the same standards of accuracy and construction as moulds for precast concrete, i.e. without the need to produce extremely high quality moulds which are essential for the plastic product.

Glass reinforced concrete has been mainly used for soffit formation sheathing between bridge beams. But as designers have begun to realize its potential, it is becoming more universally adopted for column and beam sheathing and in locations where an accurate line is required, or some high quality surface finish.

One problem which the manufacturers have experienced is that of mistrust because many potential users think of the material as being similar to the asbestos cement sheeting which is sometimes friable and easily damaged. The grc sheet is usually denser and its impact resistance is higher. When properly supported, it is capable of containing concrete pressures while imparting the appropriate surface finish.

As it is composed of a cement rich mixture, there has been some evidence of crazing and problems of durability have occurred where the material has been used as a structural material. When utilized as a permanent form however, grc can be assumed to contribute to the 'cover' of concrete required for durability and fire resistant qualities of a structure. In normal conditions, i.e. for bridge soffits, there should be little problem with differentials between the characteristics of the grc and the concrete which constitutes the structural element.

Many manufacturers now produce grc solely for formwork purposes, and the material is thus rapidly gaining general acceptance.

The advantages of a permanent formwork material can be summarized as:

It produces guaranteed surface finishes

Problems of geometry can be resolved away from site

It eliminates the need for access scaffolding between beams in bridging

Elements can be pre-coated with finishes such as epoxies and polyurethanes

Positive arrangements can be incorporated for locations of reinforcement

Modular motifs can be repeated simply and cheaply

Numbers of moulds required for a given surface area can be reduced

Form elements can be easily handled – manually if necessary.

From a study of these attributes it is evident that grc can be placed high on the list of materials available for fulfilling the requirements of both the structural designer and the formwork engineer.

Whatever material or filling is adopted the designer needs to maintain a close contact with the firms that manufacture and distribute these ancillary products and he must keep abreast of new developments and watch for new and useful accessories and treatments. Samples should be obtained and *controlled* tests carried out to assess the value of the materials in particular situations.

Ease of use and simplicity of installation are essential and these must be considered under the conditions that exist at the time of application. Access and orientation can affect the actual installation, so that the location of the application must be carefully considered.

Only proven methods should be adopted in production situations because the consequential losses from a breakdown of a system of sealing or coating application can be quite disastrous when a large quantity of concrete is being placed, especially where some load-bearing structure is involved.

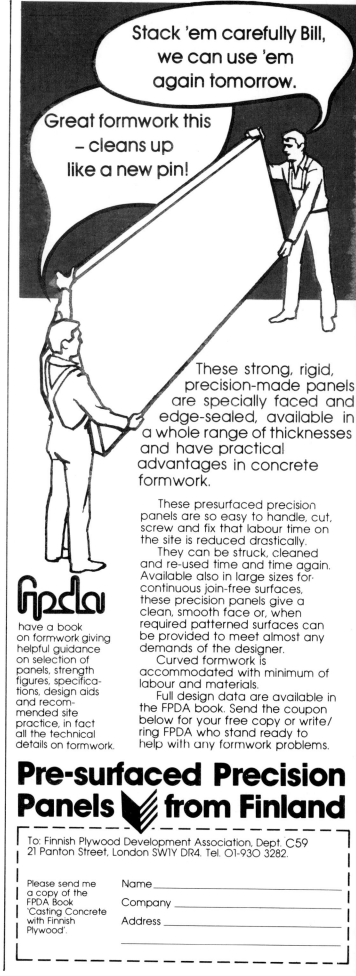

Figure 9.1 A massive form used in the construction of a dock. It was designed to resist the pressures imposed by the concrete and the pressures created by tidal flow. (Stelmo Ltd.).

9. Special formwork

General

There are a number of occasions where concreting work is sufficiently repetitious to warrant the provision of a purpose-made or special form. Of course the adoption of such formwork arrangements ensures that there are predictable outputs and that the programming of activities are more precise. Other factors will determine the demand for a special form arrangement and these include:

Degree of accuracy required

Production of exotic textures, profiles or surface finishes

Shortages of skilled labour

Exceptional costs, for example, extremely high lifts or very long or complicated bays

The need for an inbuilt means of form travel

Physical limitations on tie arrangements.

Special forms

Certain types of construction depend on the provision of special formwork or forms that have to be built for one particular application. Slipforming and tunnelling works both require special formwork arrangements. Contractors tend to use special forms that incorporate travellers and access scaffolds in connection with the construction of retaining walls, sea walls and similar structures. Special forms are probably used more in civil engineering applications than other types of construction although occasionally special forms, tunnel and purpose-made form arrangements are used in some building work.

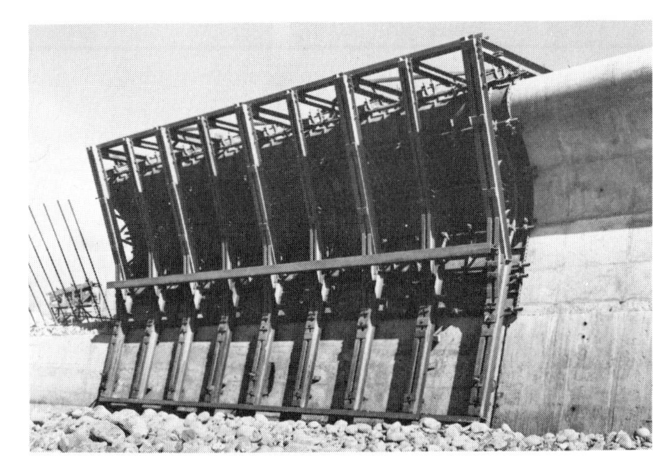

Figure 9.2 Typical application of special formwork – for the construction of a sea wall. Formwork is required to withstand wave impact forces in addition to forces set up during casting of concrete.

The fact that the form panels are larger and of proportions designed for handling reduces crane commitments on site. The massive nature of the form members simplifies the collection and transfer of forces, and for soffit work reduces the number of support towers required. The special form is generally, though not always, manufactured off site and special formwork is often sub-let to specialist suppliers. Good results therefore depend to a large extent on the proper communication of design criteria, and specification and standards requirements.

Usually, formwork manufacturers are well placed to produce technically sound formwork, they are given the essential information relating to type of concrete, admixtures, rate of fill, accuracy required, acceptable surface finish and position and cover to reinforcement, access required and means of form handling. The check list at the end of this chapter covers most of these and should serve as a useful guide to the form designer or contract manager who is concerned with ordering special formwork.

The majority of suppliers of special steel formwork have their own drawing offices staffed with suitably qualified design personnel. The status of those who process enquiries and subsequent designs must be established during early discussions with the supplier.

It is not intended here to validate the whole philosophy of special formwork design and supply, save to say that some concrete construction contracts cannot be economically and successfully completed unless special formwork is employed. Some contracts may be borderline cases because the work of forming concrete may be carried out using proprietary equipment including the utilization of some special components although with escalating labour costs the trend is towards entire special forms which in the long term are more economical.

The contractor has to decide where his money will be spent and what skills are available. What may appear initially to be an expensive form with a traveller or rig, could prove to be fairly economical once those on site overcome the preliminary problems associated with achieving the quickest way to set up the form and the simplest means of moving the equipment. Unless special formwork can be handled in large sections, either by a skidding traveller or a crane, the advantages of the special form will be lost.

Many special form systems are adopted for other than

Figure 9.3 Special formwork for tunnel work.
(a) Form retracted to pass within form.

(b) Traveller projecting to place invert.

just casting concrete, they provide reinforcing jigs, methods for handling and placing concrete and transporting forms about a site. In each case they can act as a buffer in providing slack time, or they may be used to uncouple the activities from those essential ones which govern the deviation of the complete construction process. For civil engineering work, the formwork arrangement may incorporate access, space for materials storage, and provide an interface with some falsework arrangement. It is difficult to draw the line between falsework and formwork, so that obviously there must be a high degree of co-operation between the formwork designer and those say, employed on the temporary works or falsework department, where these are separate entities.

The special formwork arrangement may be constructed in timber, steel, grp or even any combination of these materials. Whatever the material, it is essential that the design work and calculations should be carried out by an expert who is completely familiar with the particular materials selected.

Figure 9.4 Mushroom-headed columns for a reservoir. The next stage involves constructing the roof, thus providing an ideal application of system formwork arrangement.

Figure 9.5 Special formwork for casting columns of a viaduct. Note use of guy ropes, access ladders and platform. Excellent concrete finish is shown on completed column.

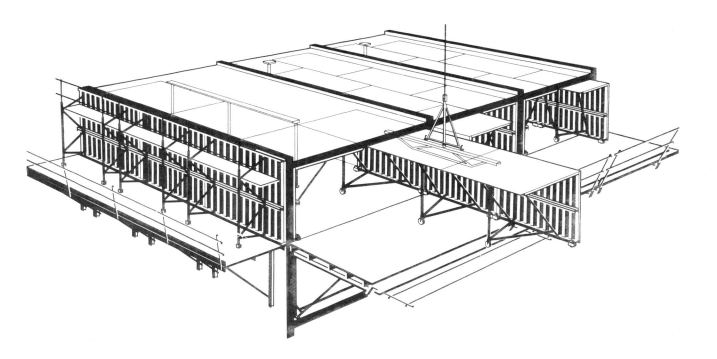

Figure 9.6 The advantages of pre-planned prefabricated formwork for sewerage work which often presents repetitious casting. All operations are clearly defined and labour and material wastage is eliminated. (Outinard).

Where more than one material is used, the connections between the materials must be carefully arranged to ensure that these are adequate fixings to transfer the stresses, this being particularly important where grp and timber are involved because the bearing and shear stresses at the connections are critical. These connections can be improved by the inclusion of embedded or attached washer plates or some other type of local strengthening or stiffeners, each connection being carefully designed and detailed. The forces imposed by handling, lifting and even some methods of stacking can be greater than those imposed during the construction process when the form is duly braced and tied at regular intervals.

The adoption of special formwork techniques translates the formwork activities into situations of mechanical handling and the designer needs to concern himself with what is essentially a machine which, to provide a satisfactory return for investment, must be used often and methodically. The essence of economy is re-use and the promotion of a continuous cycle of work for those who fix reinforcement and place concrete.

The frequency of use, which involves early removal of formwork, introduces the need for care in curing and the use of a carefully devised system of standing supports. These are essential to the complete success of the special formwork arrangement and should feature in the manual of information or on drawings prepared for the work.

Equally essential is the conveying of information to those who are to use the arrangement. Site staff need to know the safe rates of fill, what admixtures are to be used and the consequent reduction in stiffening rates. Instructions must include information of methodical filling and the type of vibration envisaged when the original design was prepared. At a more basic level the site personnel must understand the urgency of preparing mature kickers and providing suitable footings or kentledge to allow bracing and strutting during erection and concreting. The bracing and support arrangements are particularly critical when only part of the form is erected or installed. For example, for one side of a wall for the support from a kicker may only be minimal, but the support must still be capable of sustaining wind or impact loading from any direction.

An essential part of special form design is the arrangement of stopends which have to be integrated with the main construction arrangement adopted for the form panels proper. This stopend can be so arranged as to form not only jigs for continuity steel but also templates for the form proper.

Special formwork arrangements depend to a great

Figure 9.7 Purpose made formwork for columns of elevated road.

extent on the adequacy of fixings cast into previous lifts and bays of formwork, and relevant drawings should have details of the positions of such fixings whether they constitute sockets, pigtail bolts, through holes for needles, blocks for paging or packings of specially cast anchor blocks for bracing members. They must also include information on the maturity required for the cast-in component, i.e. how it can satisfactorily sustain the loads and forces to be imposed upon it.

Special formwork can accommodate the casting of larger bays and deeper lifts that can be obtained by proprietary formwork. The mechanized value of the formwork reduces the skills requirement because the process becomes a simple sequence of erect, strike and travel. Because the number of sheathing joints are reduced surface finishes tend to be of a higher standard, and the accuracy is enhanced by the improved limits of deflection and movement, and the continuity members built into the system.

Special formwork designers generally make provision in their constructional details for the attachment of external vibrators. These facilitate compaction, doors or openings inserted for placing concrete and for the provision of poker vibrators. Tamper forms incorporate facilities such as screw jacks, pneumatic or hydraulic rams which help with the striking operations. Tunnel lining formwork provides an excellent example of these various facilities which are incorporated into what is virtually a casting machine.

The particular aspects of tunnel operations require that a form should incorporate mechanical striking arrangements, some type of power unit and handling facilities, the whole lot being further complicated by the need for a form to travel through restricted spaces.

Checklist

Has contract been secured?

Does price form basis of tender?

What time is available for manufacture?

What is specification regarding surface finish?

What are stipulations regarding tie arrangements?

What degree of accuracy is required?

What are permissible bay sizes and of heights?

How are forms to be handled?

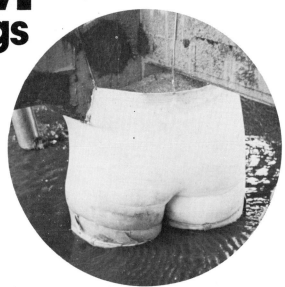

Figure 9.8 Casting in deep lifts. Note filling chutes and provision of access to critical parts of the mould.

What is crane capacity?

What is concrete specification regarding
characteristic strength?
workability?
admixtures?

How will the concrete be handled?

How is concrete to be compacted?

How many uses can be achieved?

What are preferred bay and lift sizes?

Have details of construction joints been established?

What skills are available on site?

What are details of projecting steel?

What cast-in components have to be located?

When will reinforcement drawings and concrete profile drawings be available?

What fixings, ties, holding down bolts and such like need to be incorporated into previously cast concrete?

Is kentledge available on site?

What is permissible striking time?

Is there a specified sequence of striking?

What are curing techniques?

Is accelerated curing to be employed?

Figure 10.1 An interesting solution to the Assembly Hall Column Exercise in Appendix 2. The designer has combined the attributes of ply, timber and steel to provide economic solution to a low usage problem.

10. Basic formwork construction

General

Much of the success of a formwork arrangement depends on the skill with which the forms have been constructed. To a large extent, the life and re-use value of a form or mould is determined by how well joints and fastenings have been made and on the form materials themselves.

Most concrete construction can be carried out by the utilization of simply fabricated plane panels which can provide up to 30 uses. However, where the shapes and profiles of the concrete components become more complex, then obviously less uses can be obtained from the form units. Generally it can be said that an estimator will be pleased if 12 uses can be obtained from a fabricated panel. Unfortunately, however, in certain situations only four uses can be achieved. It therefore follows that the construction of a form panel or unit should be consistent with the quality of the surface finish of the concrete and the number of required uses.

It is desirable to establish certain construction standards and to use them throughout a contract. These standards depend on available labour and equipment, and policy standards with respect to the concrete, and accuracy and finish. The tradesman and trades supervisor are quite used to certain standard methods of construction, since traditional methods of form construction employ very direct ways of fixing and fastening.

Timber formwork, traditionally, has been constructed using nail fastenings and the required shapes have been achieved by means of simple tools, such as the saw and, in some instances, the axe.

Sophisticated forms of construction are now being adopted more and more by the increasing use of hand-power tools, saw benches and bandsaws. This equipment together with machined sections and feature formers ensures that formwork can be constructed to a very high standard.

The new generation of adhesives, pva, resins and contact adhesives allow the best advantage to be obtained from bonded construction. It cannot be over-emphasized that a form component's life depends on how it it has been made, and the form designer should study the basic and traditional methods of construction so that his details combine the best of the proven techniques with the advantages achieved by the application of modern techniques. He must also understand the techniques of form construction, especially in respect of plastics, glass reinforced resins, concrete and steel, so that his details can fully utilize these materials. He will not know everything about the various techniques, but he must be prepared to consult those experts who possess knowledge of the respective materials.

The principles of construction revolve round the following basic requirements:

Containment

Resistance to grout infiltration

Accuracy that is consistent with the specification

Surfaces which are capable of imparting the required finish to the concrete

Construction that is consistent with the number of uses required

Ease of removal from the freshly cast concrete without damage to form or concrete

Construction details

The actual construction process begins during the early stages of design, when the decisions are made with regard to material selection as discussed earlier in this book. The selected materials obviously affect the disposition and direction of the carcassing members, the sheathing spans, edge treatment and the way in which features and recesses are formed. These must be carefully considered and incorporated within the designer's construction detail.

The designer should decide, by calculation, or by some empirical means, on appropriate spans as required by the specification. He must then produce sketches and drawings from which the tradesmen can construct the forms.

Form panel construction

Ideally, there should be some space allocated in a suitable site shed or workshop for panel construction. The area should be floored to give reasonable conditions, while the regulations with regard to working conditions, space ventilation, lighting and heating must be in accordance with the Factory Acts.

Various types of construction require particular types of stooling or benches, and where geometrical forms are involved it may be advisable for the construction to be carried out over the full size setting out area. Whatever arrangements are used, a level surface will be required which has sufficient headroom to allow the panels or components to be turned during construction as required.

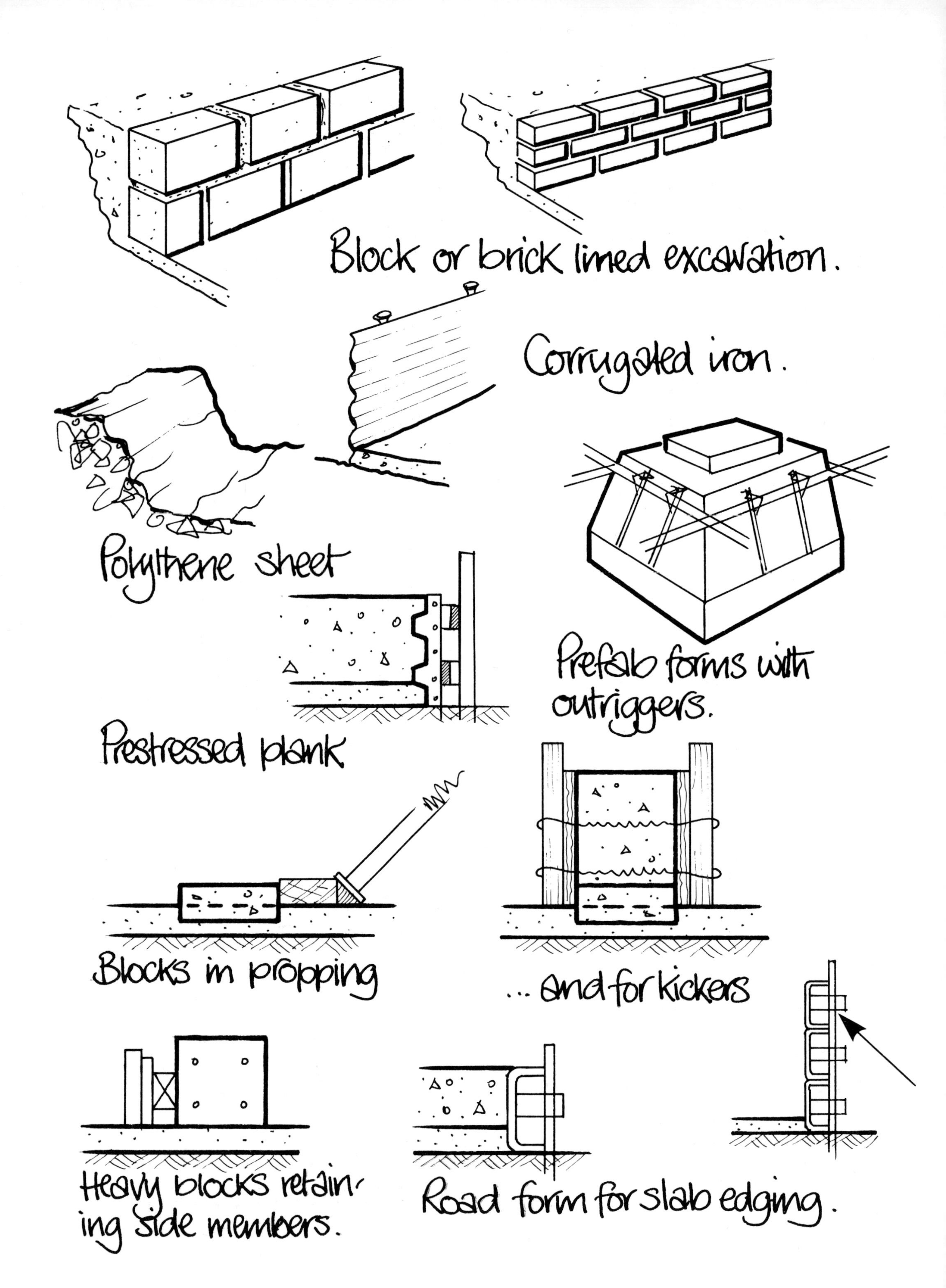
Block or brick lined excavation.
Corrugated iron.
Polythene sheet
Prefab forms with outriggers.
Prestressed plank
Blocks in propping
... and for kickers
Heavy blocks retaining side members.
Road form for slab edging.

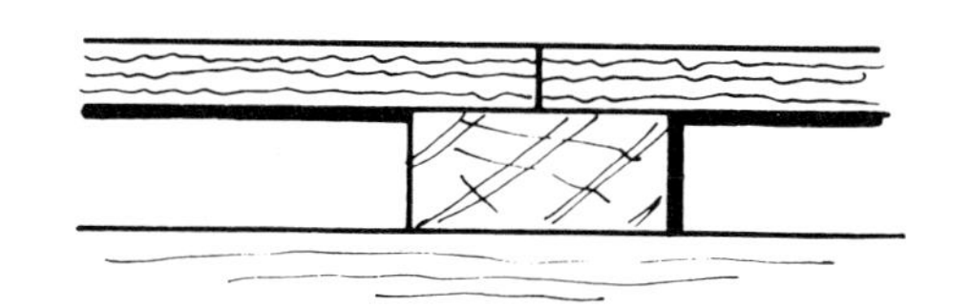

First layer carcassing laid flat provides sound bearing for sheathing.

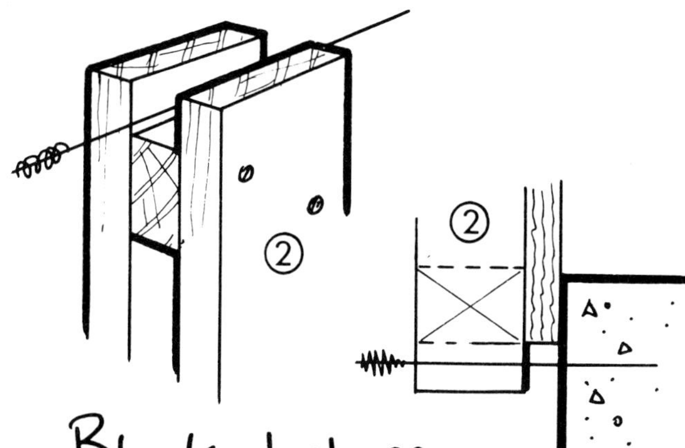

Blocks between twins allow tie or form support.

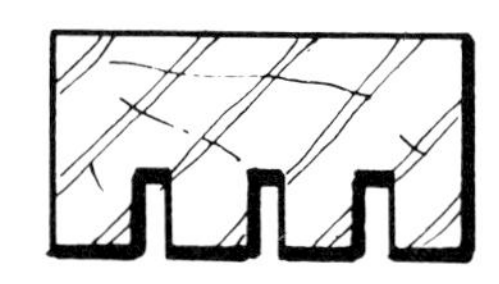

Saw kerfs in stout materials avoid cupping, bowing and distortion.

Shouldered joints improve stiffness of frame.

The slot between twins allows latitude in fixing.

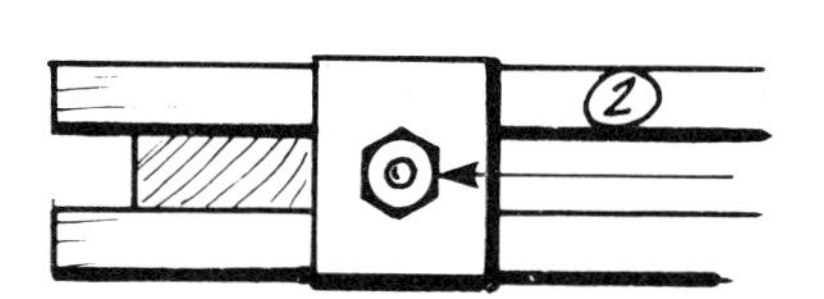

Blocks at end of twin can be used to restrain ties.

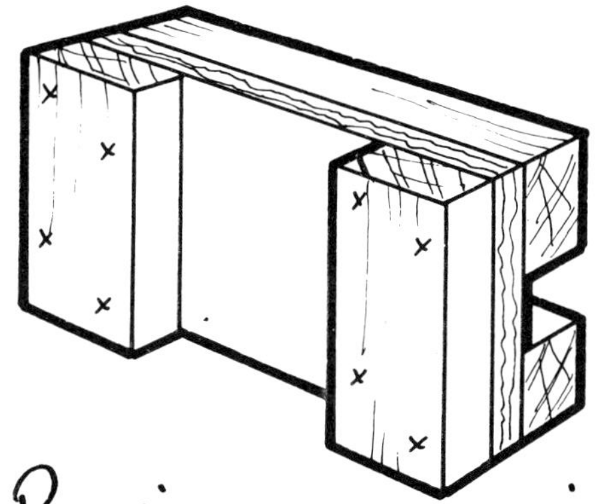

Ply diaphragms stiffen carcassing and frame.

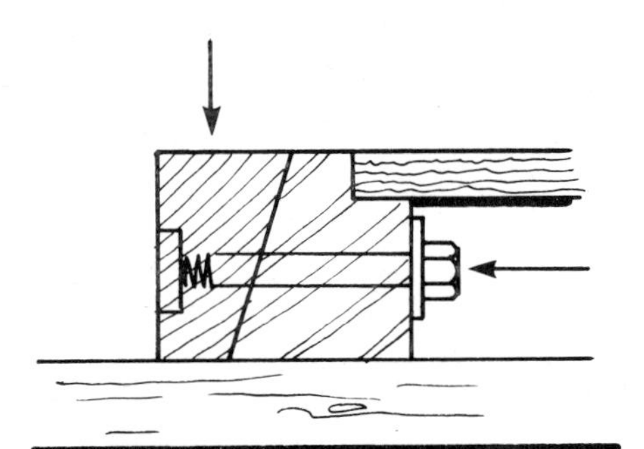

Striking fillets allow removal from between square reveals.

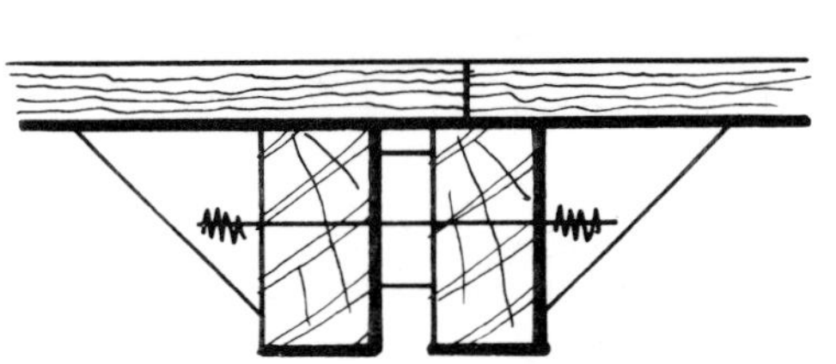

Joints should include lap and land to ensure line of face.

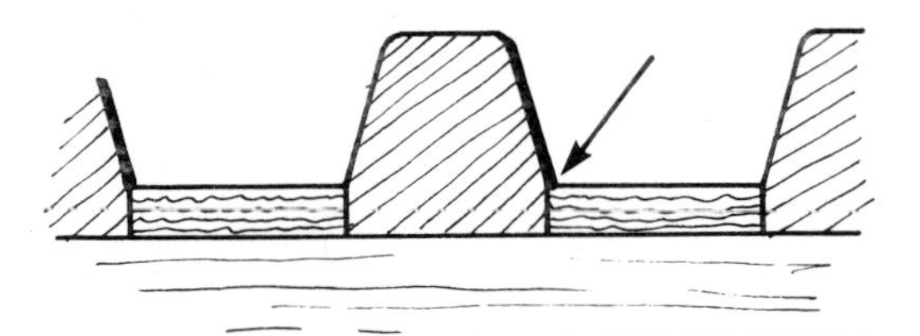

Ply fillets form rebates into which features can be mounted.

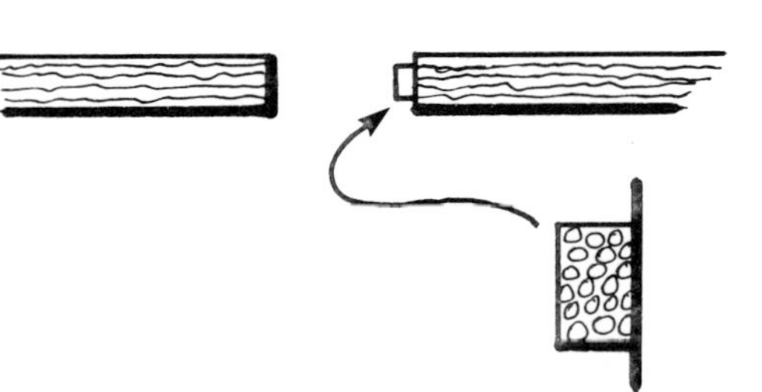

Closed cell foam prevents grout loss.

All features and fillets should have rounded arrisses.

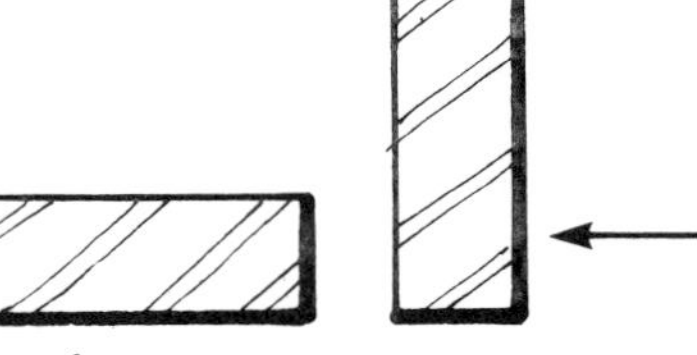

Side members should always clip soffits.

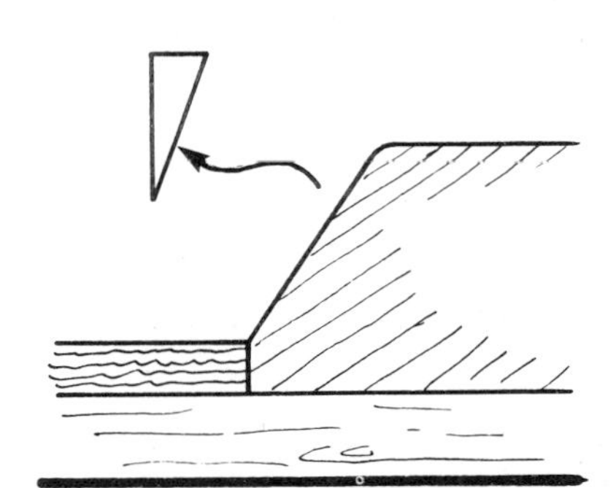

Ends of features should incorporate min. 1 in 6 draw.

Wedges provide simple fixing adjacent to soffit.

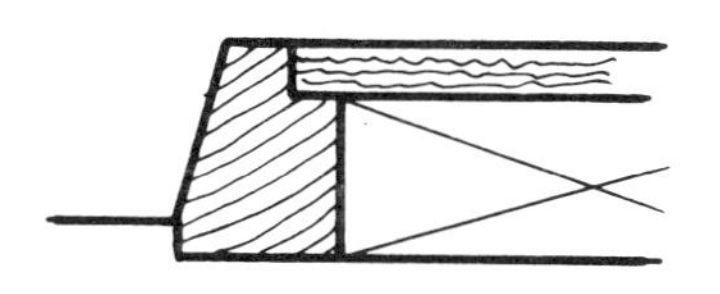

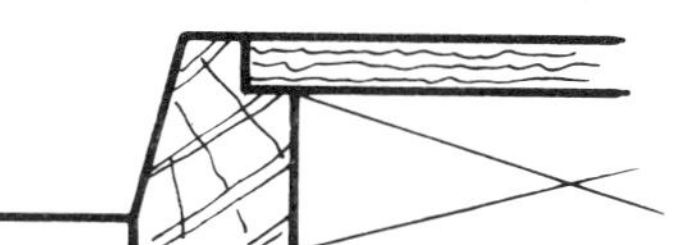

Hardwood or softwood fillets form feature and mask end grain.

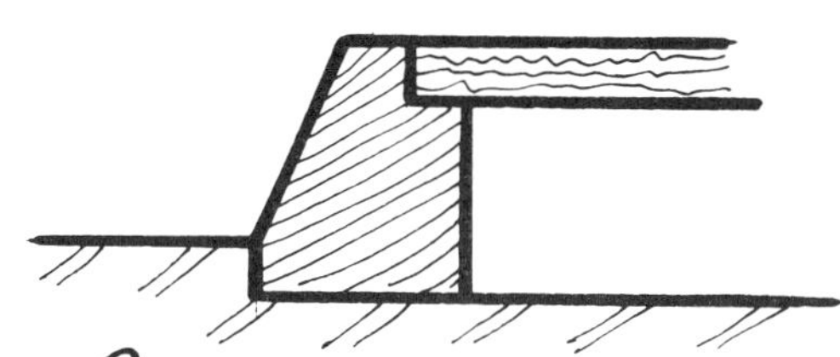

Rebates resist grout infiltration.

Staggered joints in carcass maintain line of face.

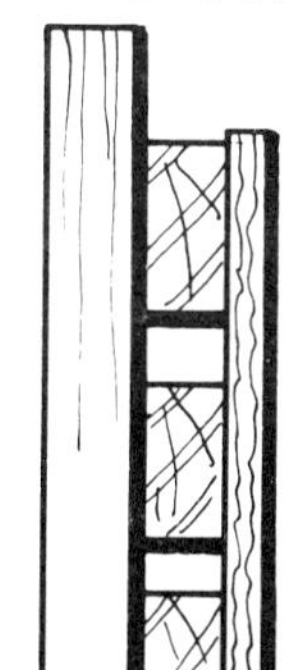

Open spaced carcassing economises in material.

Folding wedges provide cramping action.

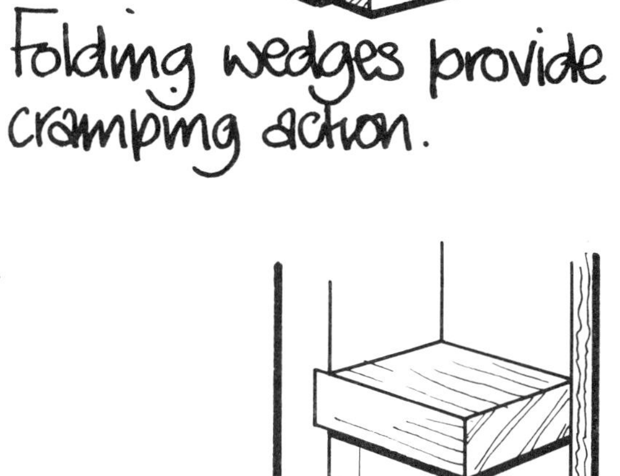

Blocks and housed joints resist turning and stiffen backings.

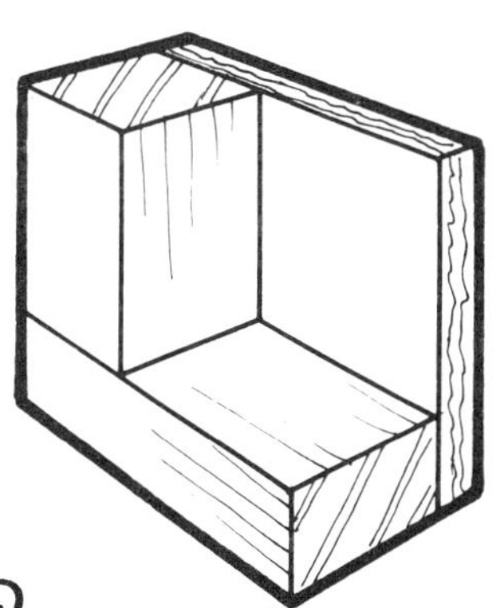

Ply mounted on timber frames lasts longer.

Modular panels increase re-use potential.

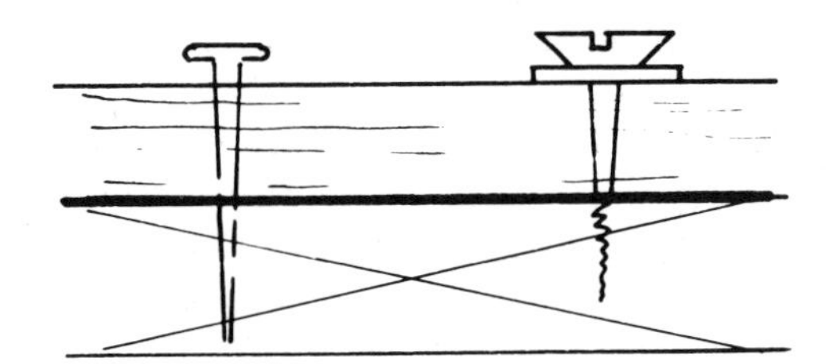

Nail heads left protruding and screws with washers ease striking.

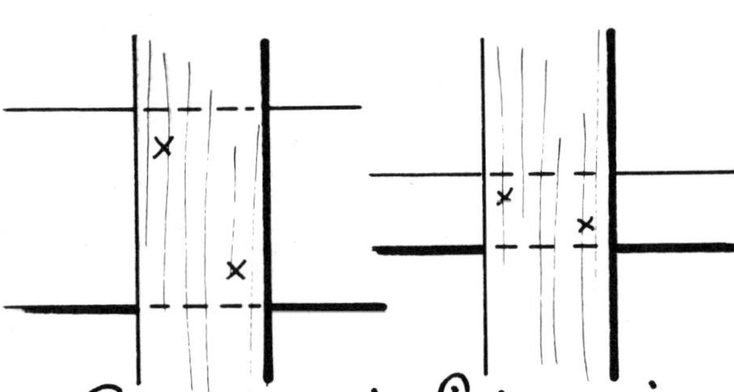

Staggered fixings improve joints. Allow 3mm at each for bearing.

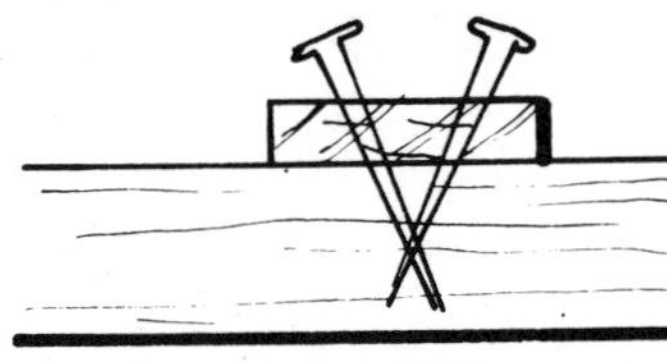

'Dovetailing' improves resistance to pull out.

Always withdraw nails at time of striking.

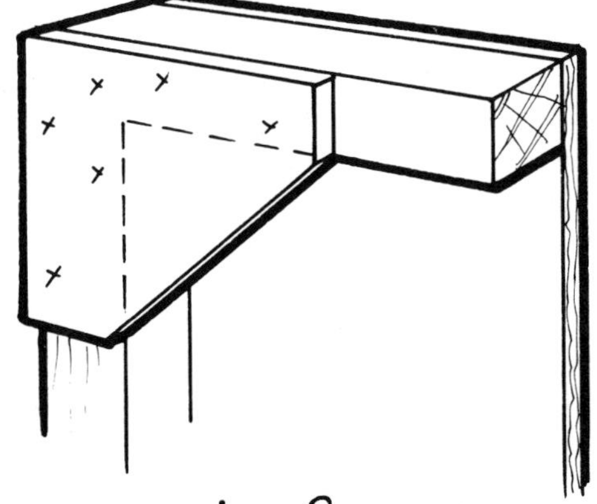

Ply gussets from offcuts improve simple joints.

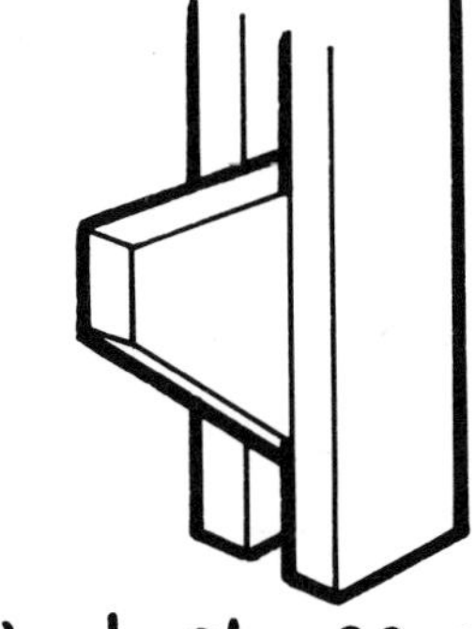

Blocks between twins support walings.

Always use substantial washers.

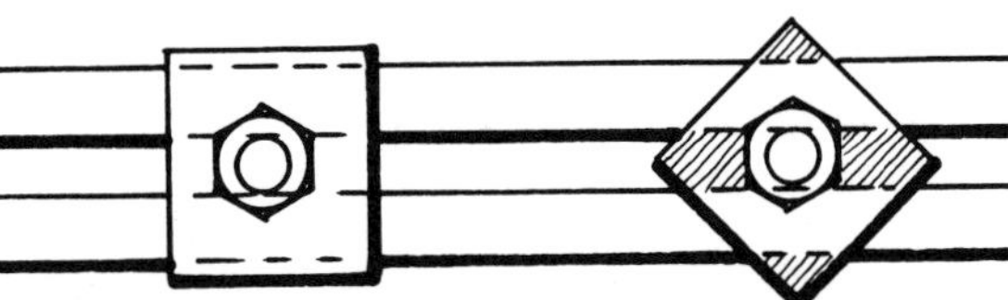

Ensure that washers are of sufficient area and are placed to maintain bearing area.

It is useful if standard bench heights or trestles can be used in the workshop, placed side by side or end to end, to accommodate the largest panels to be constructed. The provision of some selected long members of timber of 225 mm × 100 mm nominal section in lengths of say 6 m may prove helpful. These can be laid over the trestles to provide a grillage on which the form can be constructed.

When considerable repetition becomes necessary, either with form panels or component parts, various jigs and templates should be built. These can be used to provide uniformity and accuracy of location of the members within the full construction, or for accurate spacing of drillings for through ties. Certain hand power tools such as electric drills, electric screwdrivers, nail and staple guns are indispensible since they reduce the actual time of manufacture.

The tradesmen fix deals onto the timbers or benching because these locate the main carcassing members and soldier members or bearers. Succeeding layers of carcassing are then securely fastened with coach screws, wood screws, bolts or nails. Where there is a feature or some arrangement for forming a concrete profile, it is necessary to build up a suitable framing of members which are stiffened as required, with struts or ply diaphragms, to provide an accurate and rigid backing to the selected sheathing material.

To achieve the best returns from a panel it should be constructed in such a way that the basic plane carcass can be re-used on a variety of jobs. It should also be such that the sheathing, and particularly any special feature, can be removed and the whole panel re-sheeted for further use.

Attention must be paid to the standardization of the panels and joints while other panels should be made either by lapping and bolting, or by lapping only, the connection being maintained by the form tie which must be square and accurately handed.

For changes of plane, arrangements must be made for masking the end grain of timber or ply, and where returns and striking pieces are incorporated close joints must be achieved and possibly gaskets inserted to resist the infiltration of grout.

Fillets and features which provide details in the concrete face should be firmly attached to the form face and, where possible, let into the sheathing face either into rebates in softwood sheathing or, with ply faced panels, inlaid within the thickness of the sheathing. All corners should be treated to generous pencil rounds which prevent spalling of the timber at the corners arrises. Particular care must be taken in achieving smooth grain-free surfaces at re-entrants and the sides of features to avoid the forms becoming keyed into the concrete.

Where pairs of 'handed' panels are produced, each should be built against either the matching panel or the backing members. This ensures that the backing members will register when the opposing panels are erected, and through ties can be inserted, normal to the concrete face. It is essential that the ties are used in conditions of pure tension.

Carcassing members which are used in twin format should have blocks inserted between them and should be bolted or screwed to form a rigid assembly. They are thus restrained from rotating under load, while the blocks provide what is virtually a slot through which tie bolts or rods can be inserted at virtually any position.

Sheathing may be of board or ply. For ply, a thickness of 19–25 mm is normal; this needs little support for ordinary form purposes. However, when the thickness is 10 mm an open boarded sub-sheathing arrangement becomes necessary. Thin ply on a sub-sheathing allows the sheet to be replaced after several uses, especially when the face shows some sign of deterioration.

Figure 10.2 Shaft lining being cast by formwork which is handled by cable winch.

Figure 10.3 Groundwork and foundations accurately executed provide the basis for successful operations carried out above. (Amey-Fairclough.).

When geometrical forms are manufactured the arrangement at the face may comprise either shaped laths or cams fixed to shaped ribs, or battens sheeted with ply of such a thickness that they can be bent to the required radius or form. Thin 'aeroply' can be used for tight radii. Where a lot of geometrical work is involved, or where features extend for a considerable distance along a beam or wall, or perhaps where forms are being constructed for the casting of an artist-designed feature, the forms should be built for the whole feature or face in one piece, the backing members being arranged so that the large form can be cut to suitable, sized panels which fit tightly together when erected. This technique is valuable for the manufacture of forms for casting cambered soffits. Each panel, as it is constructed, should be indelibly marked with a number and hand to the schedule, and then, as prescribed, treated with a sealer, lacquer or paint. This should be carried out when there is little likelihood of dust settling on the freshly coated surface. Some of the new coatings give off heavy fumes which can be easily ignited, so that every precaution should be taken to ventilate the workshop when a coating is being applied.

Stopend cleats must be firmly bolted or screwed into position, preferably through the sheathing and into at least one layer of the carcass.

Mould panel construction

The basic principles of mould construction are similar to those employed in form construction. However the tradesman or mould maker fabricates a complete container which comprises base, sides, features (if incorporated) and stopends. Construction normally begins with the fixing of the sheathing to soffit bearers which are either set into a suitable jig, or fastened to the trestles or bench on which they are being made. Next, plates or wedging blocks, which are used to clamp the form during casting, are fixed so that the base forms a jig upon which the side bearers can be located for the construction of the side panels. The sheathing can be either ply or timber boarding, although for moulds thin ply and runners on a sub-sheathing are usually adopted.

For simple moulds, used for small products, bearers of say 75 mm × 50 mm in section, slotted at the top for the tie rod and splayed at the bottom to allow a seating for the wedge clamp are sufficient.

Large moulds for bridge beams and heavy structural components demand something more substantial, and the soldiers may be of twin 150 mm × 50 mm members blocked and fastened as for forms employed on in situ construction.

Large moulds are best constructed as whole panels and then cut into smaller components; this ensures the line of the features and details. Where geometrical forms and heavy features are concerned, they should be built onto framed supports that are stiffened by diaphragms. In each case the joints should be staggered to ensure maintenance of line when the mould is re-assembled.

For precast concrete production, moulds are generally required to cast 30 or 40 components and these are grouped in families of similar section or form. The face formers or features are usually arranged so that they can be moved about on the face of the main panel to accommodate various profiles.

Skew-bridge components must be constructed so that the correct skew or pitch is achieved and any formers or ducts for transverse post-tensioning cables should be capable of producing a fair line.

Blocks or clips should be arranged to ensure that:

(a) The side members do not lift during concreting, particularly where I-beams or inverted T-beams are being manufactured

(b) The sides remain correctly located longitudinally with regard to the base so that the stopends located by the cleats on the sides remain square to the beam proper.

The way in which precast moulds are tied is critical in maintaining the correct section. Spacer blocks, tube spacers and various arrangements of combined spacers and ties can be incorporated. Whatever arrangement is employed it must be capable of maintaining the correct location of side members yet allow speedy removal at the time of demoulding. Threaded components should be avoided at the foot of soldier members, while over-lever clamps or simple wedges are excellent for tying mould sides and the bases. Adequate space should be left at the top of the mould to allow trowelling and finishing to be carried out unhindered; clamps or ties which incorporate barrel spacers provide a simple tie arrangement.

System moulds and gang moulds

In precasting work considerable use is made of steel carcasses which can be clad or suitably modified by the insertion of blocks or formers to achieve the required profile.

With tilting tables or frames the profiles of particular panels can be generated by ply which is set onto various shaped wedges as formers. Features are set into the ply facings and the usual precautions taken with respect to possible grout leakage. Where heavy vibration is applied precautions must be taken to avoid infiltration of grout at the ply sheathing joints, and mastic or resin adhesives should be applied to fully seal the joints.

With battery moulds for walls the height and profile of a panel are achieved by inserting timber end formers, base stooling and opening and door formers. These are generally constructed in softwood with ply gusset pieces to maintain squareness. They should be made up in jigs on plane surfaces to ensure accuracy, and gaskets inserted to take account of the shrinkage which occurs during the early stages of curing.

Erection and striking

General foundations and beams

It is essential that some arrangement be made to drain or dewater the workplace. This may be effected by full scale dewatering arrangements or by the provision of sumps and pumps. The workplace must be made safe by the use

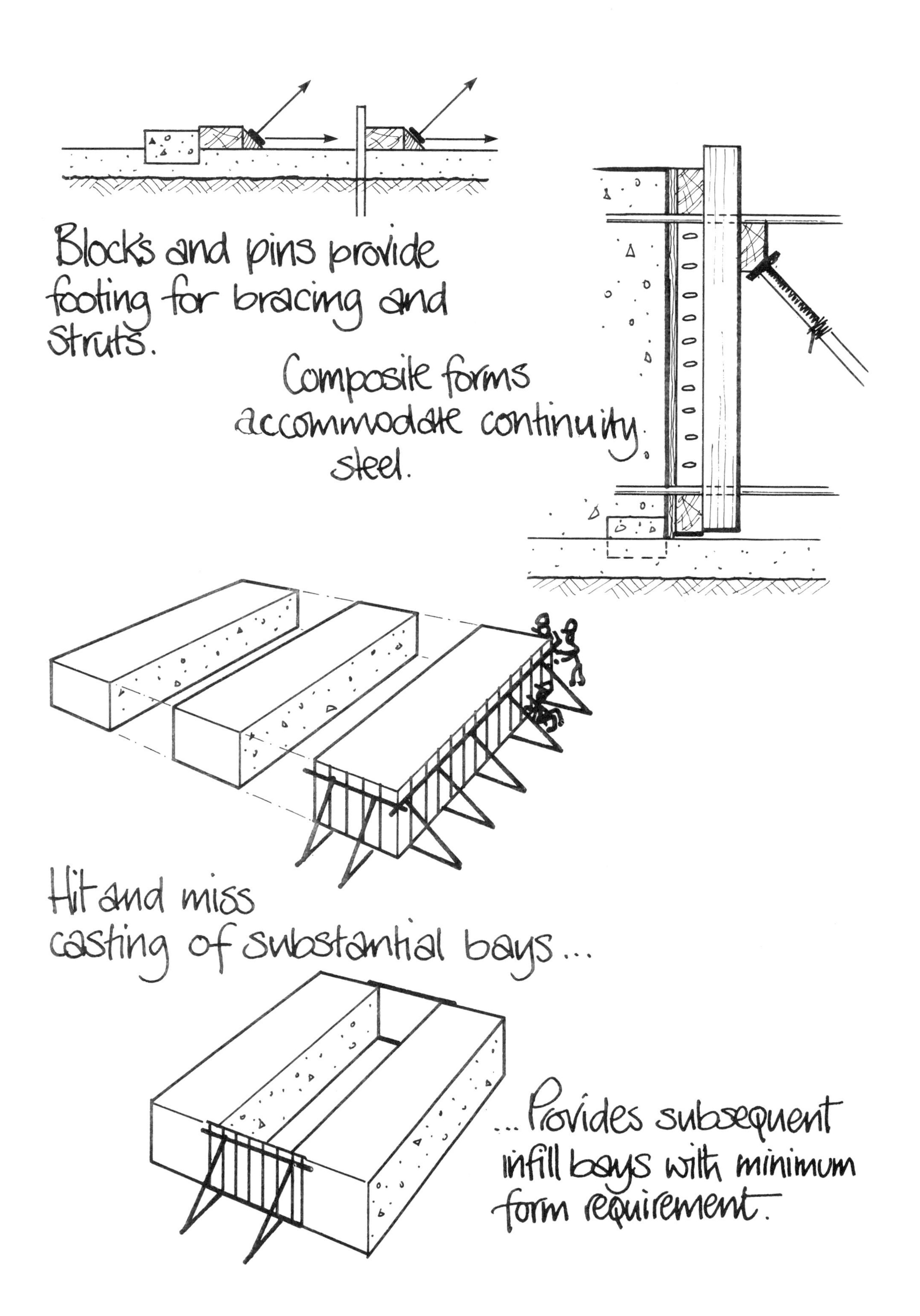

Raft construction

of sheet piling or normal timbering and strutting. Often the workplace is a congested area and this can present a danger because operatives who work in confined spaces may dislodge struts, or even temporarily remove them to gain access or head room. All timbering and strutting must therefore be carefully checked to avoid any possible accident.

Where access is particularly difficult, it may be necessary to construct the forms piece by piece away from the workplace and to treat the formwork as lost formwork and simply backfill on completion without removing the equipment, or at least, the sheathing. Corrugated sheathing, concrete planks and similar materials are excellent in such circumstances and are cheap enough to be considered expendable.

Work in the ground can prove to be difficult with regard to achieving suitable supports, and while timber struts can be utilized it is necessary, where substantial forces are involved, to cast concrete sleepers, drop precast kentledge units into the ground or to take definite steps to avoid movement from waterlogged ground or even from what may appear to be apparently quite sound ground.

It is often convenient and profitable to place a blinding on the ground in addition to those areas which are immediately adjacent to structural components. The blinding, which virtually can be any scrap concrete, provides clean working conditions upon which men can operate safely and speedily. Where props or supports are to be founded on the blinding the thickness should be increased beyond the 75 mm normally employed.

Apart from access, the clearance between the form and the excavation timbers or sheet piling should be checked to ensure that there is sufficient space for the forms to be inserted and enough room to allow withdrawal of the ties. This is not a problem where deep breakback ties are used although some of the larger she-bolts and especially one-piece through ties can present difficulties.

All formwork should be prefabricated and then lowered into position taking account of headroom or access. On foundation work or where work is carried out above foundation level accidents can occur as a result of falling debris. This usually occurs where items have been stacked adjacent to the edge of an excavation and therefore these should be kept clear at all times.

Figure 10.4 Care in the provision of accurate lead-ups and kickers will facilitate further stages of construction.

Single-sided work

Where single-sided work is being executed some support may be taken from the sheet piling by means of hook bolts or welded nuts, although if the sheet piles are to be withdrawn welded attachments must be avoided.

Anchors should be cast into either the foundation slab or retaining wall kicker, and where necessary into purposely thickened areas of the blinding. This provides a sound support for the struts and braces which are used to locate accurately the forms for the first lift.

The problems associated with the strutting of single-sided work can be eliminated once the first lift has been placed because pigtails, loops or sockets can be incorporated with this lift to provide secure fixing of subsequent formwork. An allowance should be made for some degree of inward displacement of the top of the forms throughout succeeding lifts to take account of the deflection that is bound to take place in the cantilever members or soldiers.

Because of the substantial loading that develops from the cantilevers, the anchorage should be embedded in sound, mature concrete to provide a firm fixing.

Pits and tanks

Special requirements are necessary with regard to the construction sequence, treatment of construction joints and the arrangement for tanking and waterproofing. For formwork the key points relate to achieving substantial fixings with secure fastening of formwork to previously cast concrete.

The scale of the operation will dictate the sizes of bays and location of joints. The structural engineer is concerned with avoiding rotation or some other disturbance of a freshly cast tank structure. The formwork must provide a watertight container so that the concrete can be adequately compacted without any loss of paste. Sound formwork should allow the full effect of the vibrators to be utilized thus facilitating the placing of concrete in difficult sections.

Where conical hoppers or tanks are to be cast, a central pad should be cast initially to secure the first former. Concrete cast around this former should incorporate loops or pigtails for securing subsequent lifts. It is debatable whether construction joints need be formed in tapered hoppers, and details invariably show the joints between adjacent lifts as being normal to the concrete face. In practice, the best joint can be formed horizontally, because attempts to form the joint normal to the face can be upset by the application of the normal compactive effort, whether applied by external or internal vibration. A joint normal to the surface of sloping work may indeed result in poor compaction being achieved. For most purposes a grout check or indent batten avoids a feather edge joint at the form face and the concreter can then apply the necessary compactive effort to achieve good sound concrete at the critical interface between lifts.

The casting of concrete to sloping or battered faces must be carried out fairly slowly to prevent air becoming trapped under the top form, while for deep lifts, doors may be provided to allow continued placement of concrete. It should be remembered that whatever precautions are taken, it is impossible to avoid the position of these doors being evident on the finished concrete, although it is unlikely to be critical. The real problem is the fit of a door within a form system that is under load, since it is extremely difficult to achieve a good seal, and line the faces.

Regarding the fit of individual panels in the form system for tapered tanks, these should not be as a nett fit from one splayed end to another because it is almost impossible to ensure that the panel fits exactly as designed within the given height. Vertical movement obviously affects the length of the panel and end splays as can any deviation from the correct pitch. Where conical faces are sheathed these problems can become excessive. In most locations packing should be inserted adjacent to a corner, or it should be suitably supported at mid-panel position to avoid any difficulty with the fit and to permit levelling of the form panel.

Tie rods, bolts or other holding down arrangements used to resist uplift in hoppers and on sloping surfaces should be taken right through the formwork to the carcassing members and the carcassing or supports. The formwork is thus put into compression and strain or distortion is avoided. For massive forces caused by uplift, any bolt or tie which is secured to sheathing, or even the first line framing only causes the form panel to be torn apart because of the stresses which develop during construction.

The form designer should discuss the actual concreting technique with the section foreman so that the correct placing arrangements are maintained. For sloping sections the concrete should be placed evenly in layers, and the fill regulated to prevent air becoming trapped. Any technique used for filling the forms must ensure a progressive uniform fill all round the form or core, otherwise the pressure of concrete developed either by the locally increased height of lift or as a result of vibration, may cause lateral displacement and uplift.

It cannot be over-emphasized that vibrators are used primarily to apply compactive effort – not, as is often the case, for distributing the concrete. Concrete should be placed in its final position before any compactive effort is applied.

To this end the new super plasticizers now commercially available can be used where there is top formwork or overhanging forms, since they promote flow and help with the placing of concrete. However they increase pressures so much that the design needs to be based on full hydrostatic pressures which can develop within a form.

Very fluid concretes present further problems since they search out joints and openings in the system and 'key' the forms into position. This is particularly applicable to fibre reinforced concretes where the fibres become trapped in hardened fins of concrete between the form joints and lock even the most carefully constructed forms. Where this is liable to occur all joints between the panels and sheathing members should be taped using one of the specially developed form tapes that adhere to used formwork.

When considering the concreting method for a hopper or tank located in the ground, there are two construction techniques which help to prevent flotation of the whole concrete structure. The first employs a sump arrangement which, in conjunction with dewatering, allows water to be pumped away in a controlled way, while the second consists of placing concrete in lifts large enough to avoid uplift. This particular technique often requires that the concrete be ready-mixed, and a full set of forms readily available.

To summarize the main points of concern to the designer for work in the ground:

Possibility of prefabrication

Working conditions

Safety, especially with respect to excavations

Shoring and reshoring arrangements in excavations

Work in confined places.

Formwork for columns

The most critical aspect of column casting involves the achievement of good, dense and well compacted concrete. Access for the vibrating and compacting operations is often congested either by projecting starter bars or by connections into the beams that comprise the reinforced concrete structure.

While the mechanics of pressure are often understood with respect to concreting columns, the considerable pressures generated within tall slender columns are often overlooked. Fortunately, the versatile column clamp which is used to fix the column sides is generally available on a site so that there have not been too many failures.

Because plywood requires secondary support by means of soldiers, the original problems associated with forms that were constructed with individual boards attached to cleats have now been more or less eliminated. The use of plywood has also done away with the board joint which always presented problems of appearance. Where boards are used for surface finish purposes careful attention to the sealing of the joints is necessary. Where board markings are to be expressed, only thin boards mounted onto a plywood form panel should be used. A kicker is essential with regard to position and for producing a grout-tight seal at the foot of the column. It gives a check on accuracy and provides a means of avoiding the column box being rotated by the props used in the plumbing operation.

The column form should be erected as accurately as possible and rigidly fixed for the concreting operation. Particular care is required during concreting to reduce the amount of impact incurred and the dumping of concrete from excessive heights into the form. Equally important is the need to minimize contact between the skip and the form. This can only be managed where the concreters have good access and can control the positioning of the skip.

Where possible, bracing props should be taken to previously cast concrete, or to specially positioned anchors set in kentledge blocks. The small amount of labour involved in locating some simple fixing, socket loop or dowel device for attaching the prop foot or plate, will be amply repaid in terms of accurate column erection. Final adjustment of plumb must be carried out as the concrete work nears completion, and especially while the concrete is still fresh. This helps to prevent any local damage or upset to the hardening concrete at the critical junction between floor and beams at the kicker.

Sloping columns require careful attention in that progressive, disciplined fills should be made combined with thorough compaction; where possible, mechanical ties to the kickers should be incorporated. Loops, pigtails or sockets are used to avoid the form becoming displaced upwards as the fill proceeds. Physical connections should also be made where raking props support the column member. It is worthwhile to leave the lower of the column wide members in position while the remaining three sides are being removed, and until either the concrete gains sufficient strength to be self supporting, or until the structural connection is made with the beam or slab.

Column forms can be arranged to allow the casting heights by the careful positioning of joints during their construction, although a designer will undoubtedly ensure that no single column mould is used to cast too many variations. Where forms are used to cast columns to the underside of sloping or tamping surfaces, these should be employed in a two-part casting operation, i.e. all columns are cast to one level, the columns being finally capped to the required height.

If only a small variation is involved the capping cast can be incorporated within the beam- or slab-casting operation, although this may affect the stability in that it may not be as successful as that provided where the beam or slab soffit sheathing intersects with a column.

Circular columns are usually cast from forms made out of sawn ply, or timber ribs with timber laggings, and thin ply sheathing. Alternative materials that can be used are foil, asbestos and plastics piping. One novel proprietary form incorporates modular interlocking sections and can be built to various diameters simply by adding further extruded sections.

Considerable gains in accuracy and appearance can be achieved by small details of design. The building-in of solid corner members to form rounds and chamfers eliminates much of the inaccuracy of profile and poor appearance which results from the use of planted features.

Groups of columns which repeat throughout a structure either as L-shaped members or as pairs of plain columns, can be gang-cast together. Ganging simplifies bracing and helps to maintain line and position. Where possible, adjacent side members are framed together into L-shaped assemblies; this reduces the number of pieces to be handled and the number of joints in the assembly, in addition to simplifying prop arrangements.

Considerable scope for innovation exists in such arrangements, an example of which is the use of one-piece circular forms that are struck by lifting in one piece. Another example involves the use of the landing ring technique where a small form section, fabricated to exactly match the section of the column proper, is 'leapfrogged' to provide a formwork section which is securely fastened in position and upon which the main column form can be lodged to execute line and level.

Wall casting

Continuity of line of face is an important feature of wall construction. Provided that the mechanical design has been carefully considered, and the panels have been constructed by proven construction techniques with sound fastenings and good quality materials, there should be little problem in achieving this. Because a disciplined fill is necessary, an established concreting plan should be discussed with the ganger or section foreman. It should be clearly understood that concrete must be placed in layers at some pre-determined rate of fill. The ganger must understand how he is to fill, especially around openings, and he must be careful not to choke the form. For the satisfactory placing of all concrete, an adequate access scaffold must be available, together with some means of entry to the form openings. As a result of local displacement or irregularity, forms which comprise modular components or individual panels often fail to produce the required line. This is not confined to plain concrete and is often evident where circular walls are involved. On circular work displacement of a form inwards at one point almost invariably is accompanied by an outward displacement at some point on the curve.

The key to accuracy is the incorporation of *continuous* members either as walings or ribs which satisfactorily fix all the components to the required line. Bracing must always be taken to such continuity members and in this

Figure 10.5 Stopend arrangements are seldom designed, even though the pressures which develop at the form face and the requirements for sound concrete construction are just as important as at any position within a structure.

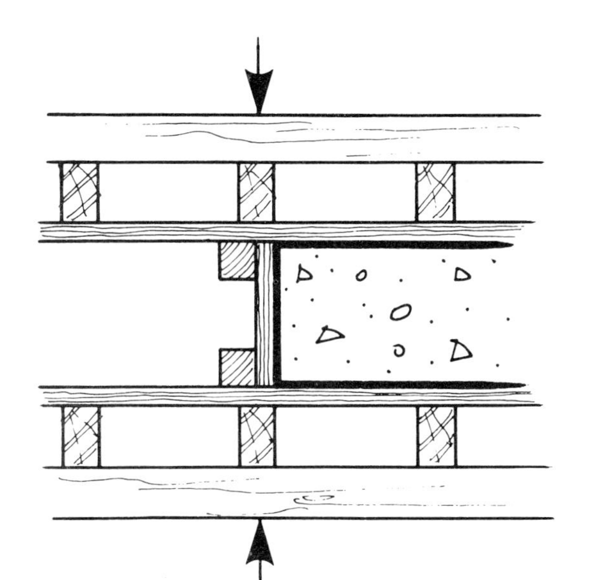

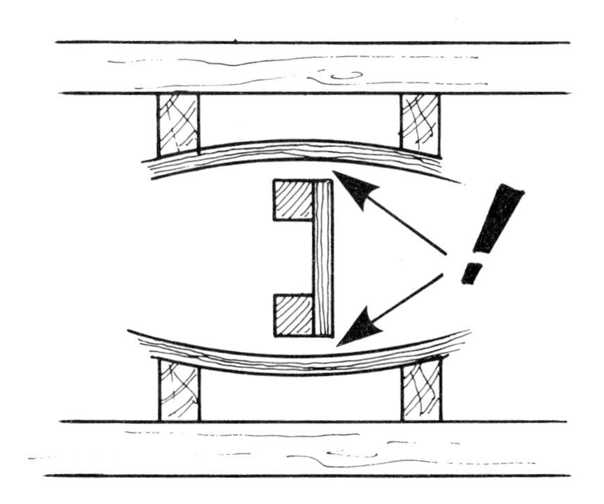

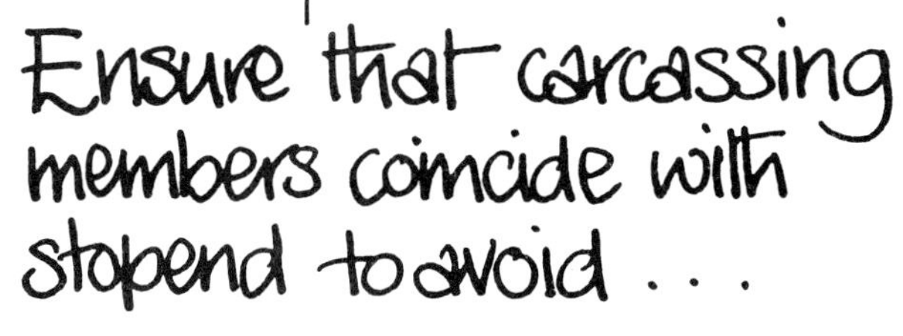

Ensure that carcassing
members coincide with
stopend to avoid . . .

. . . local deflection
and leakage.

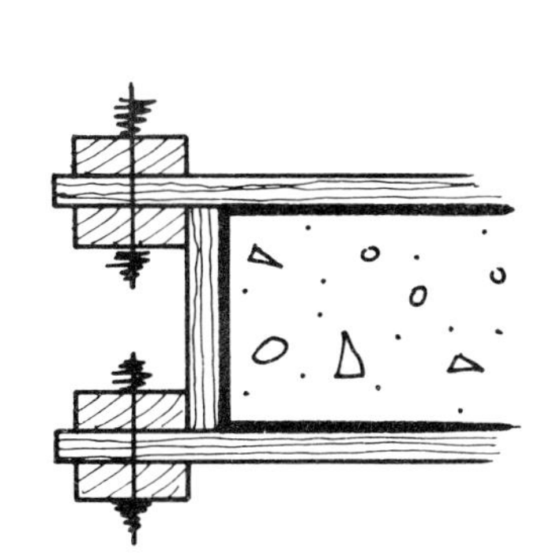

Do not rely on sheath-
ing fixings – always
bolt through plate
or carcassing.

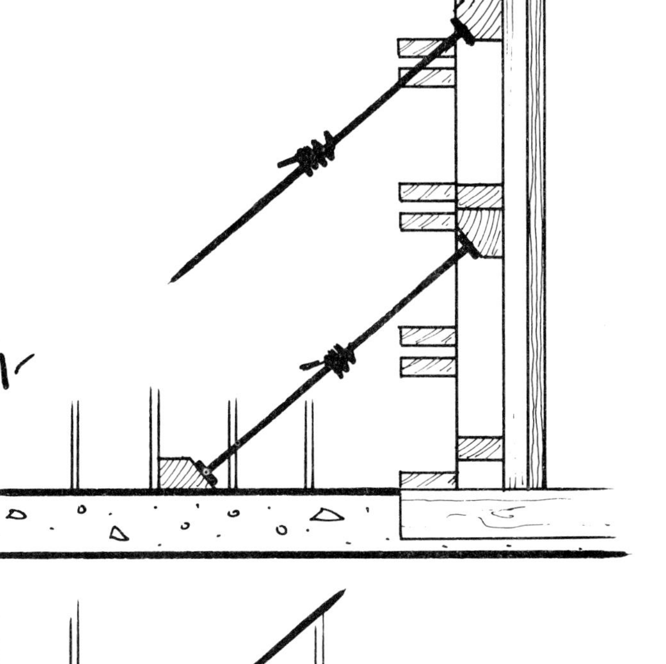

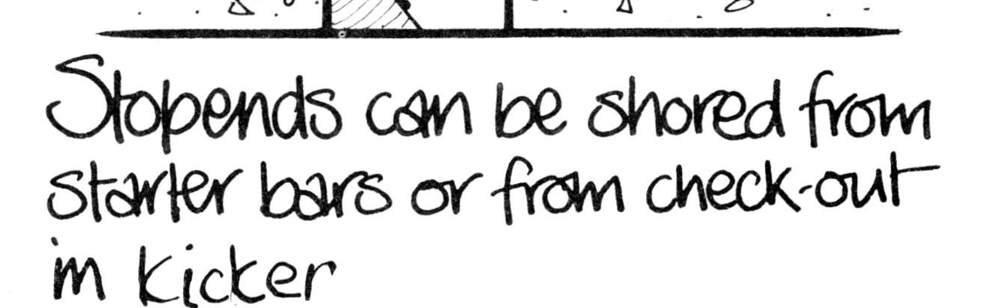

Stopends can be shored from
starter bars or from check-out
in kicker

Stopend details

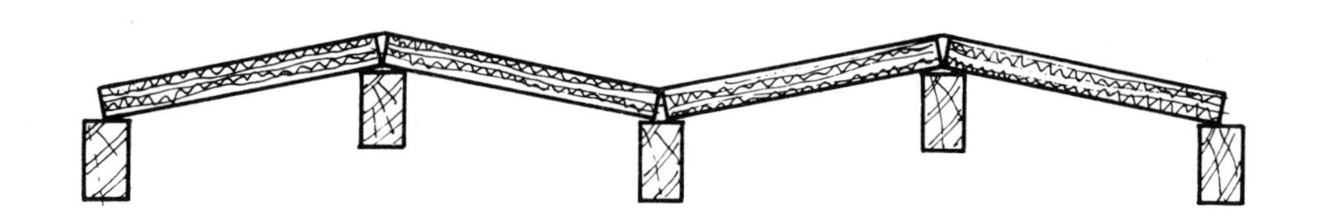

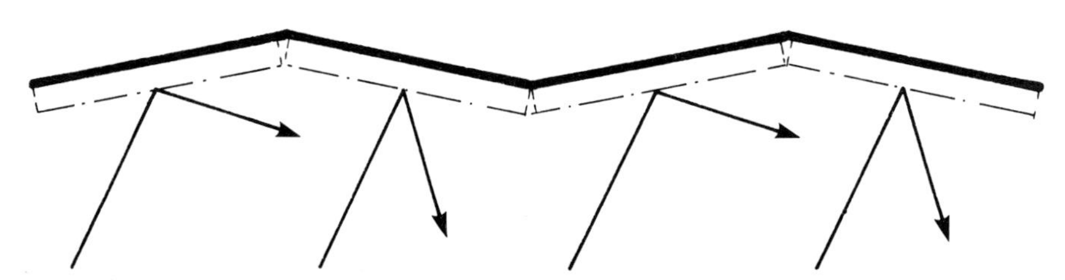

Formwork of any scale which lacks continuity members will cast surfaces which reflect light at different angles and cause variation in shading.

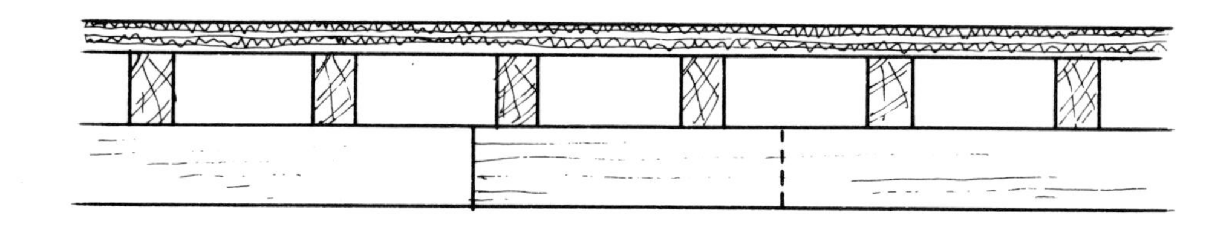

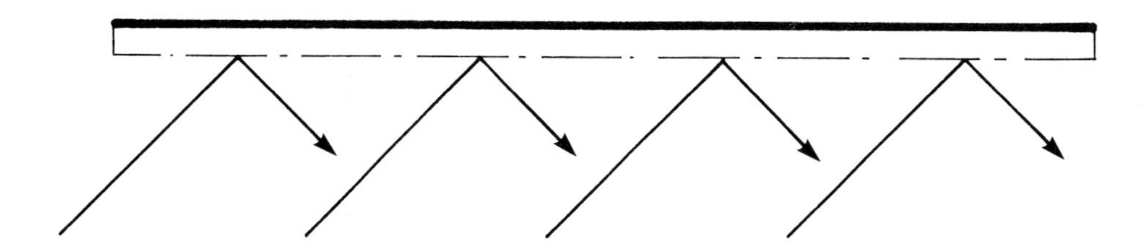

Continuity members either in long lengths or well lapped short lengths improve line of face and also allow for selectivity in tie or support position.

Continuity members

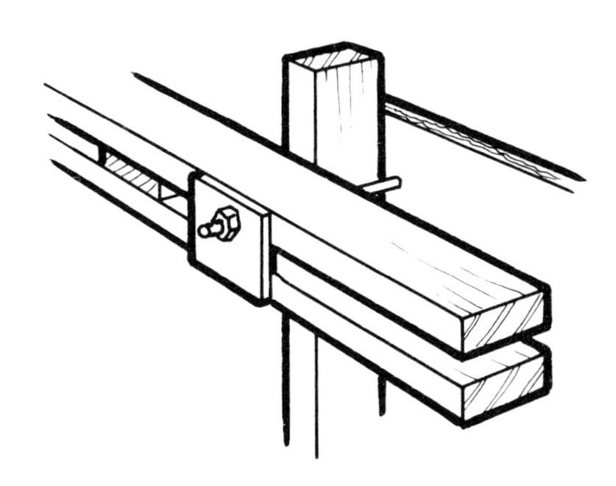

Where possible ties should be placed adjacent to studs or soldiers.

The second layer of carcassing ensures continuity of face.

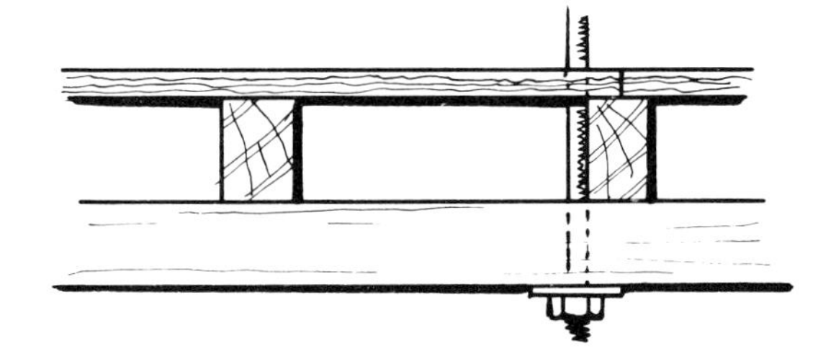

Shear rings or tooth plates improve the capacity of bolted joints.

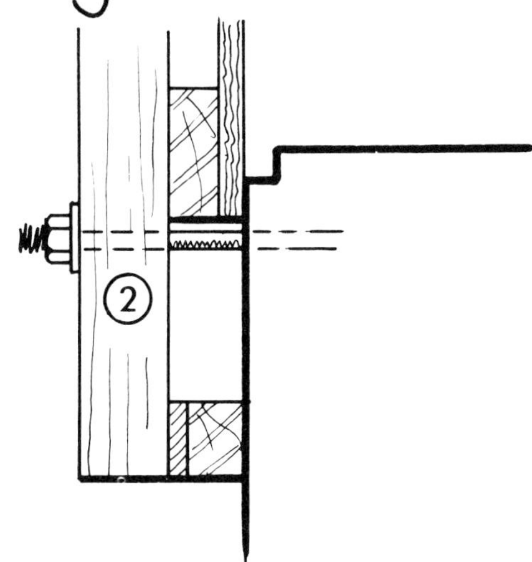

Cleats nailed to posts improve bearing.

Short projection of twins improves stability of form.

way the logical arrangement for collection and distribution of forces is maintained.

Where timber or long tubular walings are used to line formwork (and to collect the pressures and transmit them to the ties or struts) individual members should always overlap. Where there are interruptions to the line of the wall caused by openings, piers or nibs that form the starters for walls, the waling members should be so arranged as to pass uninterrupted past the projection to maintain the continuity of line.

With ply and timber forms a waling should be placed at or immediately above the kicker so that the panel does not deflect away from it, because of the concrete pressure, and thus allow grout loss. The tie arrangements should, whenever possible, be located so that the forces that result from the pressure on the form face are transmitted by the most direct route into the tie. Ties should be kept close to the point of bearing, for example between the soldier and the waling.

Large openings in walls are best treated as openings in the formwork, at least on one side of a form. Where rates of fill are high, a cill member may be inserted once the concrete reaches the level of the start of the opening. This avoids 'hungry' patches or voids which would otherwise occur with solid forms. The placing of concrete where formers are used within continuous wall forms must be carefully controlled, the concrete being placed at one side of the opening until it can be seen to flow at the other side. Once this has occurred filling can be continued at both points. During concreting, a strict watch must be kept on all bracing members to make sure that vibration or impact does not displace vital supports.

It is always desirable to spend time in discussing the joints and laps that occur in steel reinforcement. Provided this is carried out early in the structural design, continuity of work for the steelfixers and carpenters can be ensured and provision made for good access for concreting purposes so that cranked bars or floor starters do not hamper the work.

While there is continued discussion on the merits of techniques such as 'hit and miss' construction, the formation of deep lifts and other detailed considerations, it would appear that the keys to successful wall casting are:

Sound forms which incorporate continuous members

Sheathing that is capable of imparting the required surface finish

A clearly defined concreting method

Adequate access.

Formwork to walls and columns is struck at an early age and the curing of the exposed concrete is always critical. Whatever technique is employed, it is essential that moisture should be contained and the concrete protected from winds, frost or mechanical damage. To achieve maximum uniformity of appearance the striking time and the curing arrangement must be dealt with consistently, on every lift or bay. Again, early discussion with the designer can prove to be beneficial to everyone where ruled joints or battens are allowed or specified and, ideally, any large surfaces are split into almost a mosaic of smaller areas. Colour variations, which inevitably exist, are broken up by a matrix of lines or grooves that present an acceptable appearance.

Featured walls

Architects who become aware of the problems presented by large areas of plain concrete now tend to incorporate more and more features. The techniques adopted include: expressing board marking, presenting striated faces or incorporating substantial features or recesses within the face. Most contractors have had difficulty in forming such faces. The problems mainly spring from insufficient attention being given to construction details – usually that of preventing grout infiltration to the form joints and behind features.

Taking timber construction first, since it is most frequently used for specials, all feature formers should be bolted and/or bonded into position. The joints should be so arranged as to be cramped normal to the concrete face, and deep features and formers allow some form of jacking or screw withdrawal. The forms should be struck as soon after casting as possible. Glass reinforced plastics or metal provide greater latitude during striking because all the internal corners can be simply formed as 'pencil rounds'; the faces are continuous so that there is no possibility of grout ingress. With steel-faced forms care must be taken to ensure that brake pressing of formers does not result in crimps or depressions which may trap the former.

Perhaps the most recent method of forming textures is that which involves the casting of polyurethanes and rubber compounds. This is perhaps the simplest way to form featured surfaces because, depending on their formulation, they can be 'peeled' from the finished face. Preformed liners of foamed plastics have also been used to produce textured surfaces, although the economical aspects need to be carefully considered with respect to the number of uses achieved from the liner.

Floor casting

During the last ten years waffle and trough arrangements have completely changed the way in which floor designs are carried out. The economical spans that can be achieved by the use of deep waffles have been adopted for all kinds of reinforced concrete structures and the formwork suppliers have rapidly responded in producing the relevant casting equipment. Waffle and trough slabs which are cast using quick strip and double-headed prop systems provide speedy construction and help to achieve the appropriate finishes.

Traditional flat slab, and beam and slab construction however continue to provide the formwork designer with a considerable amount of work. Again, continuity of line is critical and whatever system is adopted to form the slab, continuity or the maintenance of camber, as specified, is important as is the achievement of satisfactory finishes from the form.

Table forms, system props, standard props, telescopic centres and support systems are usually well documented in the manufacturers' literature and there is little point in repeating the relevant information here. However certain essential factors should be examined which may have a major bearing on the results achieved.

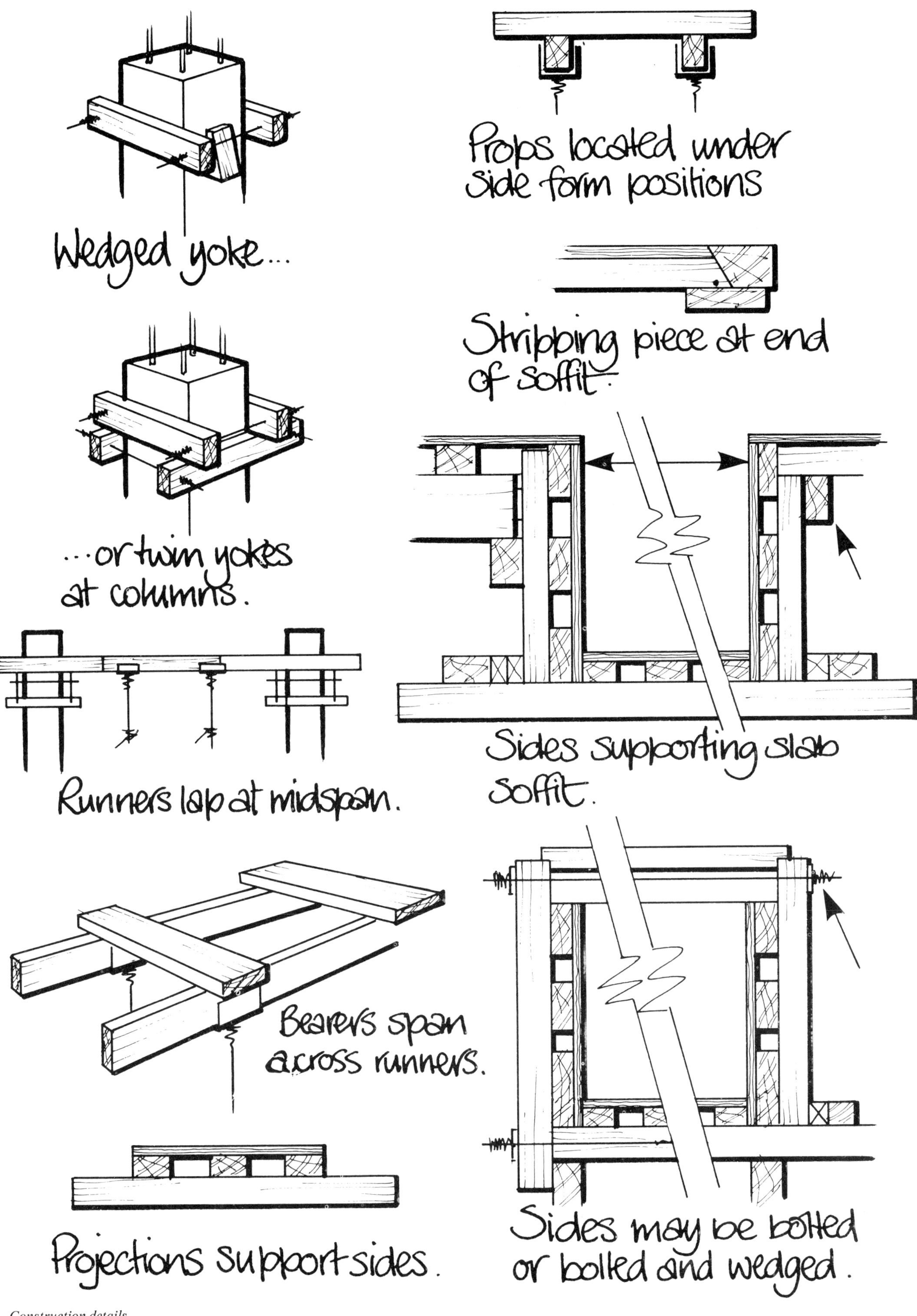

Construction details

Where ply faced forms are constructed using open jointed carcassing, blocks should be inserted to transmit wedging action.

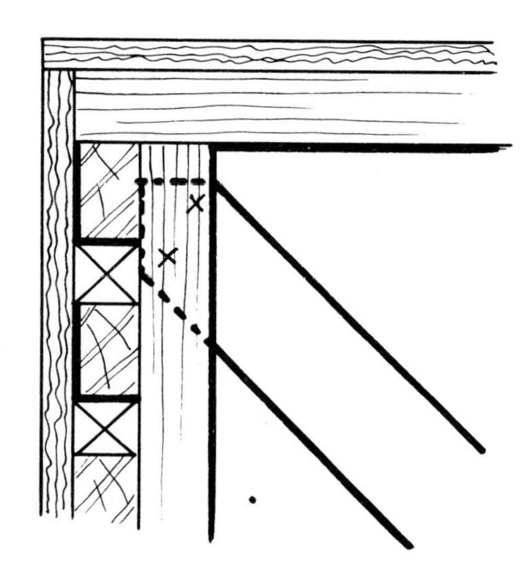

Similarly when sheathing supports bear on form side members.

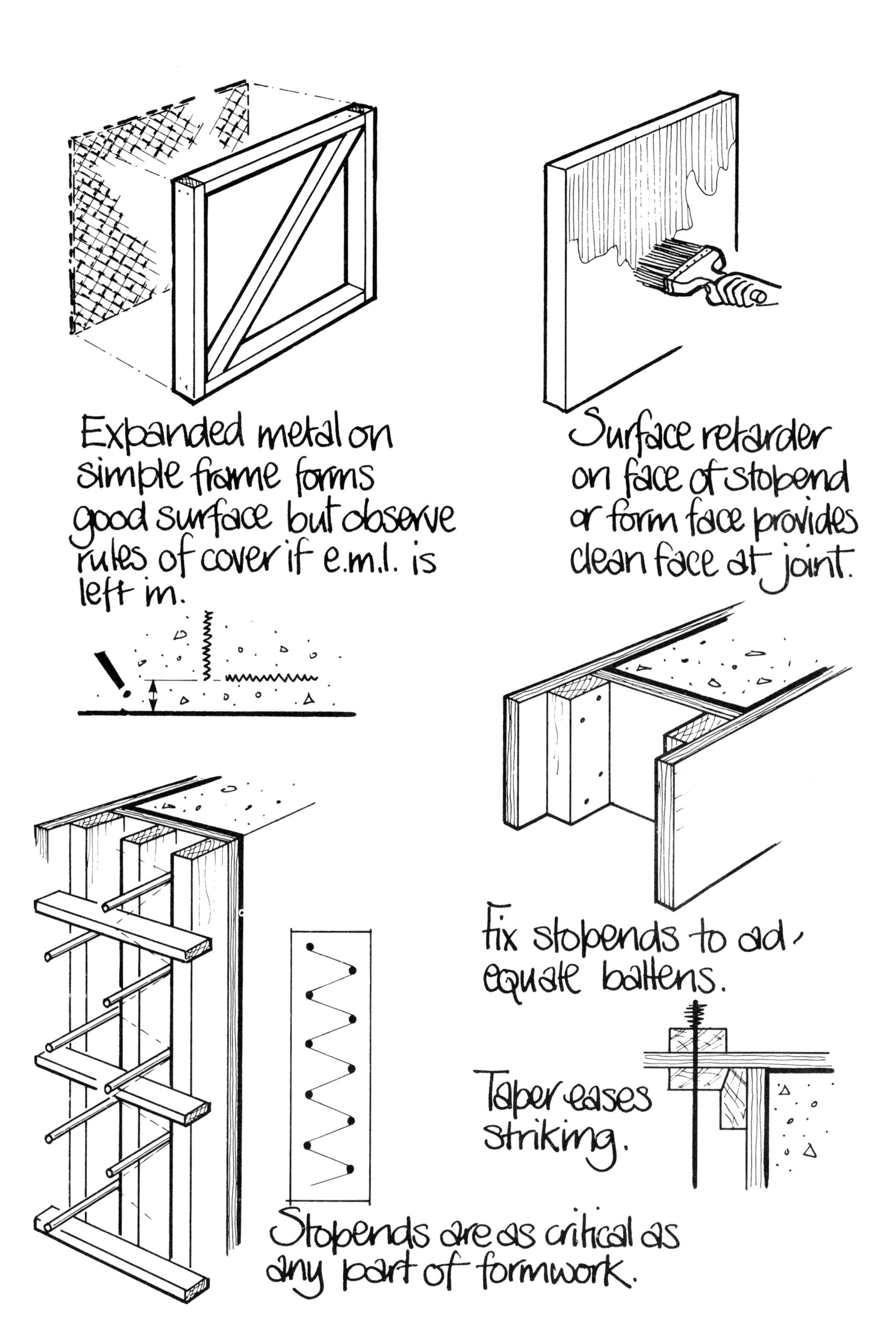

Stopend details

Joint positions

The positions of specified joints and day joints should be carefully considered and arrangements made to avoid defects of line and appearance which can result from grout infiltration caused by secondary deflections. The prop positions should be carefully detailed to provide access for construction activities, and all subsequent service installation carried out without the removal of the standing supports. These should be eased or removed only on the instructions from a competent person. The designer must always remember that once a floor has been cast onto a soffit arrangement, the whole process of striking has to be carried out under totally different circumstances from those in which it was erected, in other words the vertical egress has now been curtailed. This vitally affects table sizes and component selection. The proprietary systems present considerable assurance with regard to safety and ensure correct spacing of props and the maintenance of plumb, which is so essential to the load bearing capacity.

Openings through slabs or forms can present major accident hazards, so much so in fact that many contractors leave the formwork to soffit level at opening locations after the main formwork has been removed, so as to prevent anybody falling through. For the same reason, where trough and waffle floors are formed, many contractors provide a fully sheathed soffit upon which the formers are laid, because operatives will work more confidently in this situation when compared with those who work on skeleton systems.

Unnecessary wastage occurs, both with respect to cash and materials on the repetitious cutting-in and re-supply of formers for openings. The most profitable arrangement for repetitious work involves the use of steel or concrete formers which can give 30 or more uses, all non-standard openings and additions simply being framed in timber.

A problem often encountered with floor construction is that of the delays that can occur between the completion of the formwork and the end of the steelfixing operations, and stopend or dayjoint insertions. This happens where there are many drop beams or upstands to negotiate. The steelfixer should prefabricate the reinforcement cages and arrange the formwork in such a way that the main beams can be fabricated on their soffits before the erection of the beam side members. Where many secondary beams are to be built it is economical to provide a platform of supports at the level of the underside of the main beams, and stooling them out to provide supports for the remainder of the work.

With regard to level and camber it has been said that there can be no substitute for the experienced line and level engineer, but by using homing lasers or levels which can project a beam around some central station in the construction, the location of formwork to level becomes as simple as the traditional method of 'boning-in' a surface or a drain run.

Floor centres provide some problems with the formation of eccentric cambers. Where there are special stipulations regarding line or level, system equipment should be used in which the props can be pre-set to the approximate line, and finally adjusted using a level before concreting begins.

Centres have their uses where for example there is considerable brick-bearing work to be carried out, or on large spans, careful attention must however be given to the way in which they can be stripped from beneath the completed slab.

Stairs and ancillary units

Wherever possible stairs should be precast or at least some system designed to produce a form capable of the job. The traditional 'build it up, knock it down' approach to stair construction is expensive in terms of labour and materials and can delay the more profitable construction work by employing those skills which could be better employed elsewhere.

Stair construction should be kept in phase with the remainder of the construction work to effect easy access and simplicity of casting. The same can be said once a staircase has been cast between landings or within a core, particularly where a floor slab has been placed onto a soffit. Precasting allows the manufacture of stairs on site so that they can be rapidly placed to give immediate access when required. Should it prove undesirable to do this the staircase, even though they may be standard, will require careful designing so that they can be accommodated within the formwork arrangement.

Sockets or buried anchors accurately cast into position for supporting the carcass of the stair forms facilitate erection. Location of reinforcement laps should be such as to allow access and the panels should be carefully sized for easy handling. Prepared cut strings and riser boards of the correct profile, or timber purchased to size, help to speed up the construction, the whole being of bolted construction for maximum re-use.

Any facility which can be provided for shoring the forms will save time as much of the tradesman's efforts are spent in achieving support at mezzanine and landing levels.

Ducts, chutes and lift shafts

Because ducts, chutes and shafts are all of vertical construction they lend themselves to storey-height treatment of forms. Obviously the precasting considerations still apply, although many contractors are reluctant to hand over responsibility for major items of construction such as lift walls to a precaster, preferring to retain the initiative on such critical work. Items of building such as ancillary equipment offer considerable opportunities for innovation and a number of patent systems are available which the contractor can hire. However, design criteria are clearly defined and most designers will use some particular preferred system which they have designed, or have become acquainted with over the years. The speed of erection is a critical feature, particularly where the stair tower forms an integral core. Unfortunately this type of design often means that the stairs also are integral and this tends to make construction difficult.

The designer should discuss fully with all parties all aspects of lift and stair construction to simplify the placing of steel and formwork erection. Generally the inner forms are arranged to work from kicker to kicker while the external forms mould the concrete from the kicker to the underside of the floor. The inner forms thus provide edge formwork to the slab and form one side of the kicker above. A platform should be provided to give safe access to the inside form for erection and striking of inside forms and it can support the external forms while they are 'in store' during the floor casting operation. Where panel construction is used, corner panels may have to project and be continuous over more than one floor to help maintain line and plumb. Any platform arrangement must be foolproof and be capable of easy hoisting by crane. All connections to the concrete should be by way of bolting rather than by counterbalanced projections.

Walls, as described here, are relatively thin so that placing and compacting concrete needs to be carefully exercised. Particular attention must be given to the correct line and the positioning of lift control boxes and fixings for guides. Chutes and ducts offer considerable opportunities in the adoption of one or other of the many patent collapsible cores, although a preformed permanent liner, such as a concrete pipe, will comply with the specification on avoiding nibs and projections.

The casting of lift walls can be speeded up by carefully positioning the construction joints. The lift wall arrangement can be simplified into a two-part operation, walls with nibs being formed in one direction and simple infill panels being clipped onto nibs to form walls in the other direction. Parts of lift enclosures can be precast, namely the cill beams or the lintels, these components being inserted into the form during the steelfixing operation.

For erection, the designer should incorporate some kind of support onto which the forms for walls and cores can be located during the assembly process and before ties are inserted. This is achieved by including anchors or sockets into which a steel channel or stout horizontal timber member can be fastened. Wedges or shims can then be used to obtain the correct vertical location of the form when ties and bolts are passed through the forms. This technique is valuable even at the edge of ground floor slabs where difficulty is often experienced in supporting the form while ties are inserted.

The designer should also make certain that sockets or through holes are included in the slab adjacent to the wall starters. This allows correct location of the members from which rakers and bracings can obtain support, and avoid the somewhat risky runs of plates which are often used to transmit thrusts back to previously cast concrete.

Figure 11.1 The shaft/tunnel intersection form which is described in this chapter.

11. The geometry of formwork

General

Geometry, once a prime consideration in construction and the particular responsibility of the skilled tradesman, has during recent years become somewhat obscured. Geometrical work is now the exception, although most architects introduce some small element of geometry into their designs, whether it be a circular column, a ramping slab or a particular feature such as a dome or barrel vault. Often geometrical work is introduced as an ornamental balcony, a canopy or a particular modular feature expressed in precast concrete cladding.

A certain amount of geometrical detail is still incorporated in the structural component. Beams and walls, circular in plan are still encountered in otherwise traditional structures. Of course, some types of geometry are intrinsic to a specific structure such as camber, or splays encountered at the soffits of cantilever beams. Areas of construction where geometrical features are encountered include sewage treatment works; paper mills; tunnel and culvert schemes; siphonways and spill-ways to dams.

On many civil engineering schemes a sound appreciation of geometry and form is essential to the preparation of economical formwork arrangements, and profiles to structures such as cooling towers, wind tunnels and suchlike should be carefully examined.

Basic approach

When the actual practicalities of formwork geometry are considered a basic approach founded on first principles is bound to lead to the simplest form of construction, and of course simplicity of construction is the prime aim. Ideally the geometry of a structure which may be quite complex should be broken down in such a way that the key members in the form are generated by known dimensions. These members, framed together using simple construction techniques provide a carcass or substructure which is capable of supporting sheathing to generate the required final shape or profile.

The main techniques for generating simple form surfaces have been included elsewhere in this book, and it is not intended to deal in depth with such details of geometry as the development of a surface or the establishment of a particular splay or bevel cut. Geometry in this context is therefore considered as an integral part of the construction.

Constructional geometry

Figure 11.3 shows a prefabricated section of formwork for the intersection between a shaft and a section of tunnel, and the elements of the basic approach to geometry which can usefully be adopted by the formwork designer and manufacturer, in this case the trades supervisor on site.

The problem is fairly clearly defined and provided that the current relevant structural or civil engineering details are available the process involves an extremely practical series of steps. The operations are within the capacity of the trade supervisor or a suitably skilled tradesman. For this project the construction materials were timber and ply although similar arrangements would also apply had the material been steel.

It should be noted that the centre lines of the tunnel and the shaft do not coincide. The tunnel was offset to one side of the centre line and thus there was some small additional degree of complication as the fairing or fillet formed at the actual intersection of the faces varied throughout the circumference of the tunnel, although symmetrically placed about the horizontal plane in which the longitudinal axis of the tunnel lay.

A drawing could have readily been produced which gave the plot of the intersection or groin line, and from this, scaled components would be produced and assembled into an appropriate form. For this type of work and the required standard of accuracy necessary, this could have proved to be a costly solution. A simple line drawing produced thus can however prove invaluable as a starting point for the basic construction method.

As illustrated, a practical and direct solution was achieved on site. First a section of the actual formwork which was to be used in forming the shaft lining was set up, this being carried out in a corner of the site. The line of the intersecting tunnel was established by lines scribed on the formwork and on setting out profiles erected clear of the working space. Next, a series of shaped ribs to generate the tunnel section was prepared; these were cut to templates which had been set out full size on ply sheet, after which they were erected as dictated by the scribed lines and profiles.

Once these shaped ribs were erected, the lagging members which provided the support to the sheathing of the tunnel section were fastened in place to span between ribs. With the lagging members temporarily tied in place the splay cuts at the ends could be marked out by scribed lines made at equal distances from the shaft sheathing.

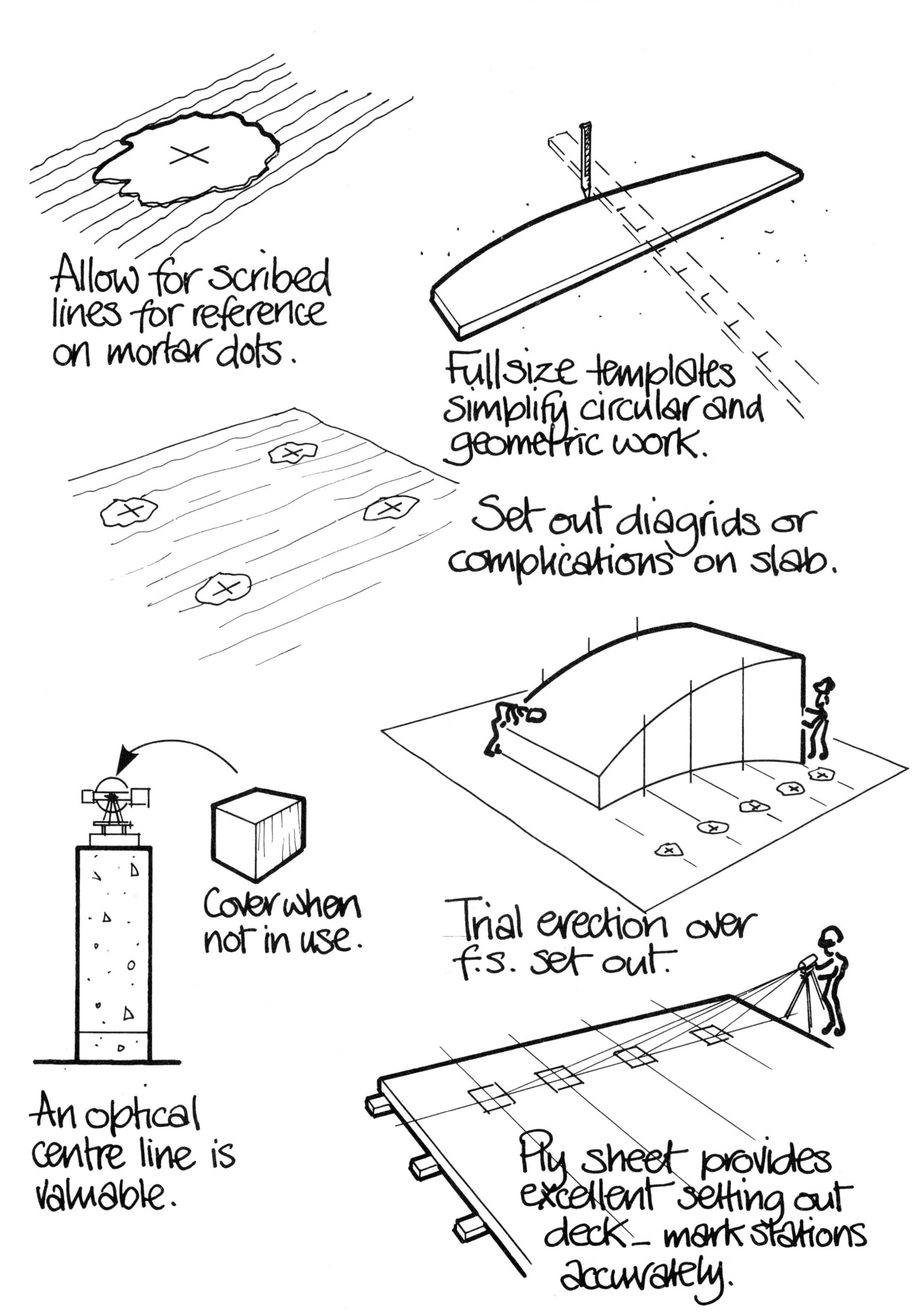

Figure 11.2 Full size setting out is essential for sound construction.

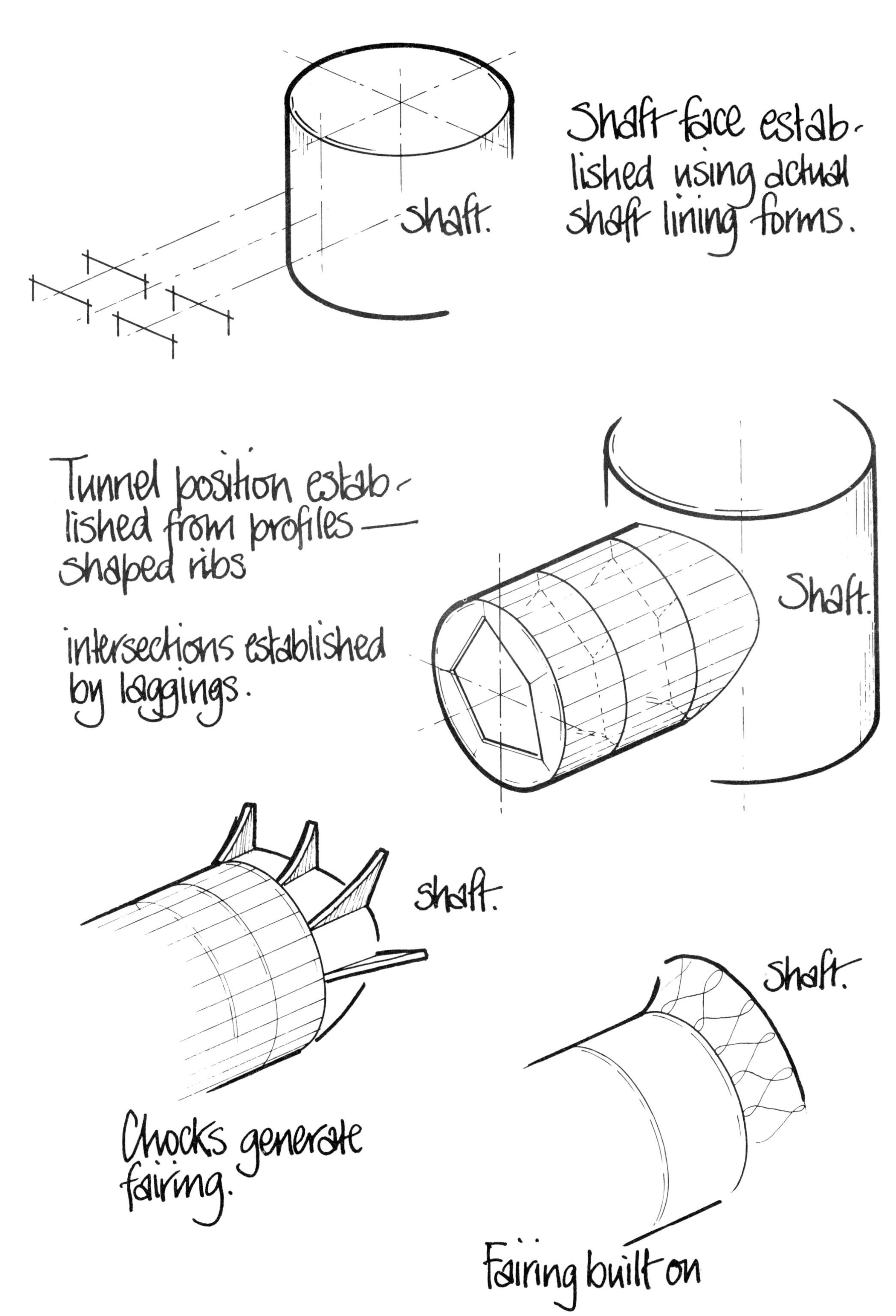

Figure 11.3 Intersection between shaft and a section of a tunnel.

Having been removed and splay cut, the lagging was then securely refixed in its final position on the ribs and the cleats planted onto the shaft sheathing face.

To achieve the correct circular form at the groin a template of the fairing curve was cut which, when positioned radially to the tunnel sheathing, was used in marking a series of chocks or cleats (subsequently used to support thin ply sprung into place), and scribed and fixed to the face of the moulded groin.

What could have been a difficult formwork problem involving considerable geometrical work and some difficult setting out had been reduced to a practical process by the application of pre-planning on the part of the section foreman. Here the geometrical problems were solved, and the form constructed well away from the eventual location of the form.

Most of the apparently complex problems of formwork can be broken down in this way, provided that the designer/constructor has a good appreciation of the geometrical principles involved. This does not need any real degree of mathematical training, merely an appreciation of shape and form. Many carpenters, joiners and mouldmakers have a good understanding of the basics, including bevel cuts and the intersection of plane surfaces. During training most joiners and mouldmakers come into contact with the development of surfaces, and what is perhaps most important they have a sound appreciation of the construction methods to use, once the elements of geometry have been identified or the main forms have been defined in a simple manner.

Making a start

As soon as the processes described earlier in this book, i.e. those of validating drawings, ensuring that the geometry indicated on the architectural or structural drawings is intended, and having established that the permissible deviation from line or the modification of profiles has been carried out, a start can be made on the necessary geometrical details. While the person responsible should avail himself of an established basic textbook on geometry for specific points, the following examples will help in the establishment of key dimensions and the location of members so that the process of installation of carcassing and sheathing generates a form or mould of the required profile.

First it is necessary to recapitulate on certain facts and simple techniques for drawing and setting out various forms. A few of the more useful ones are now included. While the definitions may not perhaps be completely to the satisfaction of a geometrician, they are intended to convey the aspects which are of most value in the *geometry* of formwork and are therefore related in terms of formwork applications.

Example 1. To fully sheath a given panel width using equal width boards (See Figure 11.4)
A line ab representing the width may be divided into equal parts by drawing a line ac from one end point at any angle to that line. Divide into a number of convenient lengths, equivalent to the required number of boards and provided that lines are then drawn through the points parallel to bc, the lines ab/x will indicate the width of the individual boards.

A similar technique is used for the rapid marking of lines of drillings for the fixing of sheathing. A suitable batten (or rule) is laid onto the sheathing in such a way that the required number of spaces are contained between the edges or outer lines of screw holes. Marks are made adjacent to the indications on the batten or rule. This is repeated as required along the length of the sheathing, and a straightedge is used to rule the lines for the boring. The same technique is used to establish the width of risers given an overall dimension over the flight, or the height of risers given an overall dimension for the height between landings.

Example 2. To determine height increments (See Figure 11.5)
Where a series of columns are to support a raking slab or beam as for a ramp, and the column centres are equally spaced, then the increments through which the column height varies will be equal. Where the column spacing varies, the increments will vary proportionally with the spacing.

In this example x is proportional to X as y is proportional to Y.

These rules apply equally to columns set on the circumference of a circle where the rake or angle of the beam or slab with the horizontal is constant.

Example 3. Bevelled cuts and developed surfaces (See Figure 11.6)
For the true shape of a surface to be seen it must be viewed from a point on a line normal (at right angles to) that surface.

In Figure 11.6 which shows a square column having a splayed head, the surface of the splayed head as viewed in the elevation is not the true surface and a sheet of ply cut to the profile indicated by points 1, 2, 3 and 4 would be of no use for sheathing the form. The sheathing area seen by the eye placed as indicated, i.e. viewed normal to the surface is however the true shape to which the sheathing should be cut.

It is evident that if the sheathing panels were cut to the net size 5, 6, 7 and 8, that when offered together, *while the true face of the column would be generated* due to the thickness of the material *only the leading corners of the sheathing panels* would touch leaving an unsatisfactory arrangement as at 'Z'.

This situation is rectified in practice by splay-cutting one pair of sides which have the face cut to the true profile, and then installing these between adjacent sides. It can be seen that if the ply sheathing (and timber backing if necessary) is brought to the correct bevel at the top edge then the necessary splay cut to achieve a neat-grout tight fit throughout the intersection between adjacent sides can be achieved by maintaining the cut to the line of the profile and to the bevel generated by the plan cut p, q, r which although it is 90° viewed in plan, of course is actually more acute than this in section normal to the sheathing face. The cut is best made by first bevelling the top and bottom of the intermediate panel and then squaring lines

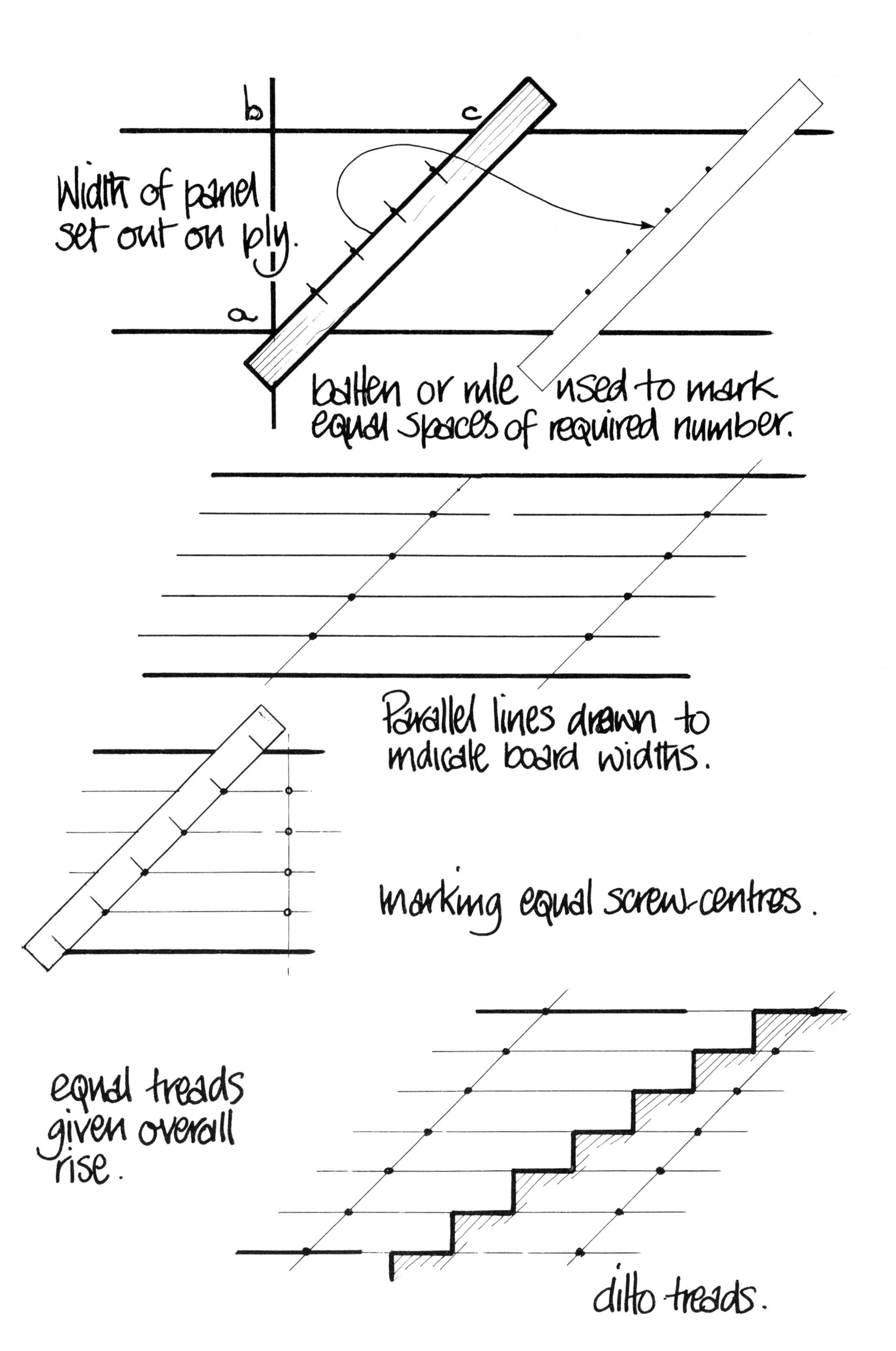

Figure 11.4 See example 1.

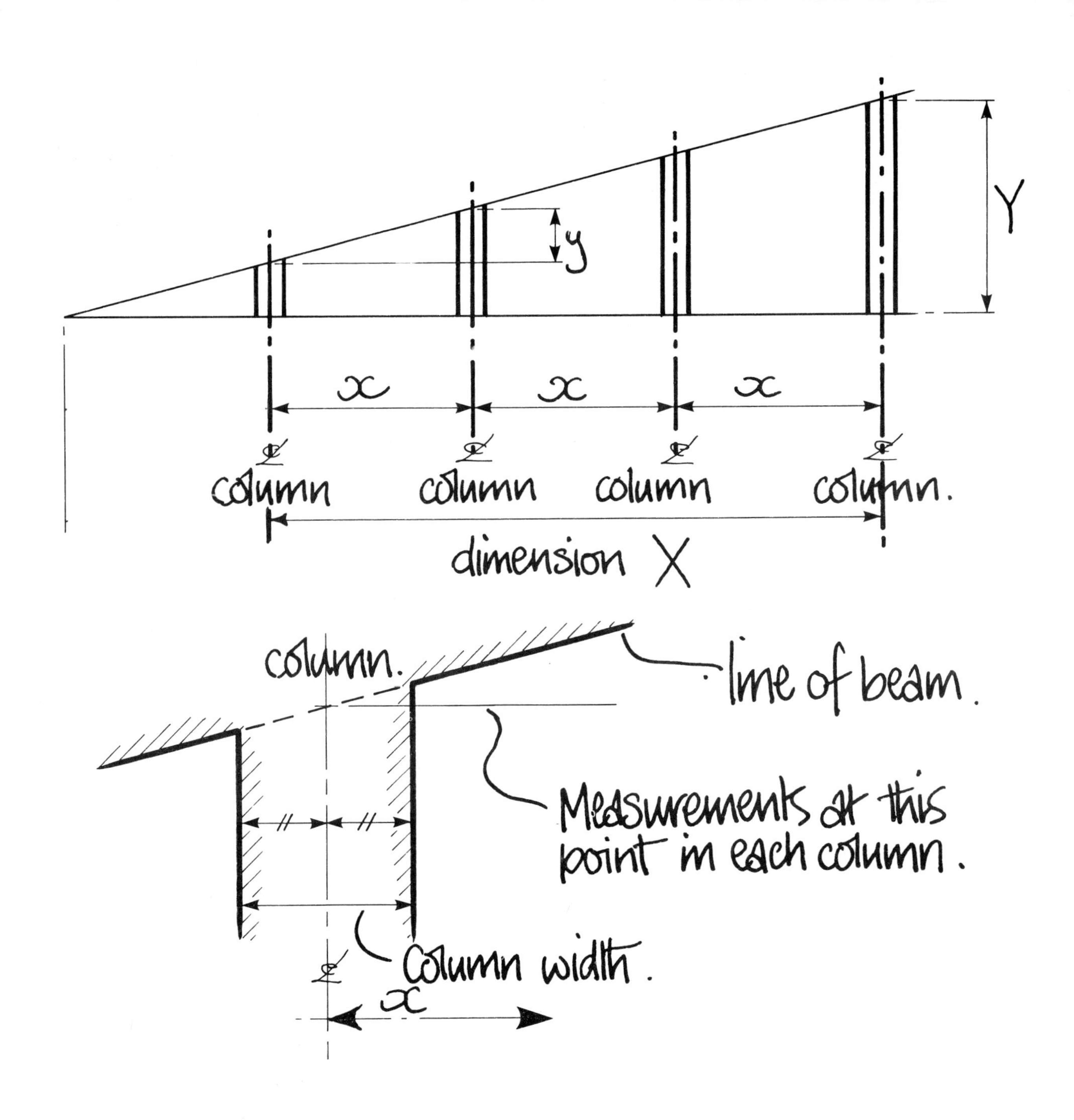

Figure 11.5 See example 2.

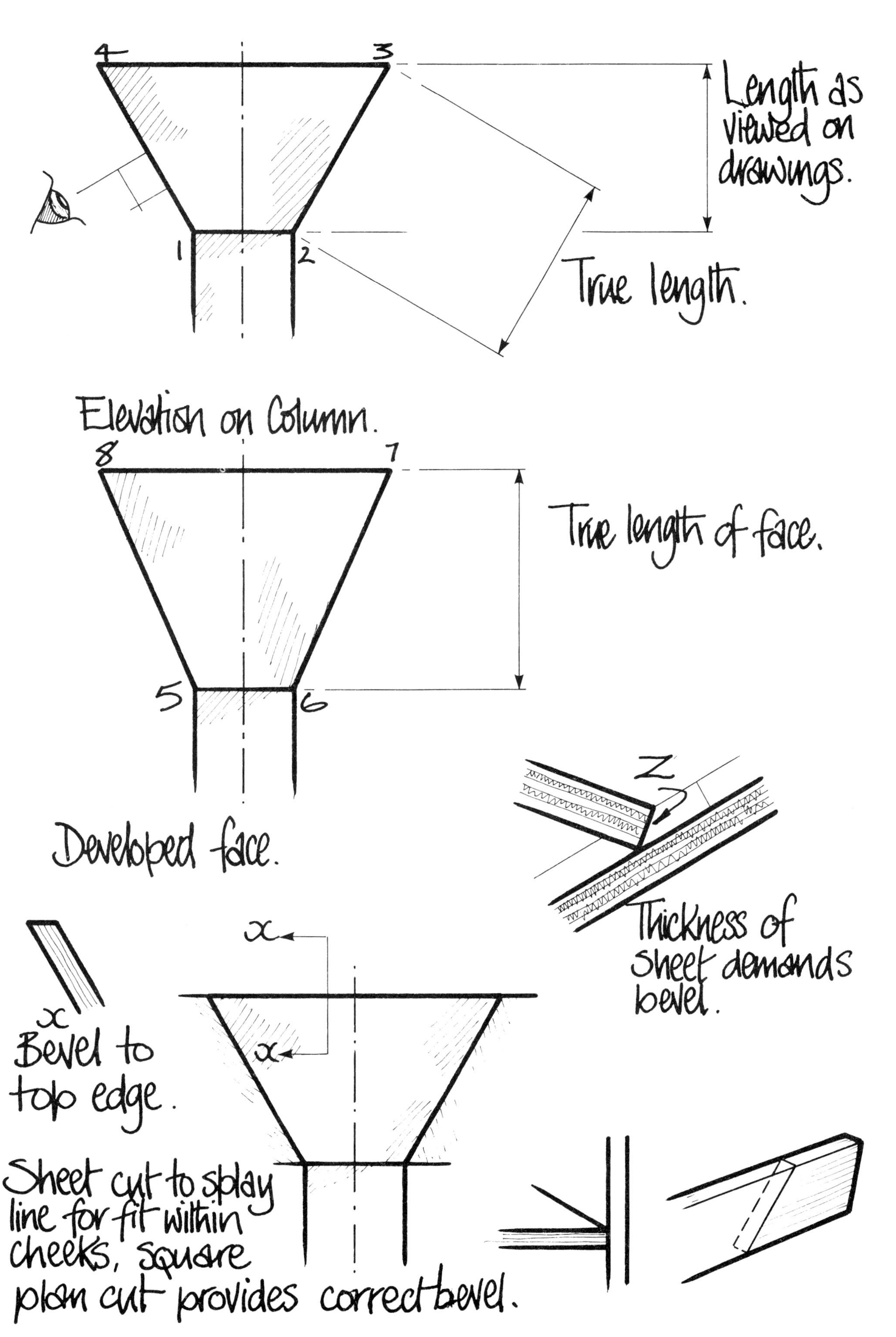

Figure 11.6 See example 3.

across the edges so obtained, these lines are then joined on the reverse face to provide a guide line for cutting purposes.

Example 4. Circular or conical surfaces (See Figure 11.7)
Development of the sheathing face becomes particularly critical when considering circular tapered columns, column caps or hopper forms. Obviously it is desirable that any sheathing joints should be vertically disposed in the plane containing the axis of the column or hopper. To achieve this the true surface of the sheathing has to be developed and the sheathing panel sizes related to the available sheet material. The surface may be developed either full size or to scale by producing the lines of the tapered part of the column face to a point of intersection x. This point is then used as the centre from which arcs a1 and b2 are drawn. Points 1 and 2 are established by means of dividers or trammel, each being spaced from a and b by the length of the circumference. The figure produced, i.e. a, 1, 2, b is the actual sheathing surface of the conical column or hopper. All that remains is the setting out of the intended joint lines determined by the size of the available sheathing material. Provided that the selected joint lines are radial to x then the sheathing joints will be vertical when the sheathing is secured back to the carcassing or supports.

Example 5. Flewing surfaces (See Figure 11.8)
Flewing surfaces are those surfaces where the pitch of a series of parallel lines in the plane of the face varies uniformly throughout the face. The under-surface of a spiral stair or a circular ramp are examples of flewing surfaces. The surface can be generated by setting up members which reproduce the angle at each end of the surface and then installing intermediate members to the positions dictated by straightedges set to points proportionally spaced on these members.

Where the surface of spiral ramps or stairs are concerned the surface can best be generated by a series of members laid radially, and each elevated to the appropriate height. Any member laid other than radially, or very nearly radially, would need some degree of curve to generate a true surface. Where the slab to a ramp is tapered in section (Figure 11.9) the supporting members should be radial, and it is valuable in the erection of the formwork for such surfaces if a series of frames of the correct pitch are constructed. These frames when complete can then be set radially as for a simple carcassing member so that the complicated form can be generated with the minimum degree of setting out.

Example 6. Dome construction (See Figure 11.10)
The principle of selecting key members to generate a sheathing form is quite clearly illustrated when dealing with domes. The dome, which was once a feature of classical architecture, continues to be used particularly with the developing technology of shell structures. Dome roofs are often employed on gymnasia and assembly halls because they provide solutions for the architectural requirement for clear space.

Generally, constructed forms for domes are onto a framed system of supports, steel or timber ribs being used to provide fixing for the sheathing. To keep the variety of the ribs to a minimum, they should radiate from the centre point of the dome or shell. Thus each rib can be cut or bent to the radius of the sphere, with of course the necessary allowance for sheathing thickness. Any member which is placed other than radially will need to be set out separately.

Members placed in planes parallel to the radius although still circular in profile will vary in diameter according to the distance from the axis.

Example 7. Circular work (See Figure 11.11)
When dealing with the carcassing members which are used to generate work that is circular on plan it is essential, for ease of construction, that the framing members are maintained in the horizontal plane. The true profile of the ribs can then be taken directly from the plan details – allowance being made for the sheathing thickness.

For ramps or sloping structures, any member set on the rake will require individual setting out because its shape will be elliptical. The actual geometry involved in setting out is simple, but any complication that might arise can be avoided by carefully siting the members horizontally.

For circular stair walls or parapets to circular ramps the tendency is to place members on the rake. With larger radii, the framing member may be of such proportions that it can be sprung into shape, but for smaller diameters this is not possible. If attempted, badly formed profiles will occur and undue stresses created within the structure.

Small diameter circular ramping beams or walls can be formed around a continuous circular core and are open boarded other than within the areas that form the concrete, so that the construction is reduced to providing accurately constructed rings which support vertical carcassing members for sheathing as required.

Example 8. Setting out circles and ellipses (See Figures 11.12 and 11.13)
When setting out circles and ellipses, this should be executed on a prepared slab or floor. Trammels and tapes can be used where there is free access, and obviously the longest possible arcs should always be laid down.

Where large radius arcs are to be set out and it is impossible to strike these with a tape, a batten can be cut to generate the rise on a given chord. This technique is based on the fact that all angles which subtend a chord are equal. The points established using the batten can be joined by using a good quality knot-free batten so that the resulting arc will be accurate enough for most formwork purposes. Ellipses can be set out by different methods some of which can give good approximations to true ellipses.

The trammels method is perhaps the simplest method. A batten marked with the major and minor axes is moved in such a way that the two marks travel along the axes so that a series of points on a segment of the ellipse are plotted. (See Figure 11.13.)

Another method which can be best used, involving a setting out table with squared edges, is that in which a circle is described to the diameter of each of the axes. These circles are divided by radial lines which serve as the points from which a series of triangles are constructed by

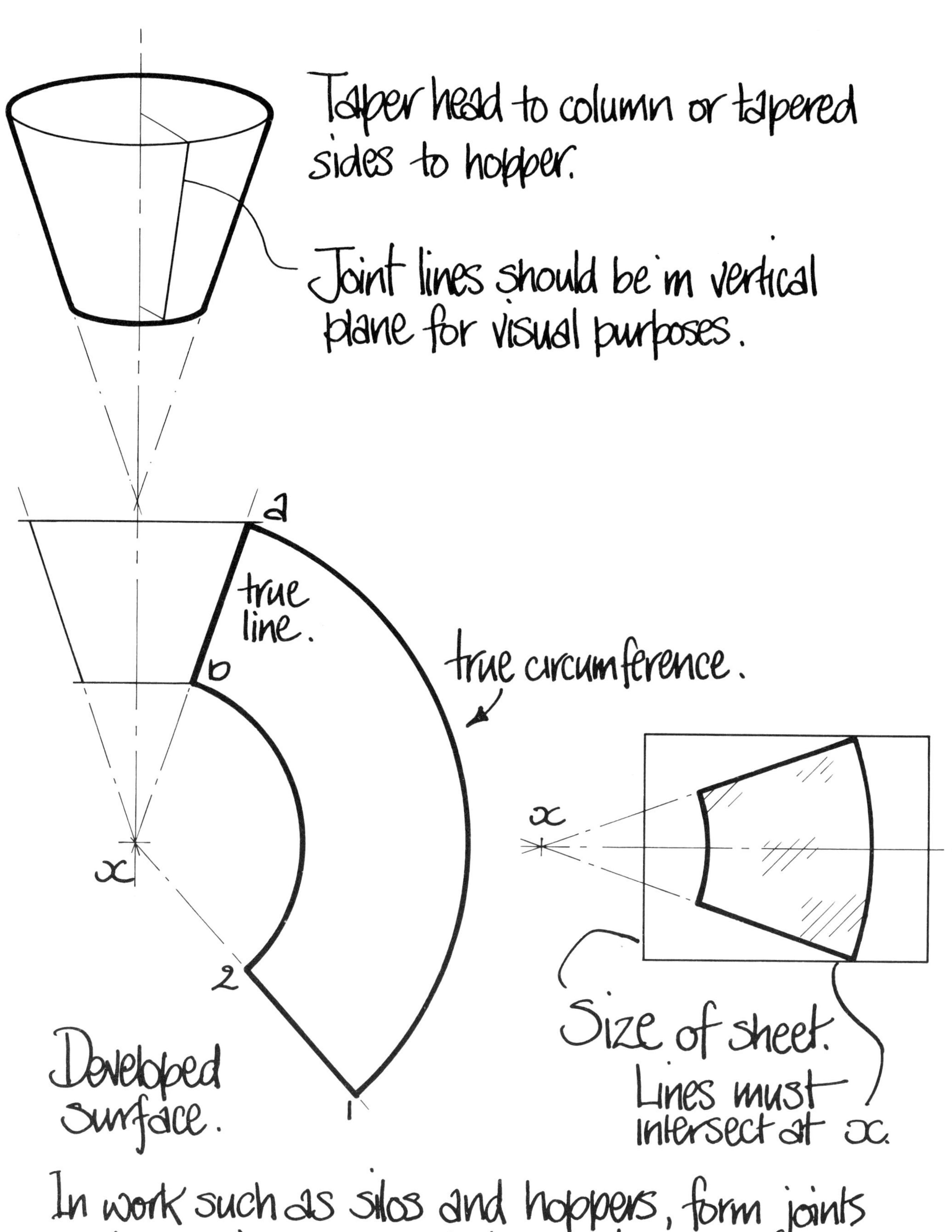

In work such as silos and hoppers, form joints must be disposed vertically to allow free flow of grain, etc..

Figure 11.7 *See example 4.*

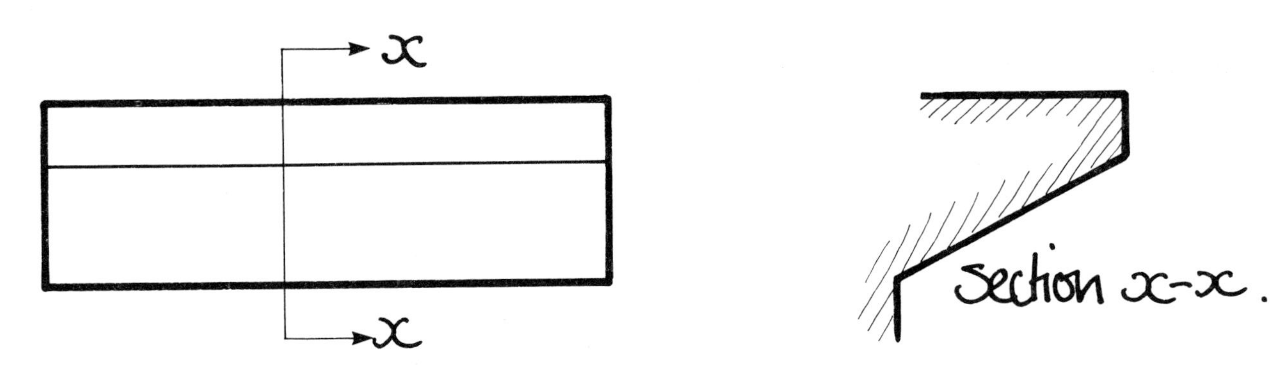

Cantilever with plane soffit, section is constant.

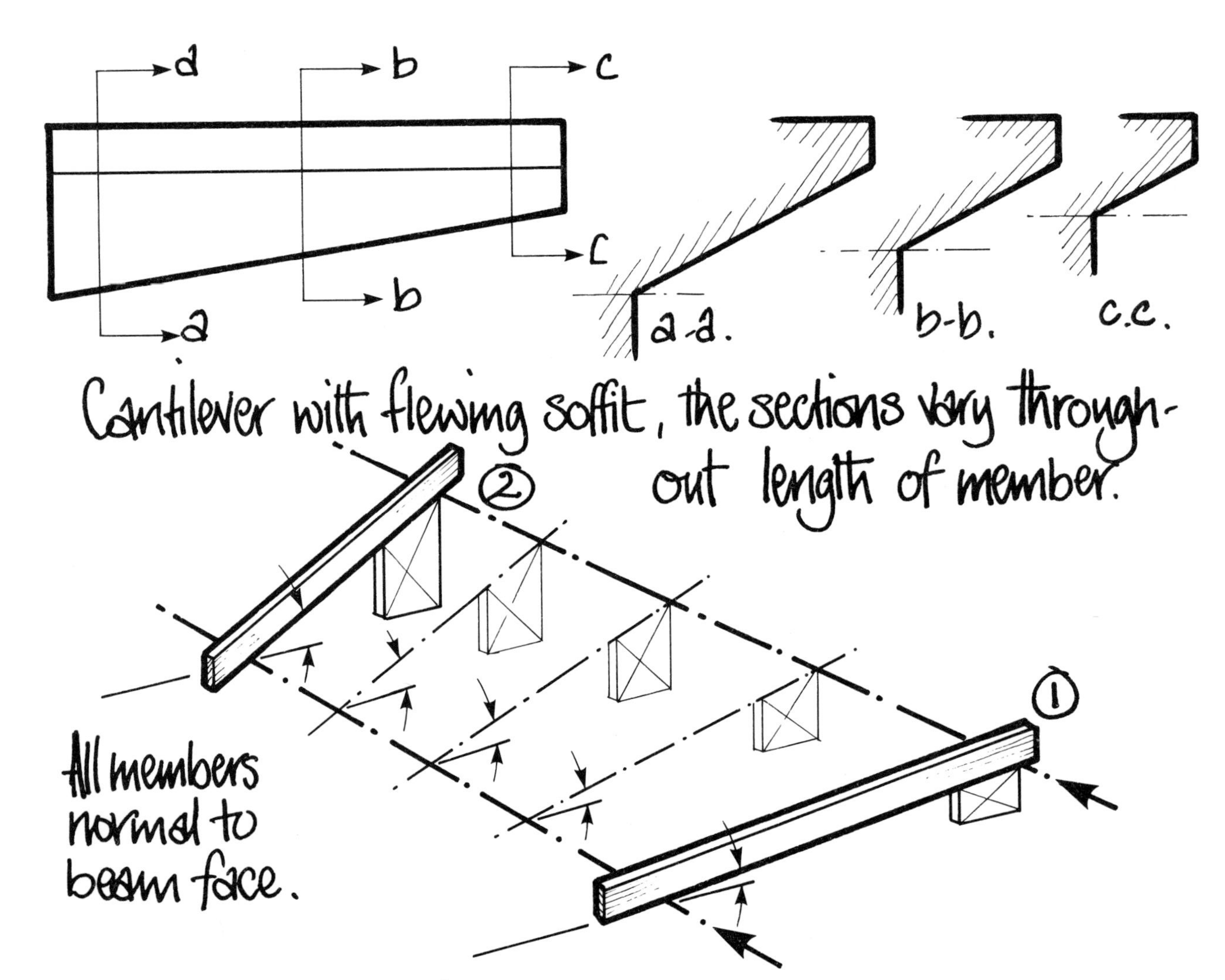

Generating a flewing soffit using members set on rake. With 1 and 2 established, intermediate members may be levelled in by eye.

Figure 11.8 See example 5.

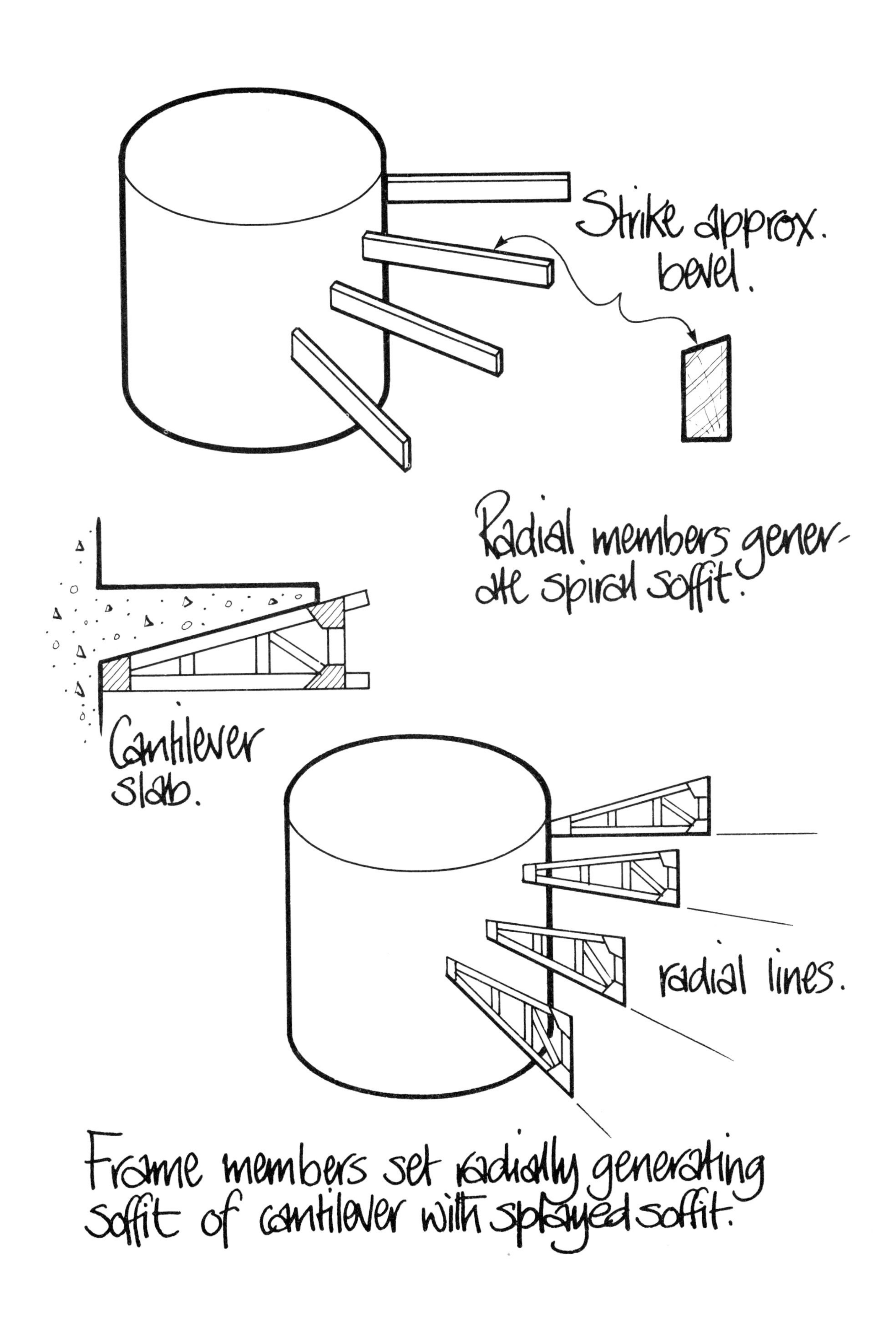

Figure 11.9 See example 5.

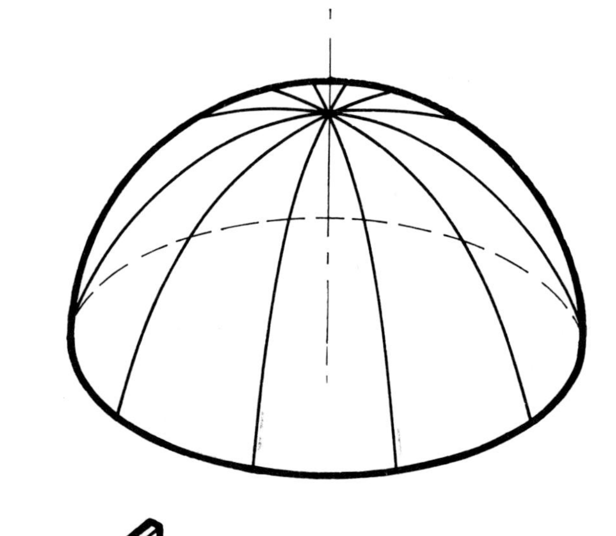

radially disposed
members are
identical

ribs may be
left square.

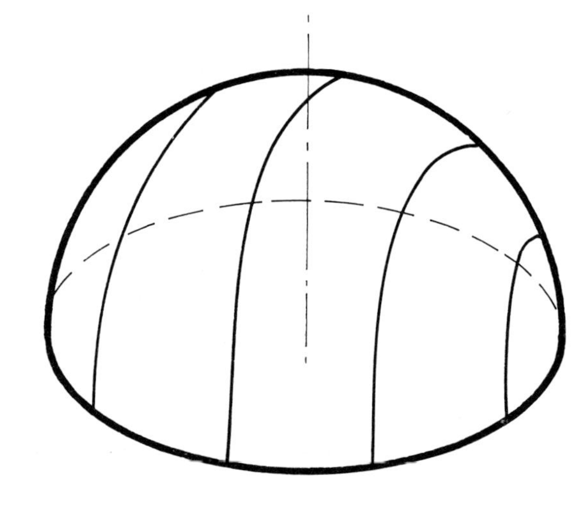

members set other than
radially will vary according
to their location.

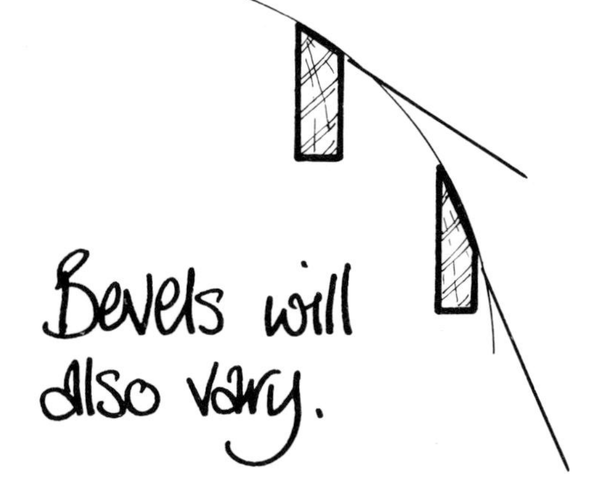

Figure 11.10 See example 6.

Plan

Shaped rib.

Shaped ribs located horizontally to generate circular work are truly circular.

Shaped ribs set on rake of ramping work become eliptical in profile and require bevel.

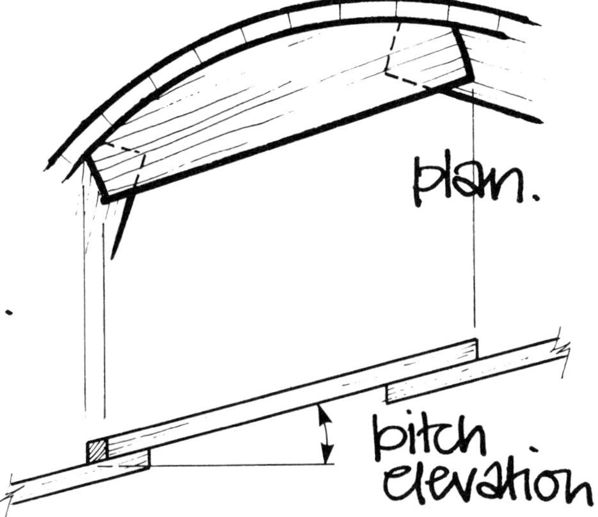

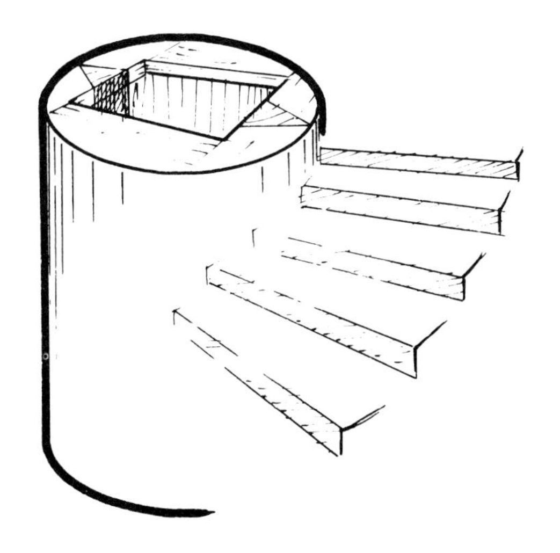

For stairs and small diameters it is advis-able to construct complete box incorporat-ing striking keys.

Figure 11.11 See example 7.

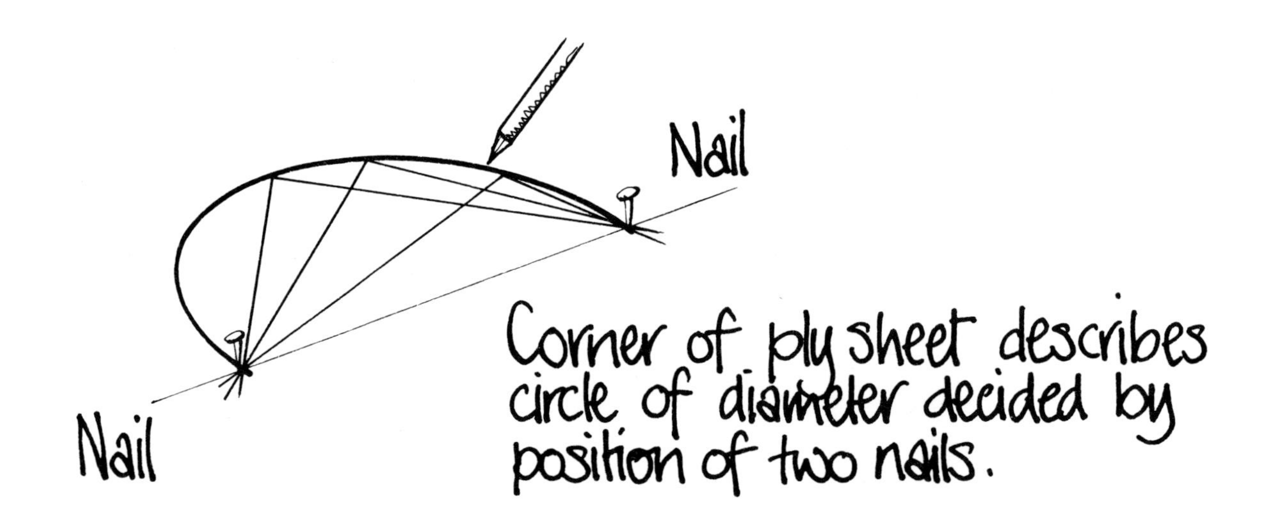

Nail
Nail
Corner of ply sheet describes circle of diameter decided by position of two nails.

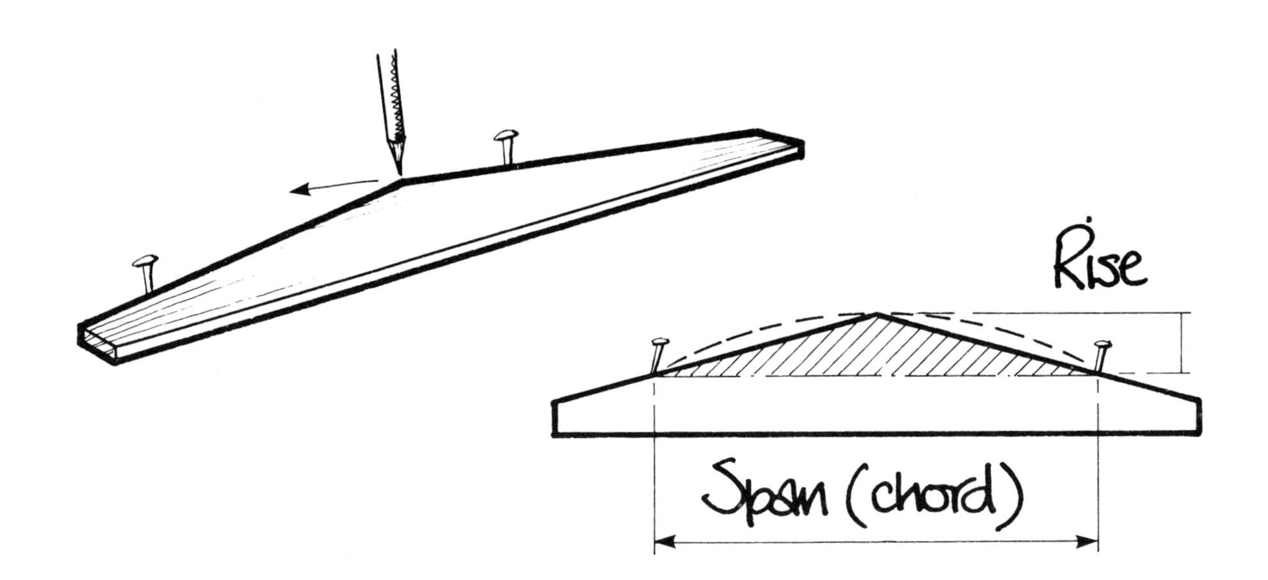

Large diameter arcs where centre is inaccessible can be set out using batten prepared from rise upon chord. Batten is run along two nails to generate required arc.

Figure 11.12 *See example 8.*

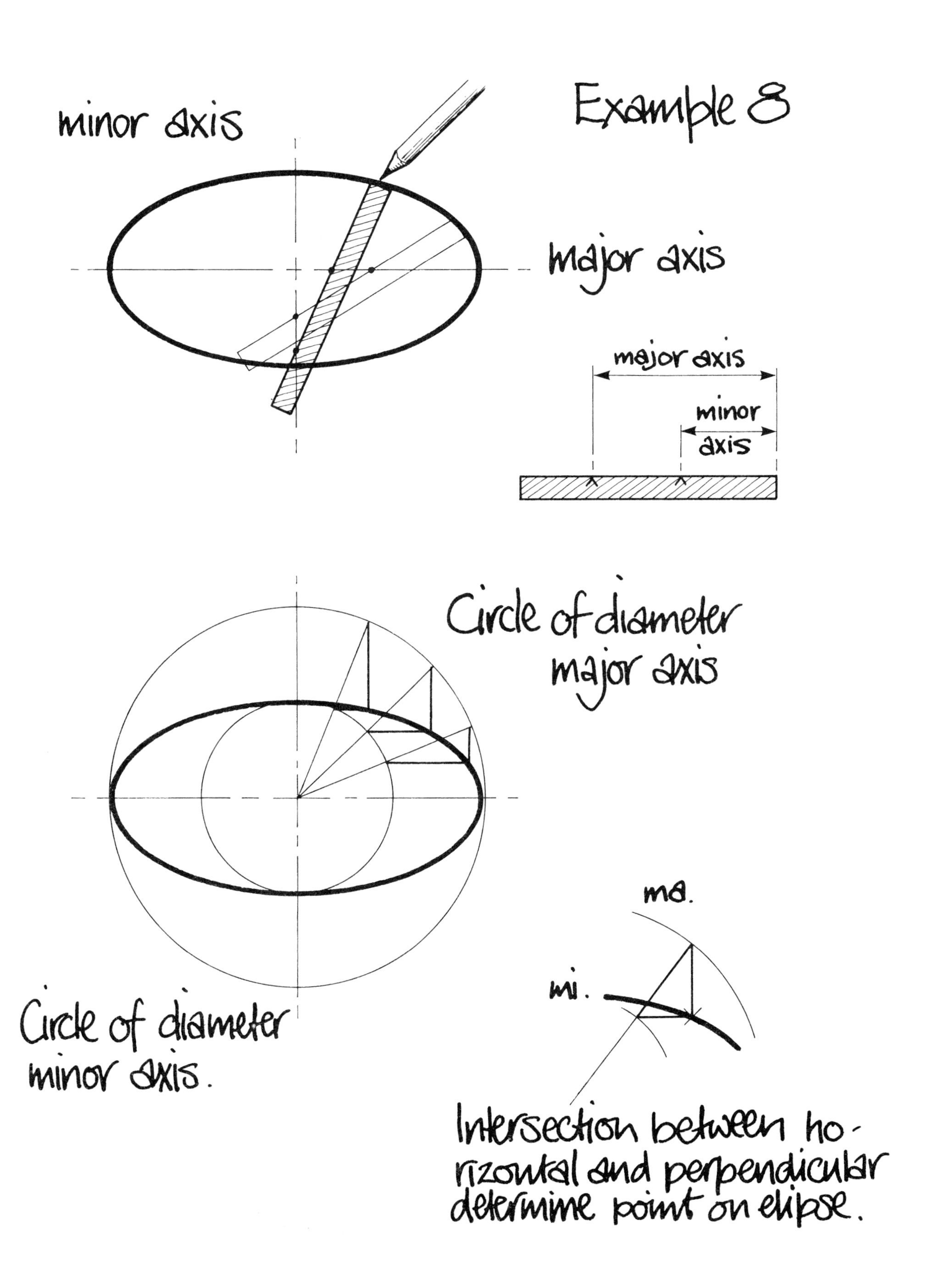

Figure 11.13 See example 8.

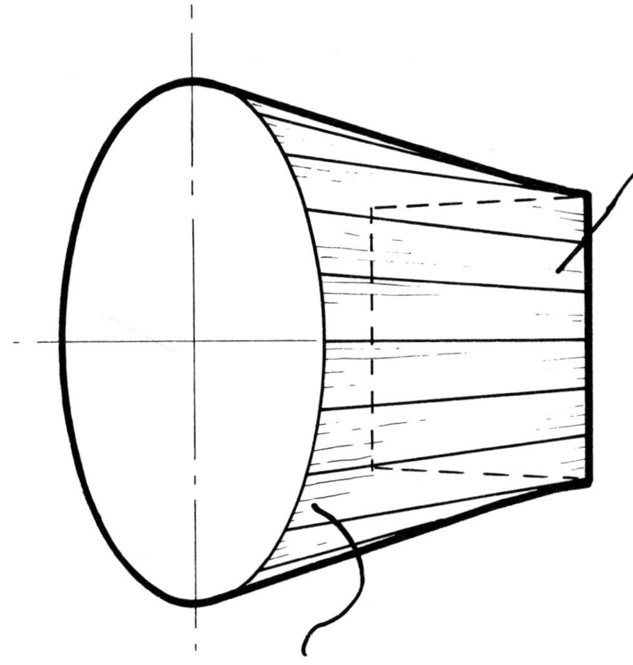

Side equally divided – each board requires to be individually prepared.

Circumference equally divided.

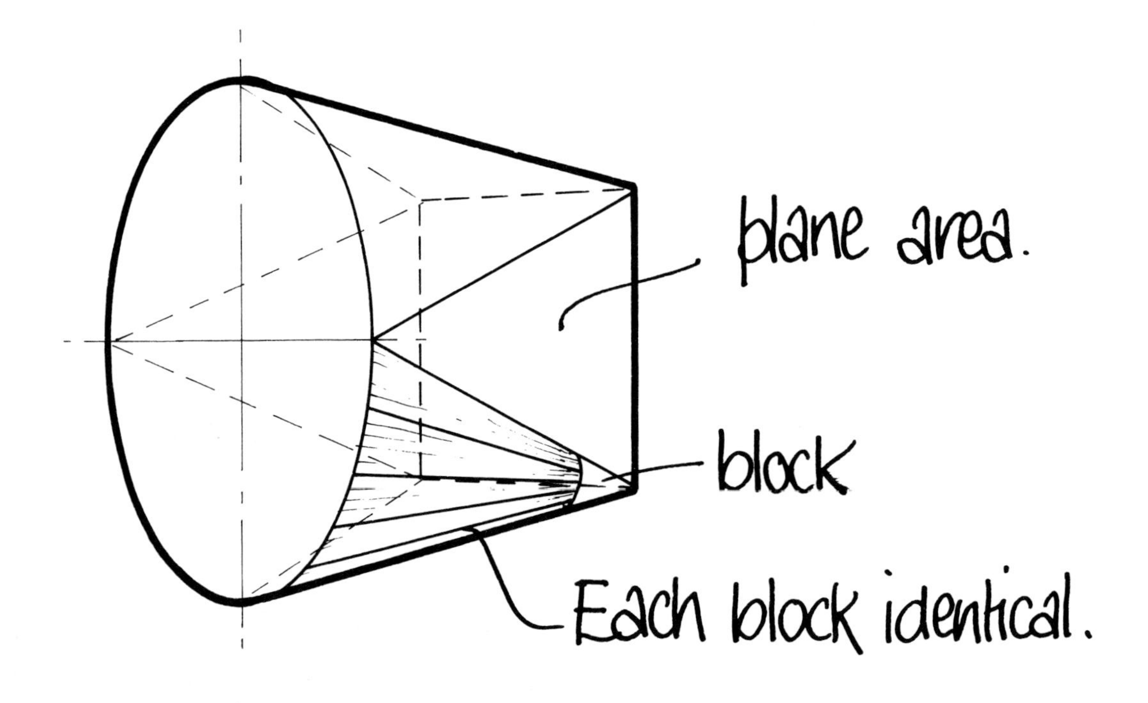

Transitions – construction is dependant on acceptability of face geometry.

Figure 11.14 See example 9.

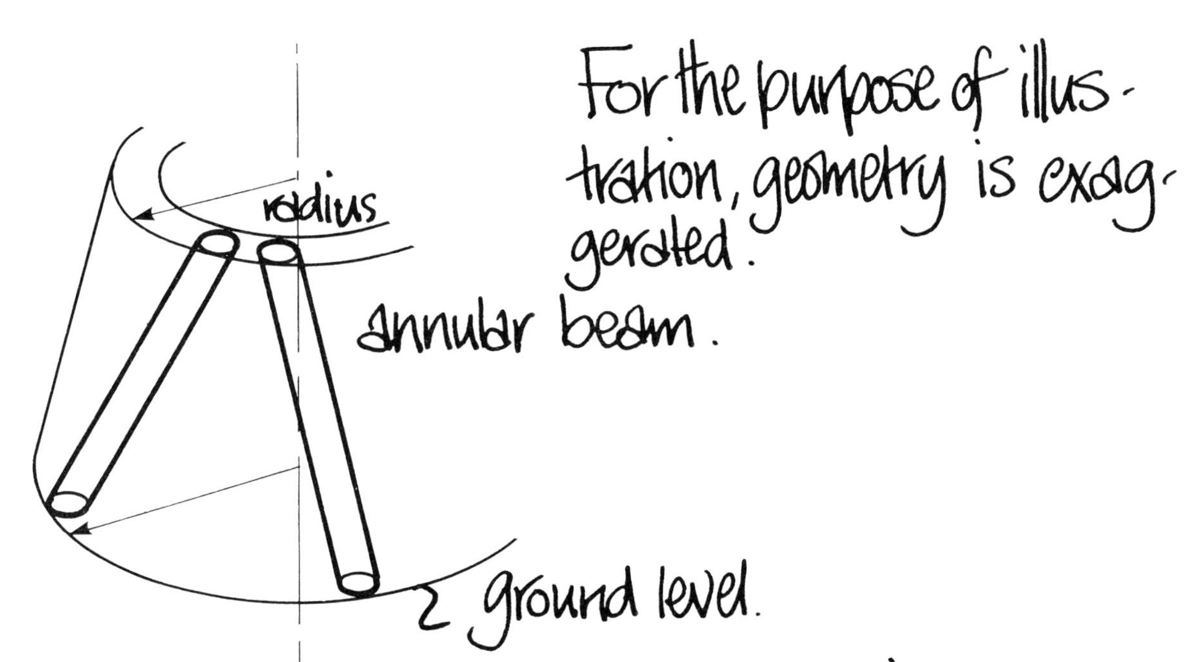

Circular columns set on rake present little problem.

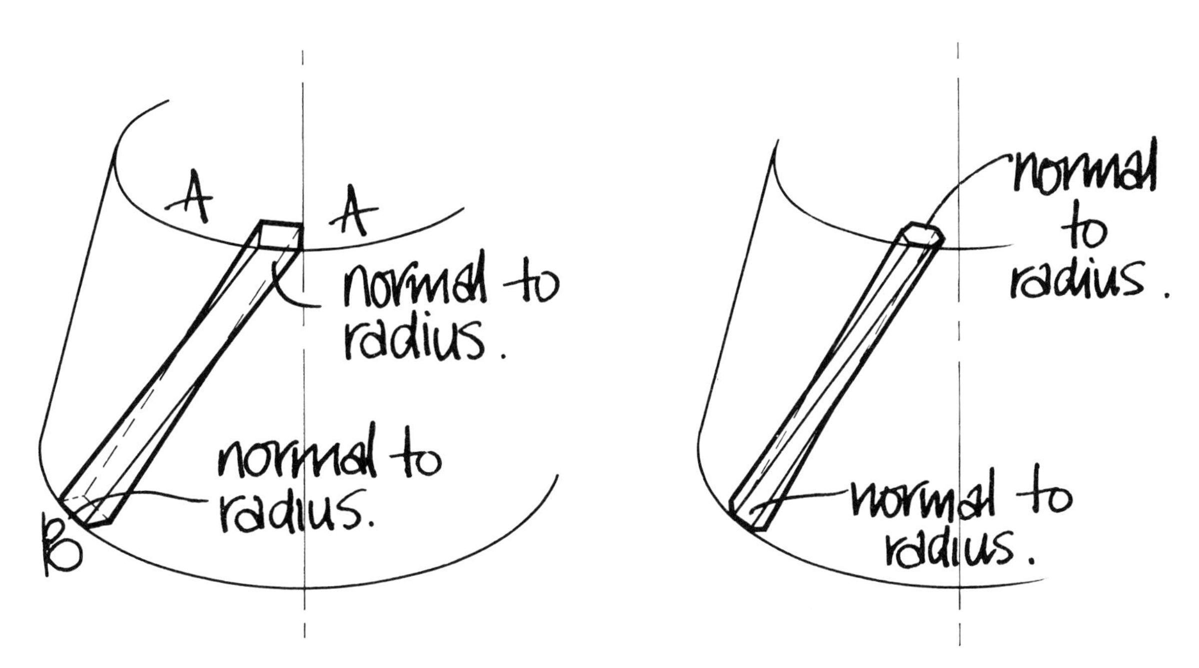

Square or hexagonal columns in, for example, a cooling tower. To provide good surface joint at AA, plane faced columns should be started out of true to ground but parallel to AA at ground.

Figure 11.15 Special problems.

lines that project vertically and horizontally. The intersecting points of the projected lines identify points on the required ellipse.

Some figures cannot be set down full size on a rod or template, although on occasions size and scale limit the practicality of full size treatment.

Example 9. Transitions (See Figure 11.14)
Formwork arrangements for the construction of tunnels, flues and culverts often include transitions from square to octagonal section and from circular to square section.

For the construction of wind tunnels transitions may require fair lines, because upsets in the line may cause upset in the boundary layer conditions. However it may be possible to simplify the form construction by careful arrangement of the sheeting material. This can be carried out so that certain clearly defined plane areas are simply sheeted to reduce the need for individually fitted or scribed boards. Figure 11.14 indicates two ways of sheathing a transition – the first demands that each board be individually fitted, while the second allows plane areas of sheathing to be placed with a reduction in the fitted panels or boards.

Information

The designer or fabricator should consult a manual on the fabrication of sheet metal where the most complicated forms are broken into a series of simply developed shapes that can be combined into a finished former. Some of the early publications on carpentry and joinery are most informative with regard to geometry because many of the illustrations can be directly translated into the sub-structure of the more exotic shapes that are devised by present-day architects who use interesting shapes and forms.

Special problems (See Figure 11.15)

Some types of construction present peculiar geometrical problems.

For example, two aspects of a shell type cooling tower where a heavy annular beam spans the raking columns: the columns may be circular, square, octagonal or hexagonal. The circular column presents little difficulty to the constructor, but the square, octagonal or hexagonal columns can provide problems which result from the rake and the inward inclination of the column. As with many geometrical situations to understand what is happening, it is worth visualising the arrangement in an exaggerated form such as is shown in Figure 11.15.

Clearly any formwork that is designed to exactly meet these requirements would be expensive and would possibly generate an unsightly concrete structure, so that a compromise must be made by providing a form which has plane sides based on the most critical intersection at AA, any discrepancy in line being taken up on the concrete ground or ring beam at level B. On a practical note this is a case where the kicker former should be a section of the form that has been suitably constructed in the carpenter's shop and set into some arrangement of yokes which relate to the setting out lines on site.

Figure 12.1 *A fine example of the use of joints.*

12. Joint considerations

General

Generally speaking structural design is based on the performance of concrete which is cast monolithically, but in practice the designer accepts that the structure is cast in a series of lifts or bays. The object of any concrete construction must be to achieve sound, and where specified, watertight joints which are consistent with the function of the structure. As noted in Chapter 2 the formwork designer must discuss the joint requirements and arrange his formwork so that the joints between lifts and bays meet the specification requirements and the needs of the construction team.

Special cases

Rarely does a concrete component have to be cast completely monolithic, although for massive foundations or certain configurations of heavy structural slabs it is desirable that particular sections, especially large sections, of the work should be cast at the same time. Some years ago, large casts caused difficulties in providing good quality concrete during protracted working hours, but ready-mixed concrete and the use of improved concrete handling equipment have now overcome these problems. Today it is quite normal for concrete to be cast in operations which involve the handling of something like 300 m^3 in a working shift, and for tank foundation slabs special arrangements have permitted the casting of monolithic construction of 2000 m^3 units. Such operations obviously make heavy demands on the formwork designer and call for large quantities of formwork and support systems to be ready for use during peak periods in the programme. Because the placing operations have to be carefully planned, some problems arise due to the rate of application of load. It is essential during the planning stages for the formwork designer to ascertain certain key factors, such as whether admixtures are to be used, or what the placing plan is to be, both of which determine formwork pressures. For large casts, particularly where suspended slabs and heavy structural members are concerned, it is vital that there should be a small gang of formwork tradesmen readily available, and that a careful watch be maintained on the system of supports adopted. The gang must be under the supervision of a person who is sufficiently skilled to identify movements or distress occasioned by, for example, the changing load pattern when succeeding layers of concrete are placed across the form. A constant check will also be needed on the support members where, due to changes in load pattern, there are relaxations in load which may cause loosening or even displacement of the supporting props.

The success of the formwork employed depends on suitably designed and located bracing members which are based on sound foundations. During the preparation for the operation the designer must check that the ground slab or walls of the excavation are capable of sustaining the loads that are to be imposed on them as the concreting operation proceeds. It may be necessary to import on to site heavy timbers for use as grillages to distribute the loadings. Even though kentledge allows bracing to be taken from poor ground, it may even be necessary to cast in concrete sleepers to enable some benefit to be obtained from bracing or support members.

Lift and bay considerations

Apart from the massive casting operation, which is normally the domain of the foundation constructor or the civil engineering contractor, most concreting operations, through negotiation with the design engineer or the resident engineer, can be broken into small, relatively manageable lifts or bays. The formwork designer should be guided here by the relevant specifications which deal with shrinkage and movement, and which tend to restrict the length of concrete cast in any one operation.

Some problems occur with regard to height or depth of lifts, as it is generally accepted that, consistent with the thickness of the member which allows satisfactory compaction, the height of lift has little effect on the quality of the concrete or the appearance of the structure. The formwork designer may, in trying to achieve a sensible re-use from formwork in order to provide continuity of work, find himself attempting to persuade site personnel to reduce the height of the intended lift because of compaction requirements and simplicity of the erection procedure. The whole process of establishing lift and bay sizes depends on a contractor's ability to make joints which are structurally acceptable and which will reproduce the required performance of monolithic construction.

Joint techniques

A lot of research and testing has been carried out on construction methods with regard to joints, and the essence is that, provided a concrete face exhibits clean

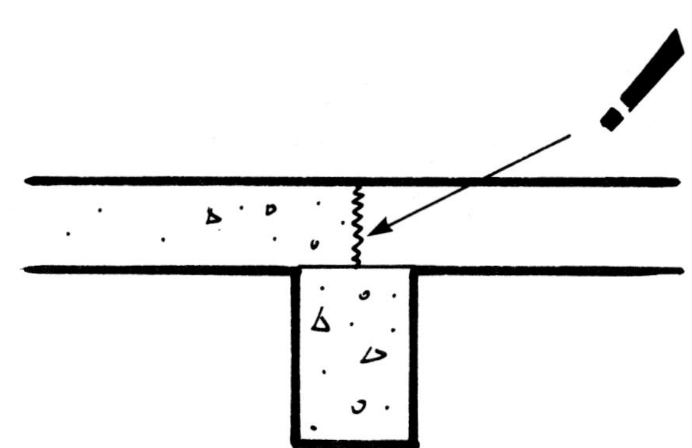

Joints at beam positions coincide with considerable quantities of steel...

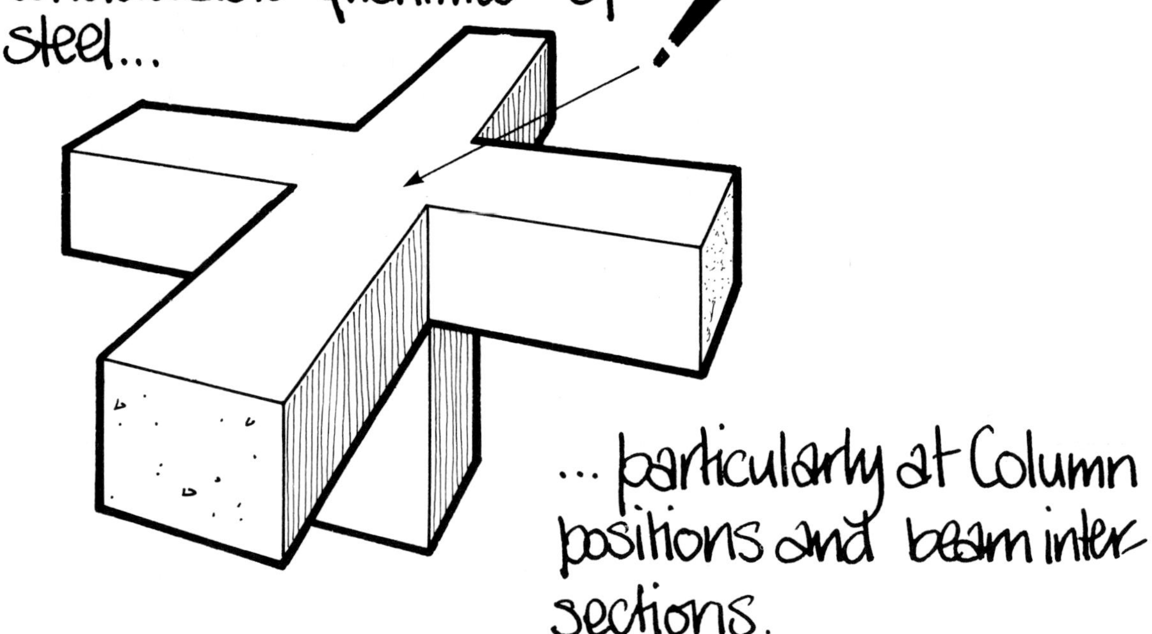

... particularly at Column positions and beam intersections.

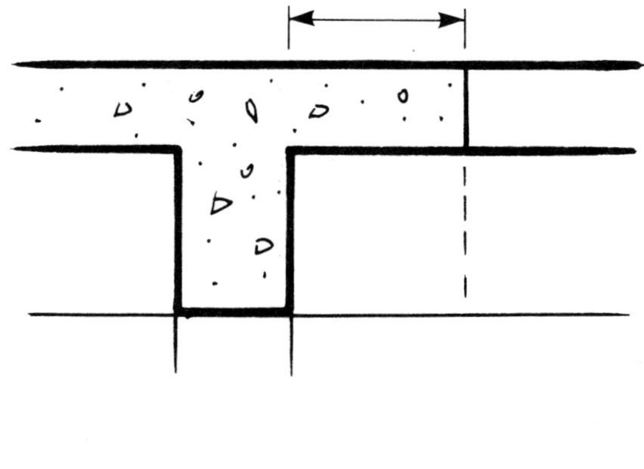

Joints at or near third points allow ease of stopend insertion...

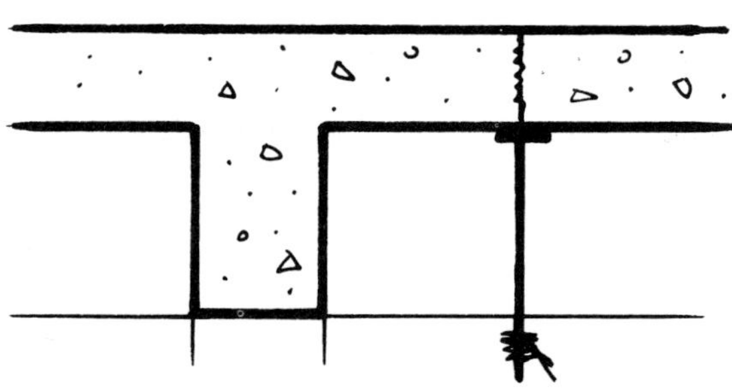

...although they demand special support if form is to be removed.

Implications of joint positions.

Shallow lifts provide repetetive work and allow forms to be man-handled.

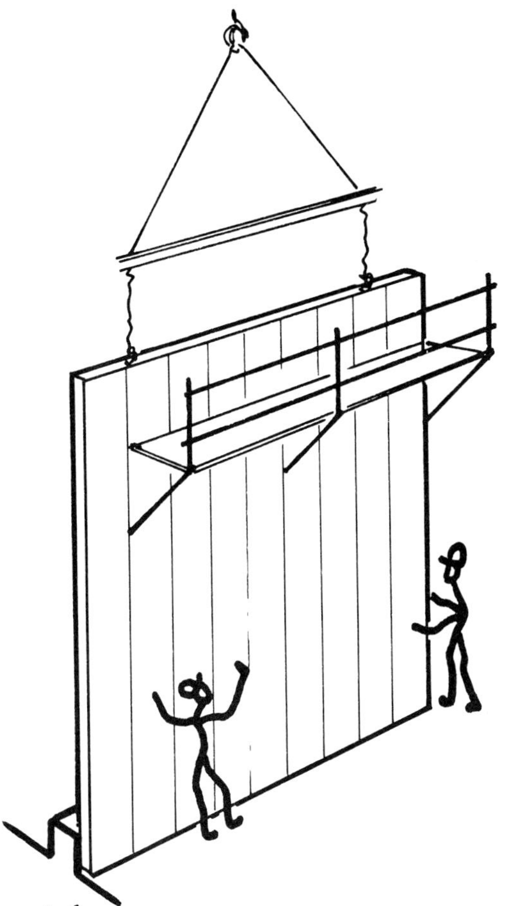

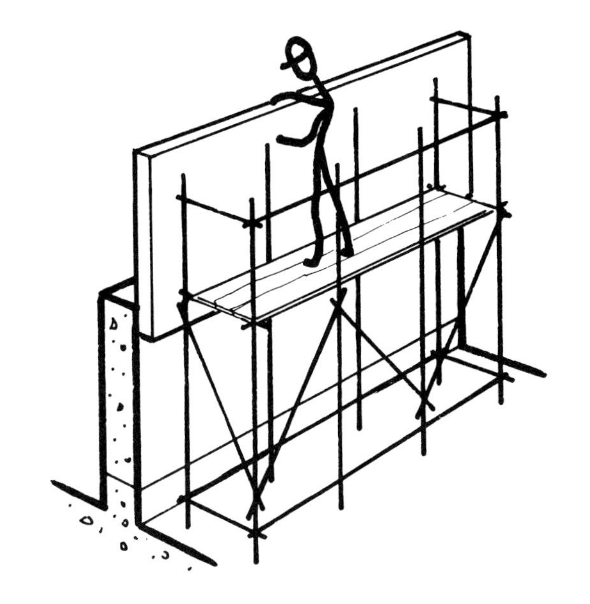

When forms can be crane handled, deep lifts avoid excessive numbers of construction joints.

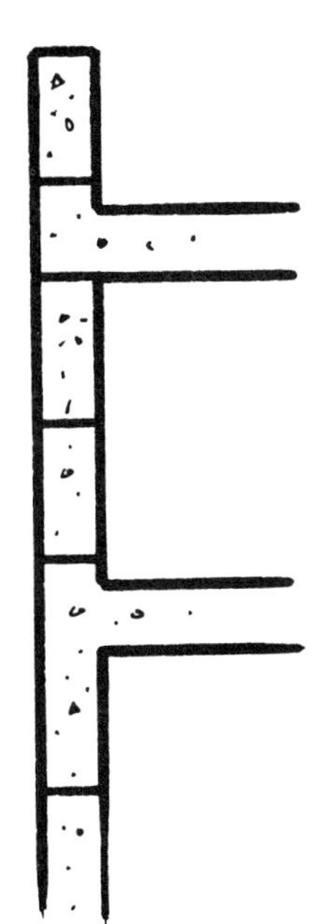

Shallow lifts simplify compaction process and allow ease of formation of adjacent slabs, features etc.

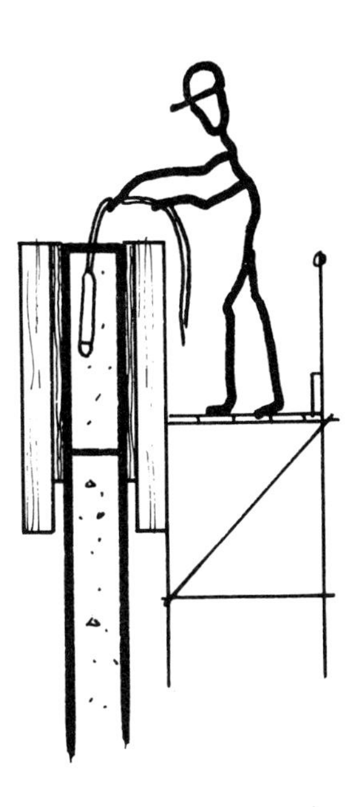

Depth of lifts.

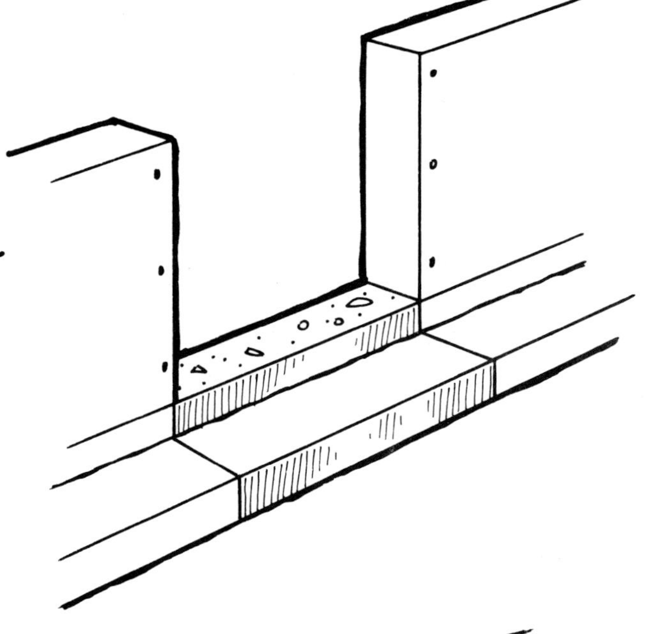

'Hit and Miss' construction allows shrinkage movement to take place before alternate panels are cast.

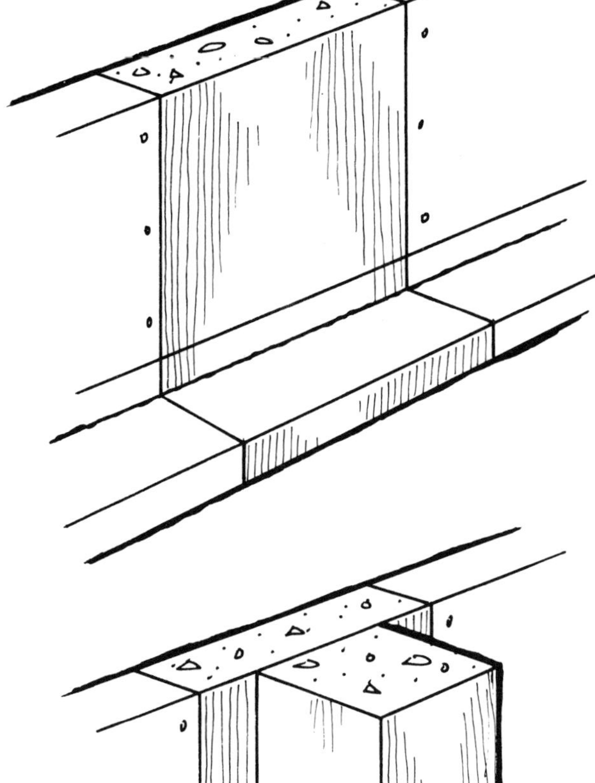

Sockets cast in first operation allow formwork to be bolted securely against original concrete.

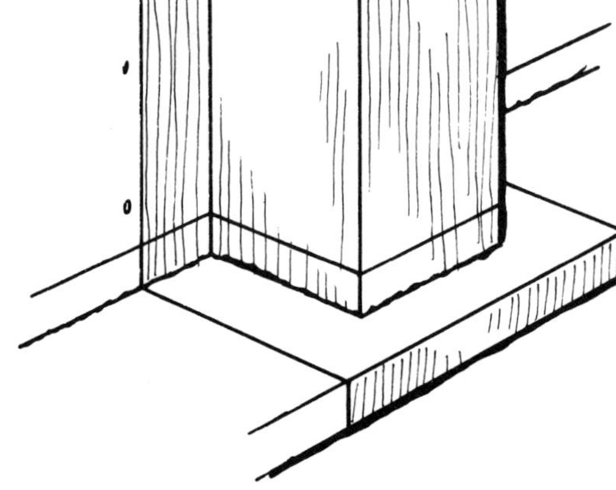

Pier positions form ideal situations for construction joints. Pier profile offers ease of filling and compaction.

Hit and miss construction.

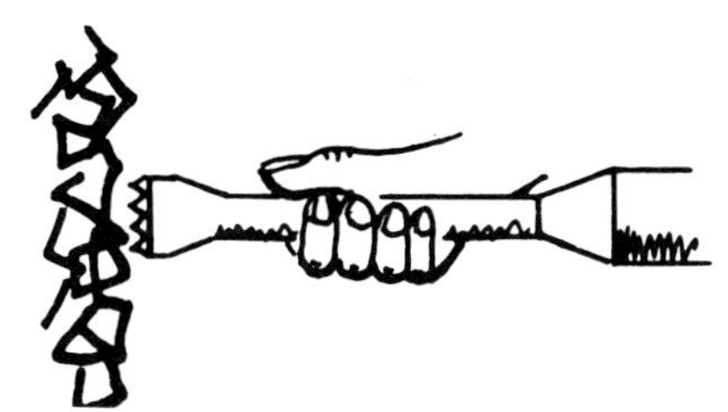

Scabbling damages aggregate — can cause failure.

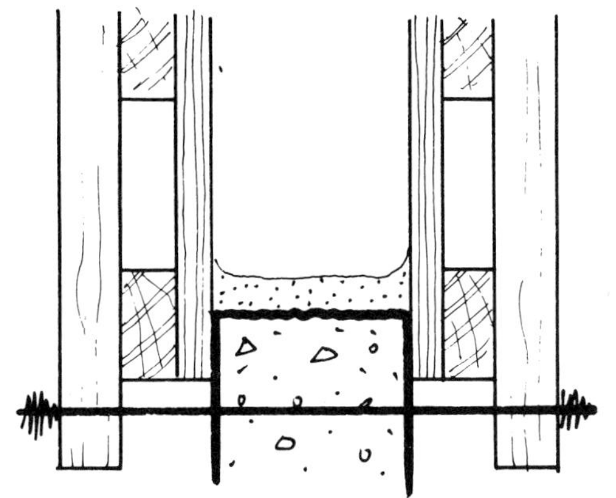

Grout or paste poured into form leakes layer of poor quality concrete.

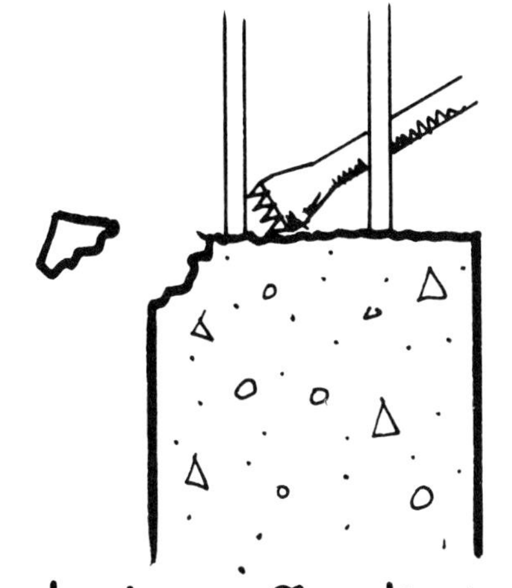

It is difficult to scabble between steel and adjacent to edges.

Simplest and best preparation is early wash and brush or high pressure jet.

Where available, grit blast provides excellent joint preparation.

Jointing techniques.

aggregate surfaces and is in a surface dry condition, a satisfactory joint can be achieved by direct placing with effective vibration of a succeeding lift or bay of concrete. Joints produced in this way are superior to joints where surfaces have been mechanically scabbled or treated with a grout or cement paste as was the practice for many years.

The methods used to achieve the required surface at an interface between the two casts of concrete vary. The most economical with respect to cost of equipment and labour requirements, is the wash and brush technique. This involves sweeping the film of fines from the top of the lift within about an hour of casting, using a soft brush and plenty of water. The problem with this method is that where the previously cast lifts are expressed visually, the cement paste may promote stains. To overcome this wire brushing with water at 12 to 15 h after casting generally provides a satisfactory surface. Surface retarders can be applied to the joint former and brushed away after the striking operation. Modern surface retarders – particularly those of the lacquer variety remain active until actually in contact with concrete, and being in the form of a drying paste will not contaminate reinforcement. Grit-blasting can be used at stopends or vertical junctions between bays if the lift is too high to permit early removal of the formwork, say within 2–3 h of completing the compaction operation. Another method of forming joints is by using expanded metal on supporting frames. The expanded metal forms a textured surface and when torn away from the surface at the joint provides an excellent key for further work.

If it is accepted that a sound joint can be achieved by one of these methods, and that the formwork designer will have the opportunity to discuss the location of construction joints and day joints with the design engineer, considerable improvements in construction practice can be achieved. The construction both of the structure, and incidentally the formwork, can be simplified by the careful location of construction joints that are relative to the physical features of the design.

Joint positions

The most critical joints in concrete construction are those formed between kickers and the structural component of which they form a part. Kickers are often badly formed, and are usually of insufficient depth to allow full compaction. The concrete used often falls outside the specification for the structure, being honeycombed and subject to leakage. They are often neglected in terms of treatment of the exposed face for achieving sound joints. Arguments put up often mention that reinforcement starter bars interfere with the process. Poor joints at kickers can be disastrous and extreme care must be taken in joint preparation. As it is not usually possible to incorporate a grout check or rule at the kicker, some trouble must be taken to ensure, that for the purposes of visual concrete, the kicker is in fact level.

On the question of visual concrete, it may be worthwhile to obtain the architect's approval for a featured or recessed kicker which allows the top edge of the kicker – the line of the construction joint – to be masked by subsequent concreting, or set in the shadow of a succeeding lift.

Rules or grout checks should be used wherever possible both for visual concrete and normal commercial situations where the concrete is to be covered. The grout check can be used to form an indent which results in a straight line or ruled joint, and to protect the top edge of the form from damage during concrete placing.

In the casting process, at any horizontal joint, it is essential that excess fines and paste be removed, these being the workability fines that are brought to the surface by compaction. For precast work, where a sound, dust-free surface has to be achieved by trowelling, it is good practice to cut away this layer of paste by screeding. For normal in situ concreting it is necessary to remove any excess from between the reinforcement with a suitably shaped trowel, and then to top up with fresh concrete. Re-vibration ensures that the surface is level and consists of sound, structural concrete.

Vertical joints in walls should be positioned where they are most likely to simplify erection of succeeding forms. Where possible tie members should be positioned in each bay where they can provide adequate restraints for the ends of the forms to the next bay. In doing this thought must be given to the lateral location of the ties for any visual concrete to meet architectural requirements, and for all concrete construction to avoid any leakage, staining and honeycombing due to normal deflections of the form.

Honeycombing gives an indication of grout leakage, and while on some contracts it is acceptable to carry out superficial local repairs to defects it is nevertheless a dangerous practice because without proper investigation which involves cutting well into the surface, or electronic testing techniques, it is virtually impossible to determine the extent of the honeycombing, and thus the extent of low durability concrete and poor protective quality for reinforcement.

Reinforcement at joints should be jigged to provide accuracy and continuity of line. Considerable damage to the concrete can occur where any attempt is made to straighten misplaced bars. The existence of a joint has a side effect in that it allows a check on the accuracy of steel location which is of course essential for protection and durability. For visual concrete and where projecting steel is to be exposed for some time before a succeeding lift or bay is cast, it is advisable, with the permission of the engineer, to coat the steel with a thin cement paste. This prevents rust from forming or rusty water from running down the face of the concrete thus causing visual defects. It has been proved that a fine coating of paste does not detract from the bond which is achieved between the steel and the concrete during a subsequent casting operation.

Joint configuration

Joggle joint forms are now being specified less and less as the problems associated with them have become apparent. Joggles create relatively small section nibs in which it is

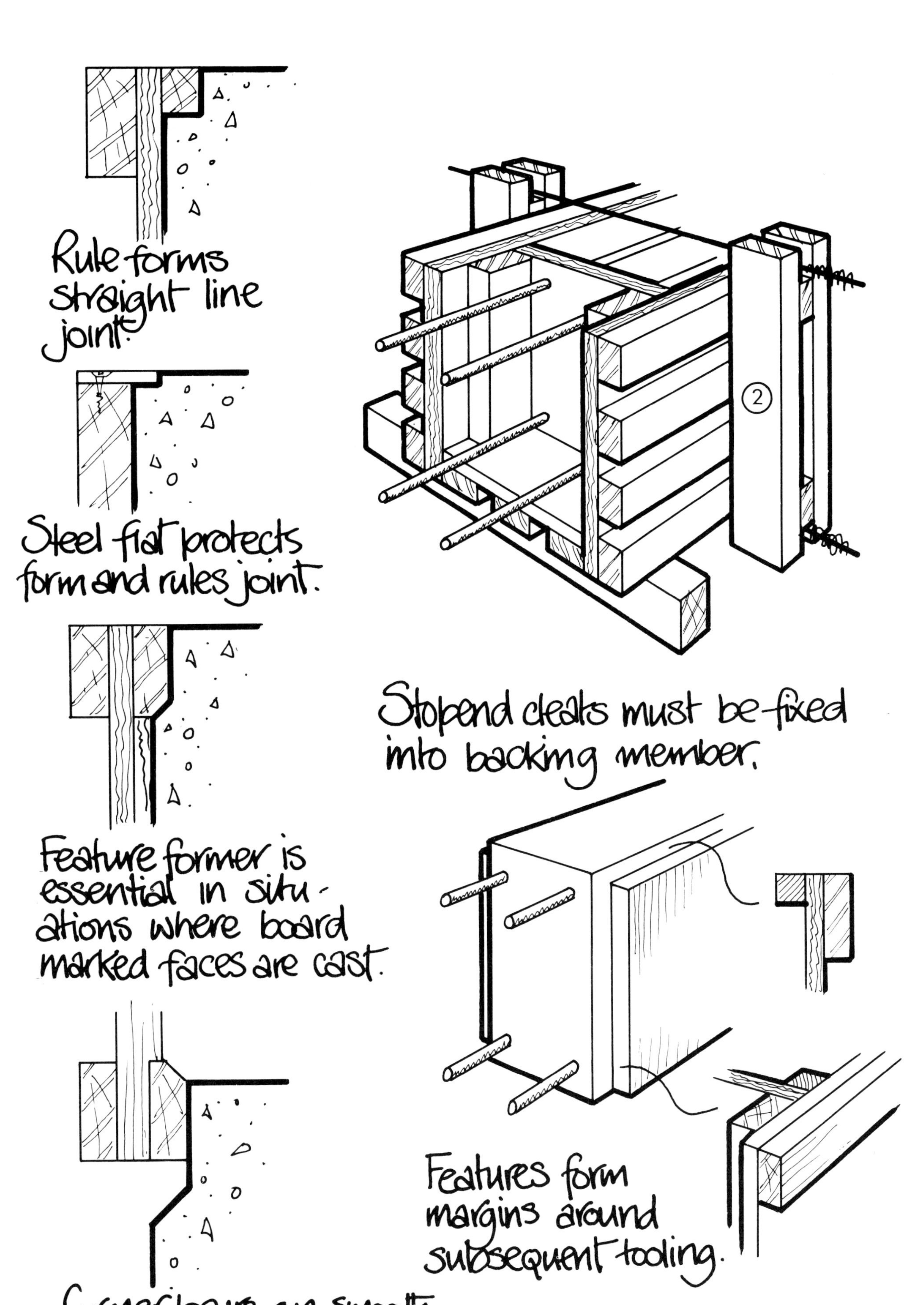

Formation of ruled joints and margins.

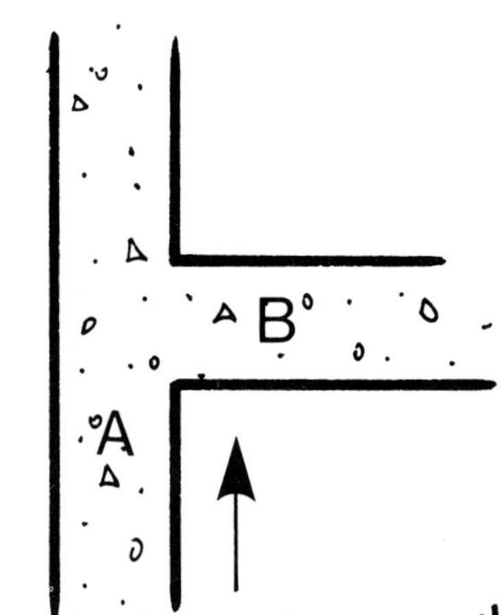

In principle wall construction should be continuous. Slabs and landings are secondary.

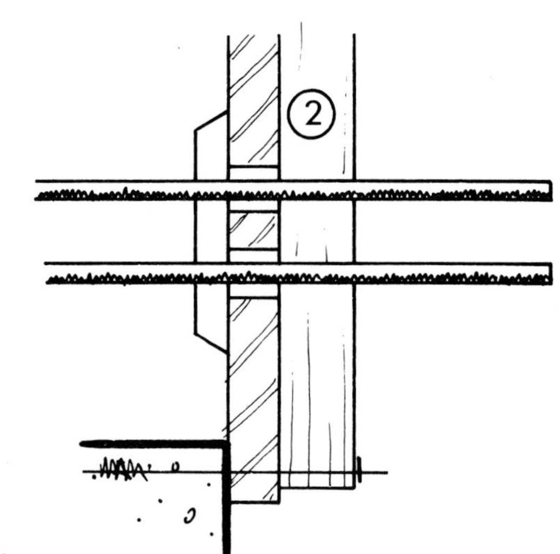

Where starters must project, lost template positions bars and masks oversize slots in form.

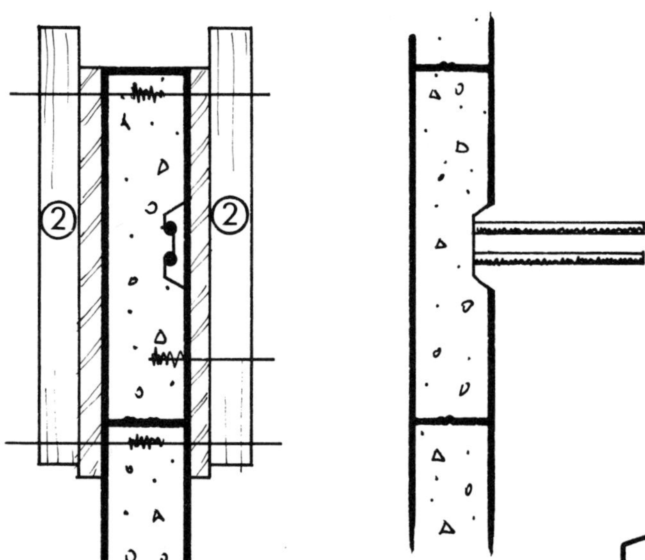

Bars set in recess later cranked straight. Chase is to locate bars not for structural reasons.

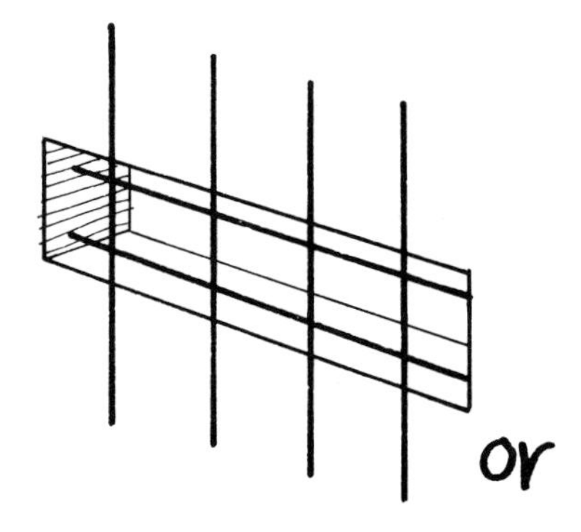

For landings, hairpins can be passed through slots or holes.

Floor/wall joints.

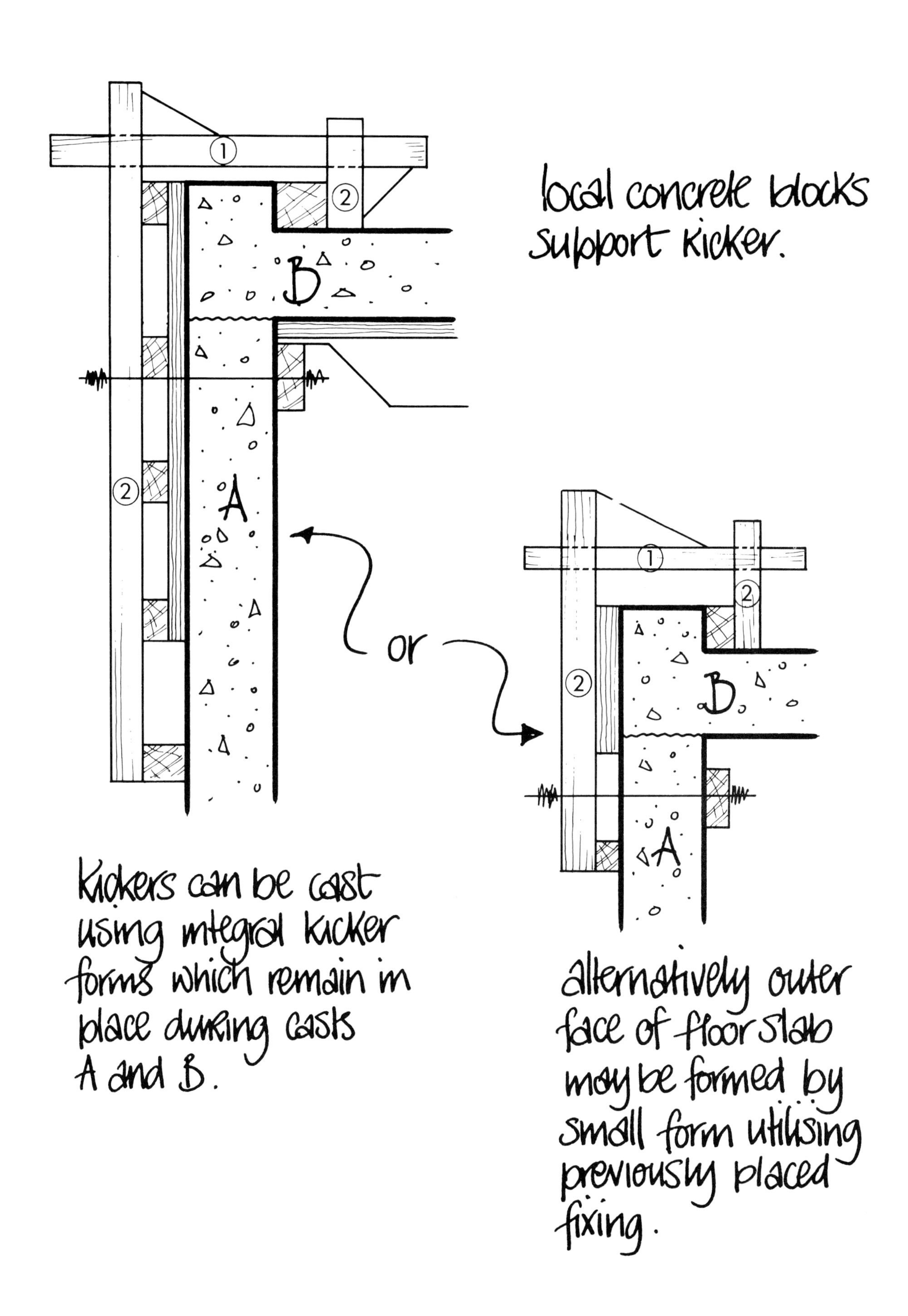

Floor/wall joints.

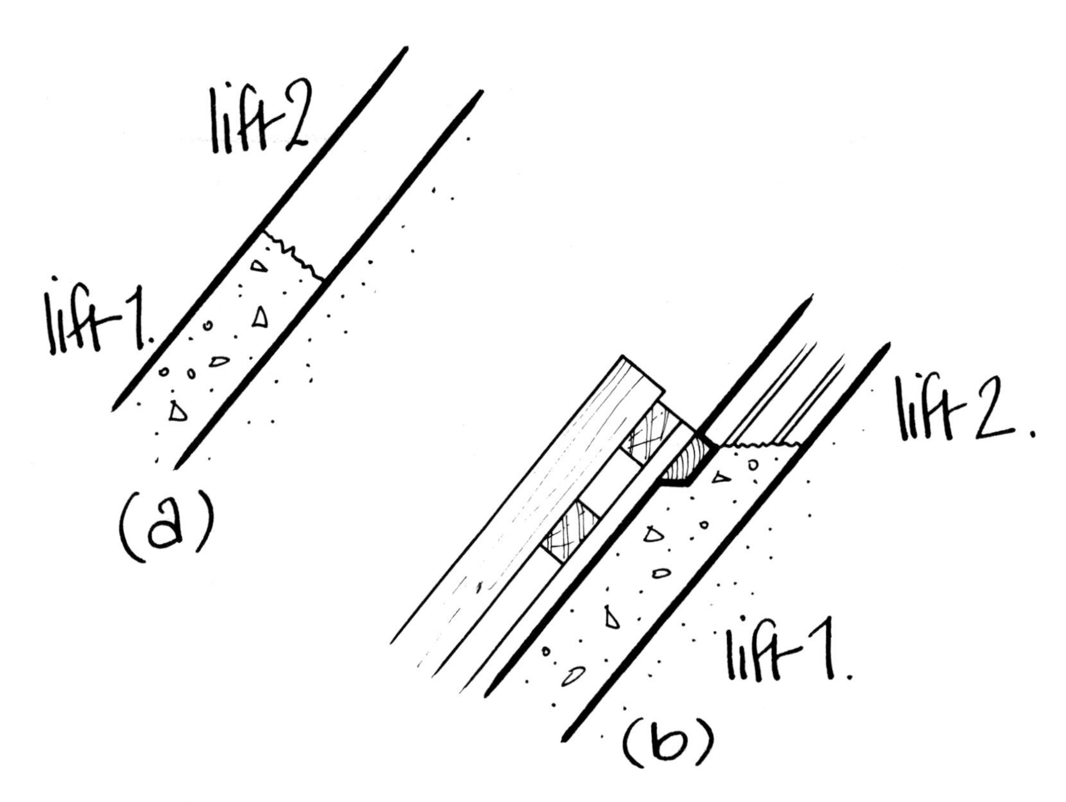

Whilst indicated as at (a), joints in sloping walls will be better formed as at (b) when concrete can be fully compacted.

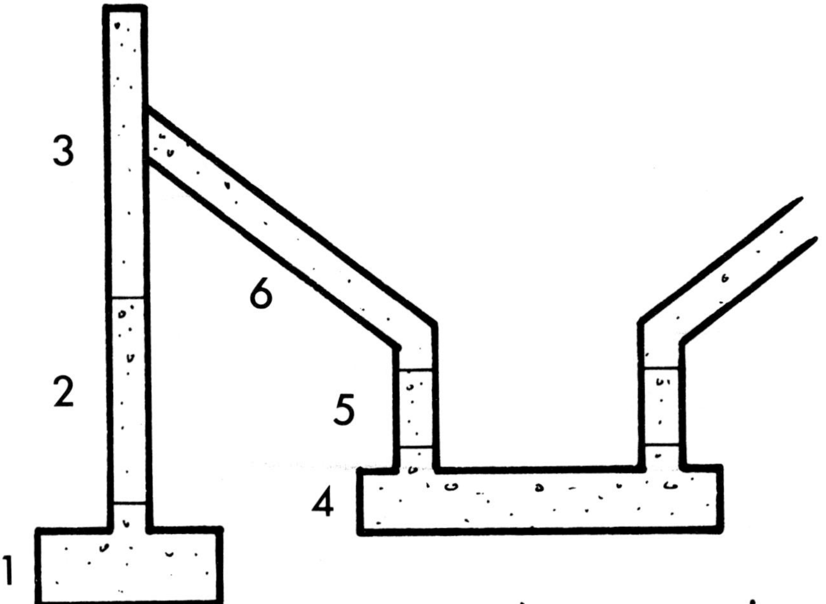

The location of construction joints will be determined by specification, structural design and reinforcement and concrete qualities, steel qualities and the dictates of construction method.

Floor/wall joints.

difficult to achieve satisfactory compaction, so that the inclusion of a joggle former within a horizontal wall joint can indeed lead to serious defects.

Where concrete has been placed beyond a previously suspended joggle former, problems of access for inspection and compaction are created. The resulting nibs at either side of the joggle former are often under-compacted and may well lack durability. The nibs can also be damaged by the force which is set up by tie tightening, and although the defect may only amount to micro-cracking there is still the possibility of the ingress of vapour or water which can corrode the reinforcement.

Joggles and shear keys increase the specific area of the joint and are desirable for certain types of construction. However the formwork designer should point out to the designer the possibility of defects being introduced by badly proportioned joggles, or by a multiplicity of shear key indents.

Water bars can present problems, although with the advent of the surface placed bar which can be simply fixed to the face of the formwork, many of the problems relating to accuracy of location, and of compaction adjacent to the inserted bar can be resolved. Where centrally located weather bars are employed, the laced variety allow for fixing to the reinforcing cage, although stopends must be split so that the projecting sections of the bars can be located in adjacent lifts or bays. If a stopend has been well designed sufficient space in the assembly will allow the weather bar to be inserted without the need of force which could cause distortion and misplacement.

Construction detail

With regard to joints between flooring bays, specifications have also changed over the years and while certain ones still stipulate chequer board casting of structural slabs, the general trend is towards strip casting of long bays. These utilize power floating techniques often in conjunction with vacuum dewatering. This trend should be welcomed, especially with respect to gangs who can be employed on gainful work thus causing less upset by teams who are working on adjacent bays. Generally the technique employs less formwork than chequer board arrangements, and the forms can be left to fly at the ends of bays thus avoiding expensive cutting or fitting of formwork within closed bays.

While considering joint positions in suspended floor slabs, the formwork designer or trades supervisor should study the formation of the formwork joist level, or the first layer of carcassing beneath the sheathing. It is advisable to ensure that the board or sheet which forms the face of the stopend either is located over a joist, or that intermediate supports are inserted from below the formwork. This provides the necessary support at the stopends in resisting the tamping action or vibrating screed used to level the concrete. More importantly where ply sheathing is adopted, it prevents local deflection and grout or paste leakage from the fresh concrete bay beneath a previously cast hardened floor bay.

The essential feature of any formwork joint is that of maintenance of line and consistency of appearance. Maintenance of line can be achieved by providing an adequate sheathing lap, the edge of the sheathing being firmly braced against previously cast concrete by lay or vertical members in the plane of the first layer of carcassing. When walling is cast vertically the form may be kept vertical by allowing the soldier members to project downwards and act as cantilevers through pads which bear on the previously cast concrete. Although this is a feature of the mechanism of simple-sided construction, this method can be used advantageously for double-sided work. The actual contact face or lap should not be excessive, something of the order of 75 mm is usual. The seal can be augmented by a strip of semi-rigid plastics or neoprene foam to produce a good grout-tight joint.

Laps of more than 150 mm can upset the line, as the smallest fin or projection can hinder the intimate contact which is required. This becomes particularly critical where closing bays or intermediate walling bays are cast by means of a form panel that is grossly oversize. The overlap would then cause considerable upset to line and result in badly defective surfaces from an aesthetic point of view, quite apart from honeycombing and the possibility of some structural defect. This can be overcome by making part of the face of the panel such that it can be removed, so that the lap is maintained at 75 mm and, for construction purposes, the backing members allowed to project. Care must be taken to make sure packers are inserted to maintain the contact between the sheathing and previously cast concrete.

Joints for producing board marked concrete can present problems because, obviously, it is impossible to achieve a perfect joint between two surfaces of varying textures. The concrete over which a form laps exhibits a natural grain pattern, the board face which imparted this grain pattern is used in the second operation at the top of the lift, and is completely different from that exhibited by the bottom end of the board which comes into contact at the point of lap. This can also occur at the edge of panels which are used to cast succeeding wall bays. The only way to avoid unsightly runs and honeycombing is by arranging a feature fillet or detail which, as it were, uncouples the two faces adjacent to the joint. A well proportioned fillet can form the grout check and joint feature and ensure that the lap is produced by a smooth area of form face.

Reinforcement at construction joints

Starter bars require particular attention during the formwork design stage so that form trapping and unnecessary drilling of formwork is avoided and the forms do not become damaged. Generally, small diameter starter bars can be bent up or along the form face and withdrawn after striking. Where starter bars have to project through the forms the sheathing should be removed locally. The prefabricated sheathing member then acts as a template for locating the bars. A cover plate left within the concrete provides a chase in addition to sealing the through holes. Long projecting bars need to be supported, the supports also being used to locate the bars at the correct centres.

All large diameter bars must be accurately positioned and to the correct projection to ensure that there is sufficient lap with the succeeding steel or cage reinforcement. Bars incorrectly positioned can damage the concrete when attempts are made to straighten them. Kickers are often damaged as a result of some straightening process, and this leads to water or vapour penetration.

Where for any reason bars need to be diverted, e.g. where they are fixed to a form face and subsequently straightened to their projecting positions, it is essential that the operation be closely supervised to ensure that the process is carried out in a workmanlike manner using the correct tools, and that *all* the diverted steel is bent into the required location.

Rebending of high yield, high bond steel is unlikely to be allowed, so that for starters it may be necessary to leave hole formers through a wall for later insertion of U-bars which, on completion of concreting, act as continuity members. This is ideal for stair wall/landing connections so that the projecting steel does not interfere with the form system. Under special circumstances it is sometimes permissible to thread the dowels and projecting bars and then insert them into cast-in sockets after the formwork has been removed. However this is an expensive technique and should be used only where forms would otherwise be trapped or excessively defaced by some other methods.

The designer can often, during early discussions, arrange the laps and projections such that the location of the steel is simplified. This is in fact considered elsewhere in the book in relation to construction and day joints.

The factors that govern joint location

As with all aspects of visual concrete, the formwork designer will need to consult with the architect to ensure that the final details of the concrete profile provide the best opportunity for practical and economical formwork treatment. In all decisions which relate to the proposed positioning of construction joints he has to take into account the following:

1 The requirements of the structural designer

2 The demands of the architect regarding aesthetics

3 The specification which governs the execution of the contract.

He will be concerned with:

1 Achieving maximum re-use

2 So arranging the joint positions as to meet the concreting capabilities on site and ensuring there is adequate access for placing and compaction

3 Achieving a form profile that is simple to construct

4 Incorporating corbels and nibs together with fixings which simplify subsequent formwork operations

5 Achieving consistency of appearance and maintaining continuity of line

6 Avoiding any trapping of the mould or difficulty of striking from between returns

7 Arranging for the progression of complementary activities, i.e. steelfixing and finishing which must continue to allow gainful work for all the trades involved.

Kickers, nibs and corbels

Two of the most vital components in concrete construction – certainly the components which contribute a great deal to the accuracy of the work – are kickers and nibs. Kickers and nibs are the small upstands formed upon horizontal and vertical surfaces respectively to which the forms for subsequent lifts and bays of walls or columns are clamped or bolted. A nib performs its function where adjacent bays are cast in wall construction, where it provides a projection against which the forms can be clamped, and thus accurately located, to line and plumb. The technique for forming nibs or projections is, however, essentially an integral part of the wall form and wall casting arrangement, while the formation of the kicker may be carried out as a separate part of the construction process.

Kicker construction

The kicker can be formed either to be integral with a slab, or as a secondary operation once the slab concrete has stiffened, or after the concrete has hardened. Kickers which are cast as a separate operation unless of considerable depth where they constitute a further lift in the sequence of operations, will result in substandard construction.

The concrete in a kicker which has been cast on to a slab once the concrete has begun to mature will inevitably move at a different rate to that of the slab, as the contraction process commences with the likelihood of formation of a shear plane at the interface between slab and kicker. While deep kickers overcome this problem, excessive height of kicker may cause the very difficulties they are intended to avoid, i.e. lack of accuracy, failure to provide a sound location for the bottom of the form, and insufficient grout seal at the joint.

Kicker forms are in fact part of the formwork arrangement, they can be compared to templates in that they govern the profile of the eventual structure and provide this opportunity for the careful and accurate positioning of walls, upstands and columns. As a secondary function a kicker provides a visual check on the accuracy of the positioning of steel starter bars and, gives an instant check on the vital aspects of cover to reinforcement.

Kicker formation integral with floor slabs or

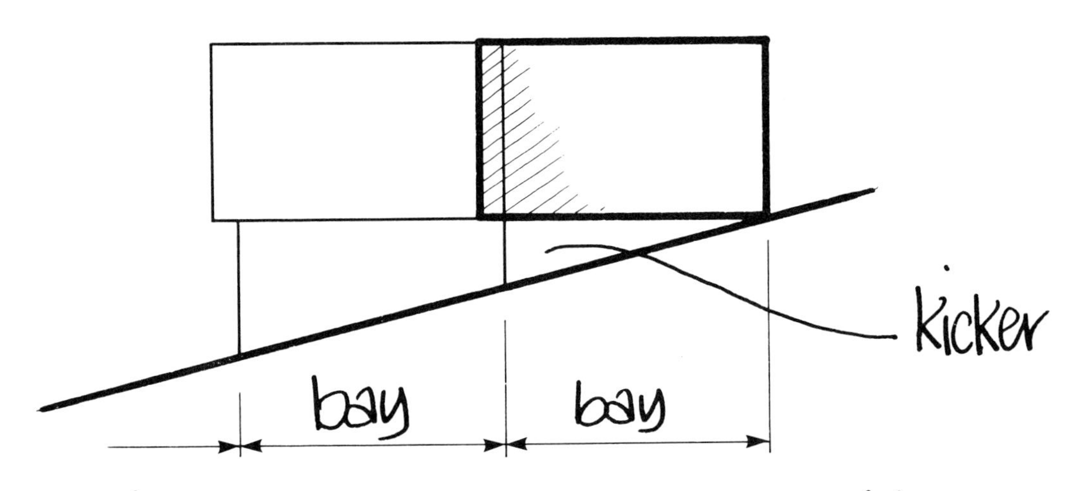

Kickers can be arranged to facilitate the use of square standard forms in ramp construction. Adjustment is carried out at time of forming kicker.

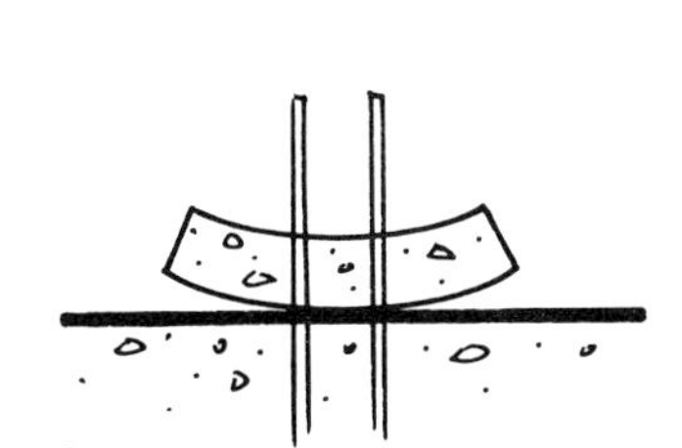

Small, shallow kickers cast other than monolithic lead to defective construction.

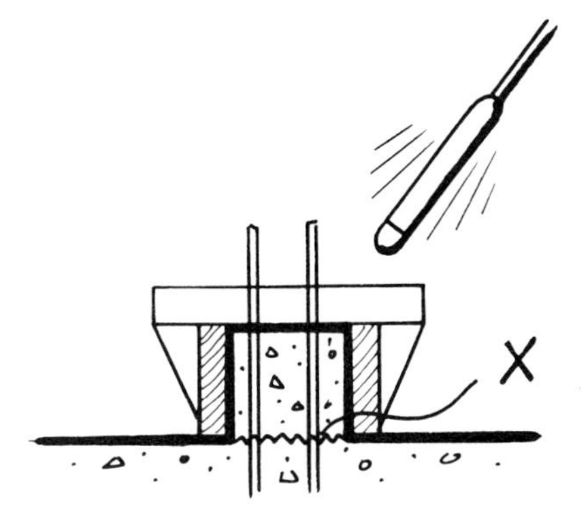

Where kickers are cast separately they should be of sufficient depth to allow correct compaction. Joint "x" should be properly prepared.

Kicker considerations.

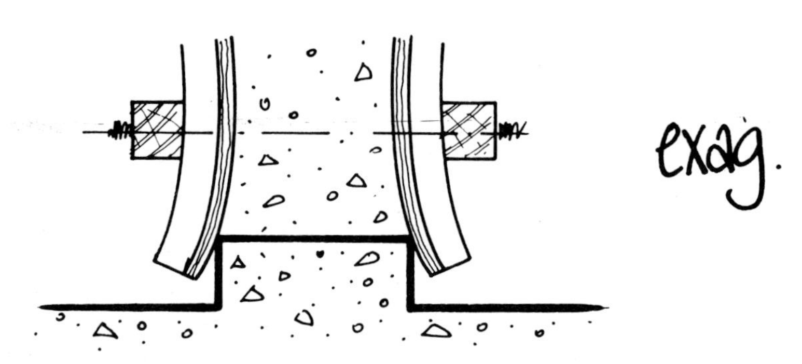

Ties positioned too far above kicker tend to allow distortion.

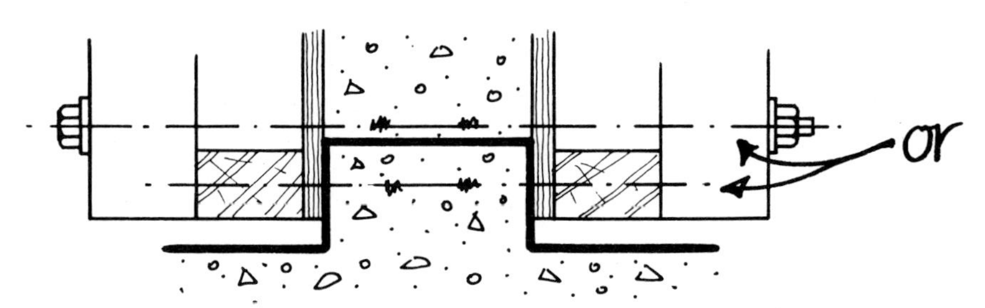

Ideally tie force should be immediately above or through kicker to ensure sound joint.

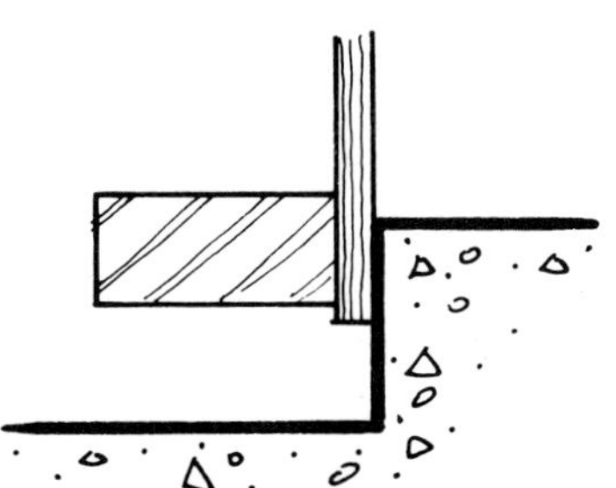

allow space for form to drop to strike.

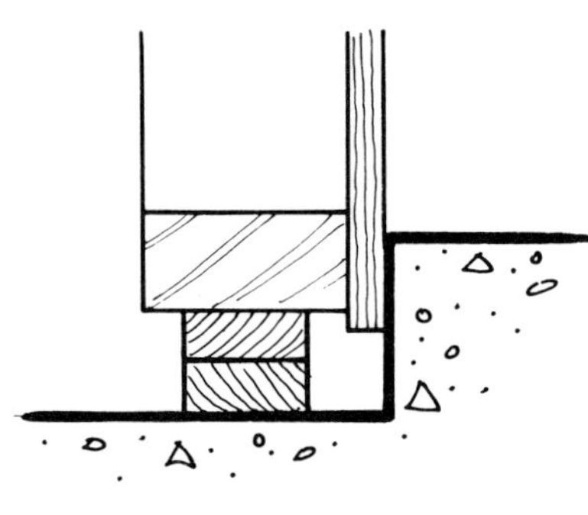

wedges ease levelling.

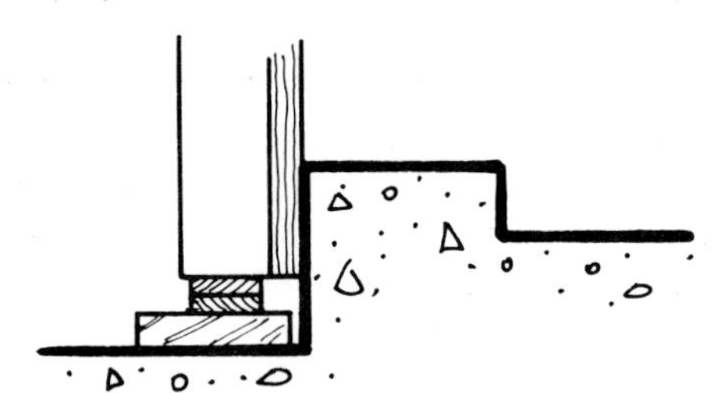

plate at ground level assists in erection.

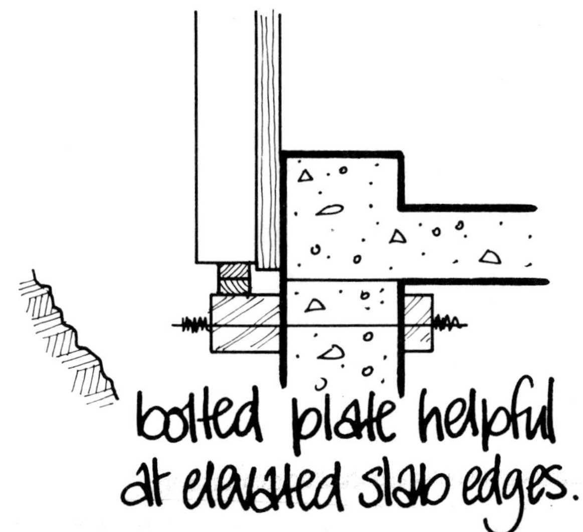

bolted plate helpful at elevated slab edges.

Kicker considerations.

foundations can be achieved by supporting kicker forms on stretchers that span the strip foundation or column base, or by simple cantilever arrangements at the perimeters of the slabs or tops of walls. However for wide foundations or where access is difficult it may be desirable for the kickers to be cast as a second part of the casting operation. On wide foundations the work should be carried out immediately access to the concrete surface is possible, if necessary using crawl boards, without defacing the finished slab surface. To facilitate kicker form fastening, suitable blocks of timber or concrete can be embedded in the concrete as it is being cast. During normal construction where surfaces such as floor toppings are to receive subsequent finishes it may be permissible to secure the kickers by steel pins driven into the previously cast, and now stiffening, concrete. Where the projecting steel is substantial enough, e.g. on columns, then the kicker formers can be strapped off or centred from the steel.

It is essential that the concrete in the kicker be properly compacted, since failure to achieve a reasonable degree of compaction will result in loss of durability or some structural weakness at a critical part of the construction. Unfortunately in reinforced concrete construction it is extremely difficult to achieve a good standard of compaction where kickers, nibs and intersections are being formed, especially as strength is necessary for sound construction at these positions. Because of this many site authorities insist on kickers with depths of about 150 mm, so that quality of compaction is ensured, and there will be less tendency for using up scrap concrete as is often the case for shallower kickers.

The kicker can be used to advantage by the formwork designer and constructor in overcoming certain problems imposed by the profile of the concrete structure. Many structures involve the formation of concrete fillets, or even rounded surfaces, at the junctions of slabs and upstands. These are imposed by the requirements of the structural design and the eventual use of the concrete structure. The kicker arrangement can be designed to incorporate such details and, by bringing the concrete joint position to a particular level, can avoid the need for the construction of complicated features at the foot of the form proper. Kickers can also be arranged to avoid the need for scribing or cutting of the main form panels. Where ramps are cast with flank walls or beams, the kicker can be arranged in such a way that the foot of the main form panel remains horizontal and yet clamps securely on the top of the kicker. Here the planning engineer must carefully relate kicker heights and lengths to the module selected for the walling operations.

Apart from positioning wall faces and allowing the forms to be clamped into position, the kicker allows the accurate placing of full height openings within the wall, and the subsequent support at the foot of the opening former. This can be achieved simply by the insertion of a timber or steel prop behind the members that form the return face of the opening. A trench prop or a timber with folding wedges will provide sufficient jacking force against a sound kicker to avoid inward or upward displacement by the very considerable pressure that develops at this point during the concreting operation.

Kickers – geometrical work

The value of kickers becomes particularly evident in the construction of circular and geometrical work. Without kickers, the forms would be free to move inwards or outwards, while joints in the forms would remain closed and the top of the form would conform to the required shape, considerable inaccuracy could exist, even to the extent of complete loss of cover at certain points either inside or outside the wall. It is extremely difficult to position accurately circular or geometric walling because of the problems encountered when using instruments once the forms are in position. The kicker allows setting out to be executed accurately, and cover on steel checked visually. Provided the forms are then clamped tightly into position and plumbed, the maintenance of a circular or ellipitical wall becomes quite a simple matter.

Just as on a drawing it is easier to draw lines tangential to a previously drawn arc than to connect the lines with an arc, so it is easier to connect straight kickers to a previously formed circular kicker and thus maintain a fair line at the intersection. Ideally, the former used for circular work should be machined to the same rod or template as the ribs for the form proper. For circular columns it is often convenient when employing timber, steel and grp for the kicker to be manufactured integral with the main form (allowance being made for the clip back or amount of overlap), the kicker section is then removed after manufacture and used separately in kicker formation. A similar section may well be used as a 'landing ring' on subsequent operations.

Where columns or walls of geometrical form are being cast this method of manufacture ensures an accurate fit between the kicker and the main form. Where architectural requirements so demand, it may indeed be necessary to ensure that the kicker and the main form are carefully marked and used in such a way that the slight differences in profile of different sides of the form and kicker are overcome, i.e. the column or wall forms are always oriented the same way on succeeding operations relative to the pre-formed kicker.

In normal construction, and essentially in splayed wall construction, the height of a kicker may be such that the wall forms are bolted back to the kicker using buried anchors or through bolts. This arrangement increases the friction at the joint and prevents the vertical displacement of forms which is often experienced in wall construction due to the difficulty of achieving purchase for bracing arrangements or to formwork pressure on a splayed face. With very pronounced splays or battens it is necessary to supplement the tie achieved with the kicker, by ties taken to anchors in previously cast concrete. Anchors in the upper face of the kicker assist in cases where there is considerable splay or overhang.

Where stepped or deep kickers are used to overcome ramping or variation in the level of the foundation or slab a timber plate, steel channel or angle member can be provided onto which the bottom of the form can be lodged during the initial form erection, and from which the form can be wedged during the process of levelling. This member can be bolted back to an anchor cast into the kicker, or where a timber plate is used, simply rest on blocks of varying heights based on the foundation or slab concrete.

With regard to the joints between the forms and the kickers it must be remembered that the only positive support occurs where backing members overlap the kickers. The joints at intermediate positions depend on the degree of deflection in the sheathing. If the sheathing overlaps too much at these points it is as bad as if there is too little sheathing; the probable optimum is about 75 mm. Too much overlap causes distortion where small particles of aggregate become entrapped, and it is therefore impossible to achieve a tight joint, while too little overlap allows little adjustment in the level of the forms and permits grout loss on minimum deflections. It is impossible to force back tightly a steel face on a rigid steel frame to previously cast concrete for the whole length of a form. Similarly, it is also impossible to clamp featured or board marked forms into intimate contact with concrete that has been formed previously by another part of the featured or board marked face, since the slightest difference in feature or marking allows a loss of grout. It is essential that the architect appreciates this, while it is the formwork designer's job to prompt the formation or an indent or feature to overcome the problem.

Ideally, the clip should be rendered more positive by the introduction of a support which can be done by positioning the backing member in such a way that it overlaps the kicker concrete. While the normal ruled joint batten or former defeats the object of the normal 75 mm kicker, it is as essential that the rule be inserted in deep kickers as it is for wall construction. Where visual concrete kickers are exposed, the ruled joint can be achieved by a minimum rule of say 4 or 6 mm, this being achieved by using a steel or alloy strip secured to the top of the kicker and allowed to project some 12 mm into the concrete.

Where it is necessary to obtain a surface finish, and essential to use a grout-tight form, it is advisable to insert a continuous member in the first layer of the carcass or within the framing system to provide a support for the sheathing over the area that needs to be grout tight. Grout seals of closed cell plastics or neoprene help to achieve the right kind of seal.

Nibs on vertical and sloping faces

Nibs or lead ups cast onto the vertical faces of beams, walls and such like are monolithically cast during the main casting operation. This obviates many of the 'quality' arguments that arise during the formation of kickers. Nibs provide a support for the succeeding bays of formwork, ensure continuity of line, and provide a grout-tight joint where a succeeding form is clamped or bolted into position. They also provide a seating against which formwork can be clamped. In fact the nib is the key to the simplicity of erection, since without one there is little to support the vertical form member until the opposing panel has been erected and some kind of tie inserted.

To be effective, the projection of a nib from an adjacent pier of a walling, beam or column, or for ducts, beams or spandrils, should be 100–150 mm. For system formwork, or arrangements that incorporate standard panels, the nib or projection can be formed within the thickness of the sheathing and first layer of carcassing or framing, so as to avoid interference with the timber or steel walings or tubular members and thus provide continuity of the support arrangements.

Where a nib is formed within the thickness of the sheathing and first layer of carcassing, the stopend arrangement can be simply fixed against the back of the framing member. The designer should watch for traps, however, where a series of panels are used between adjacent nibs, for example where duct cheeks project from the face of a wall.

Nibs at the corners of piers or columns formed to provide continuity in spandril construction present some problems to the concreter. Good compaction must be achieved and where heights above 1.2 m are being cast certain stopend boards to the nib form should be omitted to allow poker access, the boards being secured after adequate access and compaction have been achieved. This applies particularly where nibs are formed at the flanks of a column for spandrils, and that the top of the nib form should be left open for the insertion of an immersion vibrator, a small capping piece or closer being fixed to prevent surge as the concrete is placed in the upper part of the column form. The boards that form the stopends, which are face-fixed to the back of the first layer of the system, should be fixed by skew nailing since it is not unknown for the concrete pressure to overcome the fixing and cause loss of grout with subsequent honeycombing, again at a critical part of the construction.

Where possible, tie arrangements should be simplified by the insertion of some anchorage into the nib. The projection of the nib should be at least 250 mm from the adjacent concrete face. Ideally the sheathing lap onto the nib in the secondary operation should be kept to a minimum of 75 mm. As with kickers which solve problems of varying heights, or splay construction, the designer can arrange the length of the nib projection such that he overcomes variations in the size of the intermediate bays. This allows a standard panel to be used, the variation being accommodated by allowing the panel to oversail the nib to some extent at each end, it is advisable that some feature or indent be incorporated to obscure the vertical joint so formed. Where visual concrete is anticipated care should be taken to ensure that the tie system clamps the sheathing at, or immediately adjacent to, the end of the nib, otherwise the normal sheathing deflection will permit the formation of unsightly curtains of grout at the junction with the fresh concrete.

Nib and corbel construction

Horizontal nibs or corbels are often used to support cladding members or structural members for intermediate floors and mezzanines adjacent to stair walls or some other floor structure. For such cases the projecting nib will be horizontal and can upset the sheathing layer. Horizontal nibs demand careful consideration so that good compaction can be achieved at the upper face and formwork can be removed without damage to the nib at

Incorporate nibs on wall bays to facilitate location of return forms.

Kickers and nibs facilitate form location and grout retention.

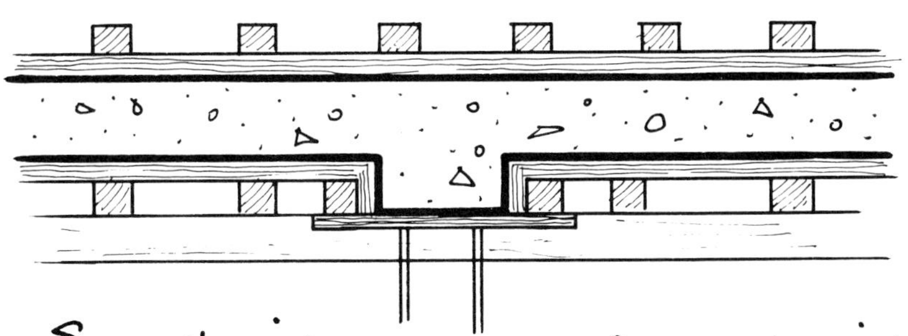

Small nibs can be formed within thickness of first layer of carcass.

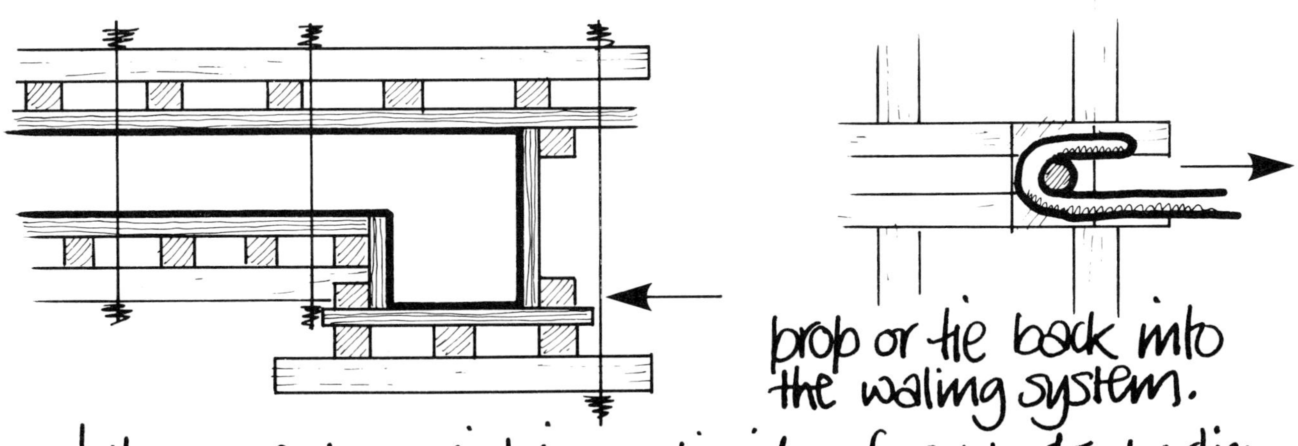
prop or tie back into the waling system.

take care to maintain continuity of support – particularly where plane of carcass changes.

Wall joints at returns.

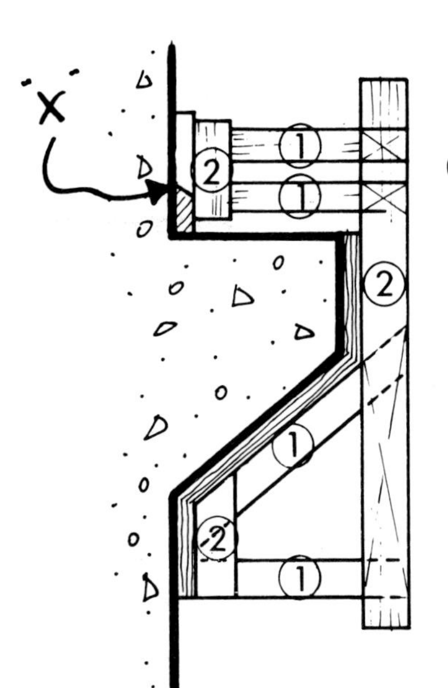

allow clearance

Fillet "x" allows form to move and clear corbel.

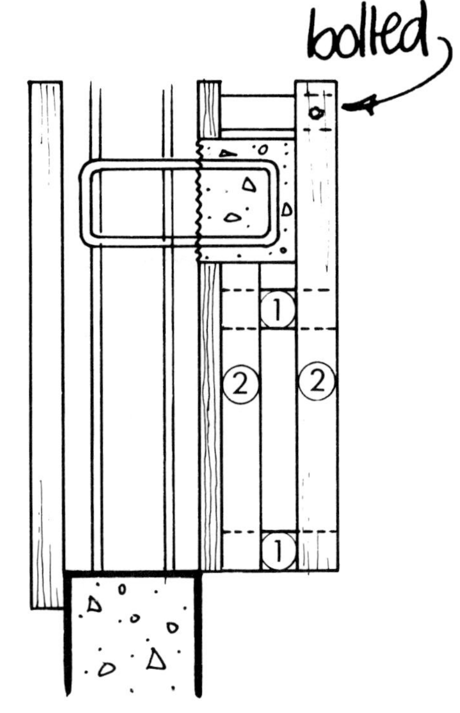

bolted

precast corbel can be incorporated into panel. Infill boards allow panel to be used for pla

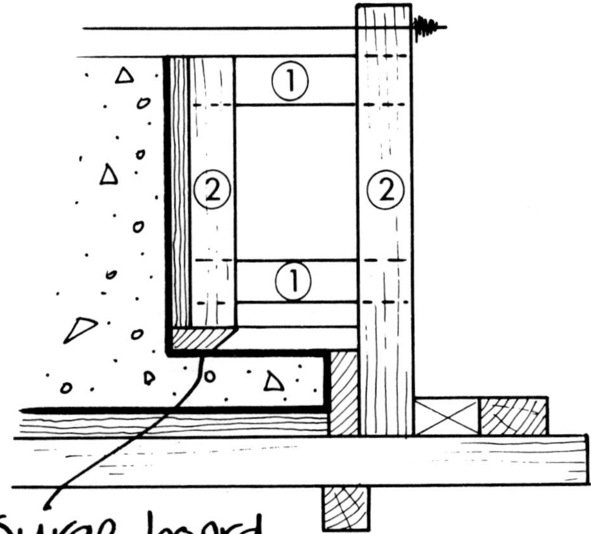

Surge board

hib or corbel with open top. Concrete placing must be timed to avoid surge or dry joint formation.

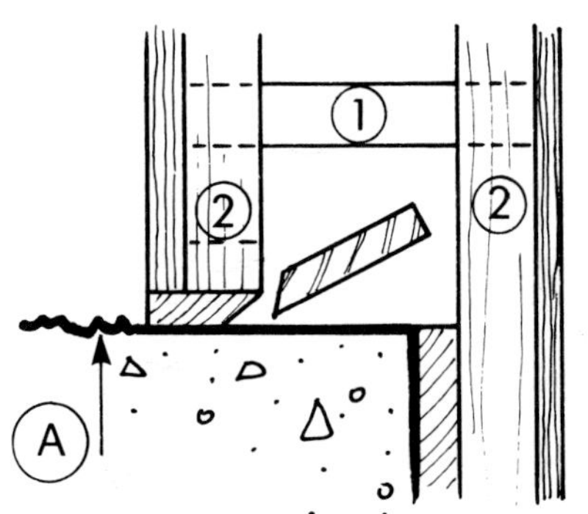

alternatively place and fix top board when concrete reaches A.

Wall joints at returns.

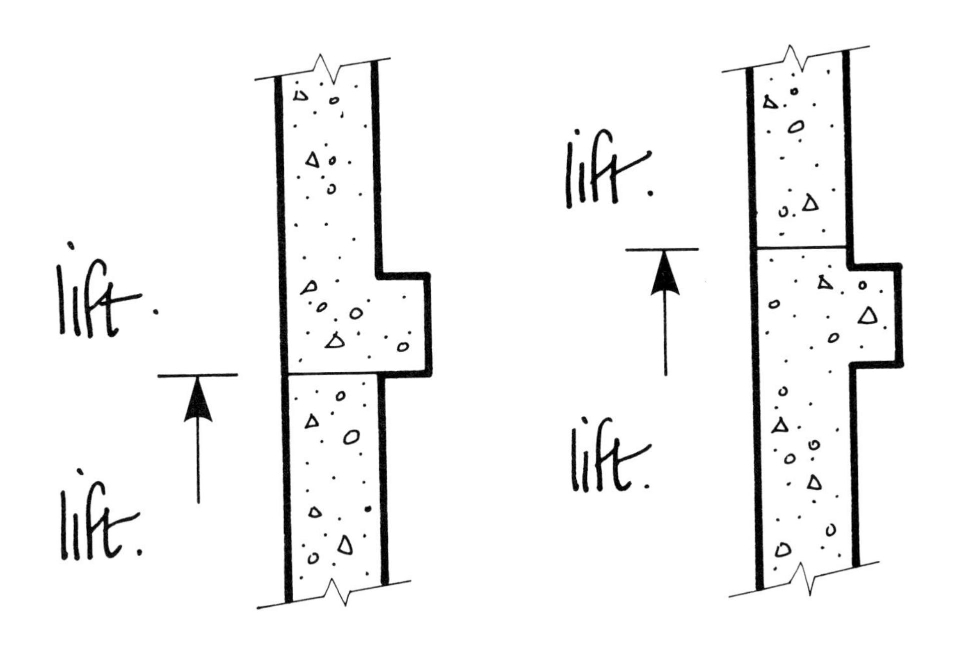

Construction joints arranged below corbels present difficulties in corbel placing and steel location.

With construction joint placed above corbel, placement and compaction, and steel location in projection are simplified.

Wall joints at returns.

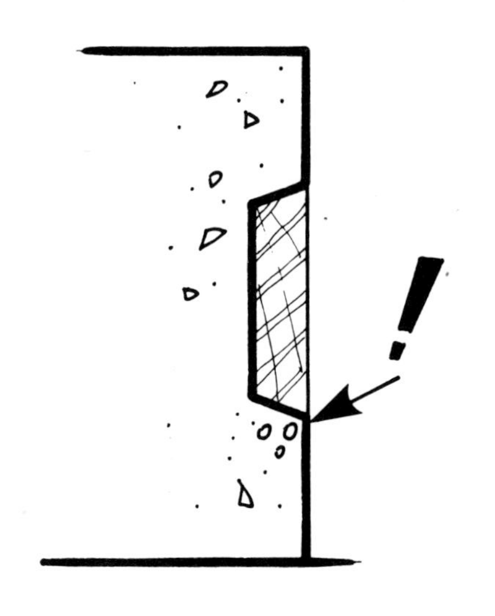

Joggles are expensive to form and can upset compaction causing defects.

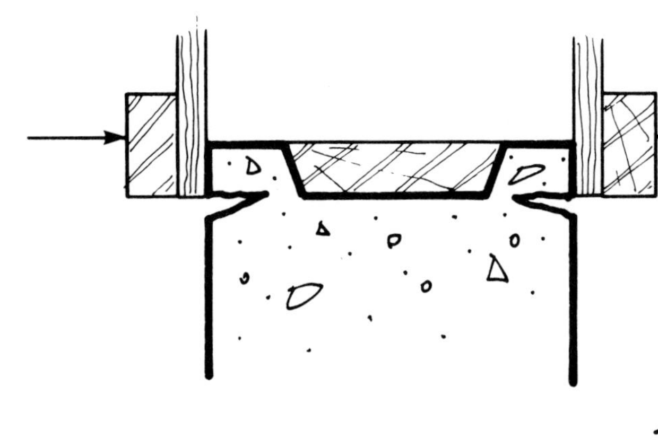

Horizontal joggles may weaken walling where strength is needed for form fixing. Compaction is difficult to achieve.

Joggles are a constant source of defective construction — joggles in a kicker, for example, detract from strength of structure and usefulness of kicker in construction.

Substandard joints.

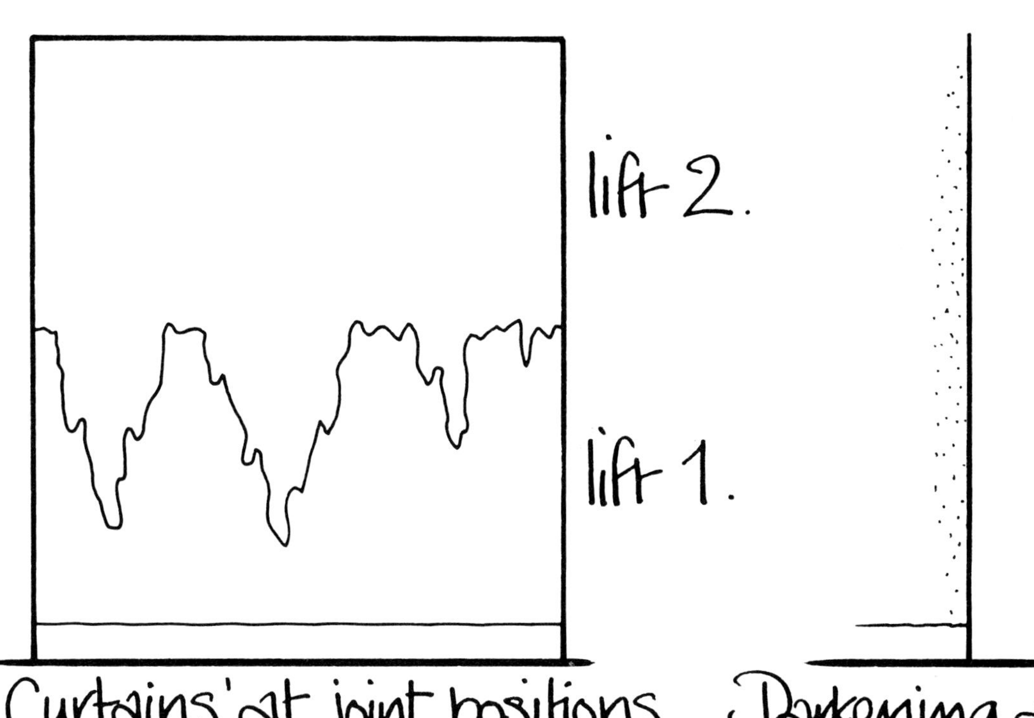

'Curtains' at joint positions indicate local leakage — probably caused by deflection of sheathing between supports.

Darkening at joint position indicates leakage of paste from form.

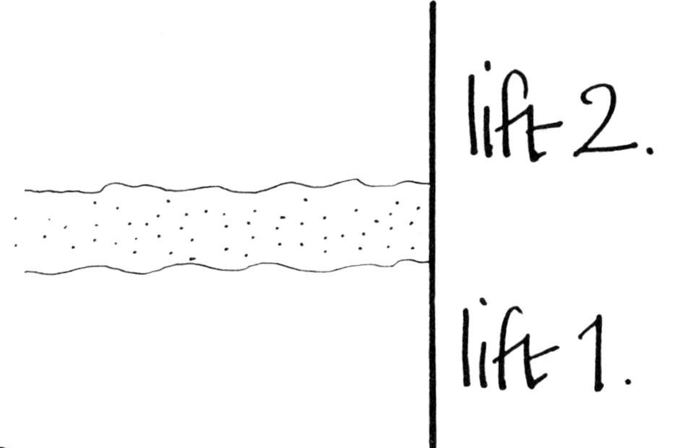

Paste or laitance at joint will be evidenced by distinct banding effect.

Darkening or staining may also occur at board or sheathing joints.

Substandard joints.

an early age. What is critical at this stage is the correct location within the nib of the reinforcement and the correct cover.

Ideally, the formwork arrangement should be such that the nib is formed at the top of the lift. This relates to the overall location of construction joints, however, and is extremely difficult to achieve, for example, in the stair tower situation. With the corbel or nib formed at the top of the lift it is necessary to ensure that the form containing the nib arrangement can be lowered away from the underside of the nib so as to avoid trapping of the panel. Alternatively a stripping joint can be inserted, or the sheathing for the nib arranged to remain in place while the main panel is being struck.

In the stair tower situation where considerable projecting steel is involved, it is especially necessary to insert fillets adjacent to the nib, to allow the form to be removed. The face of the nib will be drilled for projecting steel and it may be that this is best achieved by framing the nib sheathing separately from the main form and allowing it to remain in place while the main panel is stripped. There remains the question as to whether a top member should bc fitted to the corbel former or whether it should remain open. Although, to avoid entrapped air, it is best to dispense with the former to the top surface of a nib or corbel, it is more difficult to fill than where a top or surge board is provided. Where no top board exists the filling operation must be timed so that the concrete begins to stiffen prior to the start of the placing and compaction of the concrete in the wall above.

An arrangement where a top board is inserted after the concrete nib has been placed and compacted avoids the problems of entrapped air and difficulty in placing concrete and dispels doubt with regard to the actual position of the steel relative to the corbel profile. The cover board is not easy to position quickly, however, and where there are projecting dowel bars it is likely that the fixed upper sheathing member will provide the best results. The designer must again ensure that the formwork is not trapped by the dowel bars, and a stripping fillet will be required to enable the top sheathing member to be removed.

Where corbels are to be formed at the mid-height of a lift the best results can be achieved by providing stoolings at the face of the sheathing which generates the corbel form. These stoolings are secured to the main sheathing or carcassing in such a way that they can be released but remain in position on the concrete while the main panels are removed. Care is required to ensure that any such stooling which is used to form return faces is split in such a way as to facilitate removal from within any adjacent return faces.

It is essential that a grout-tight seal should be achieved at the junction between the corbel sheathing and the main panel sheathing. This seal apart from preventing grout ingress, will also serve to prevent the formation of fins and the trapping of the stooling member.

In certain situations, for example where there is any proliferation of corbels or nibs, it may be permissible to precast the projecting portion complete with tails or projecting bars. The precast nibs are then positioned in the form, preferably by bolting to the carcass, and then transferred into the final location in the structure. This technique will need to be discussed with the engineer, although a major point in favour of its adoption is that the compaction of the nib, and the steel position within the nib, can be carefully regulated during the casting process. This technique has particular applications in the manufacture of precast components, such as columns and cladding panels where there are often projecting nibs the accuracy of which are critical to the performance of the component.

Figure 13.1. Some examples of cast-in fixings.

13. Cast-in fittings

General

Because concrete is mainly used as a functional material that constitutes a carcass or superstructure to which may be attached cladding units, structural steelwork, curtain walling, window frames and such like, it is necessary to embed or attach various types of fixings.

The installation of these fastenings or fixings is mainly the responsibility of the formwork supervisor and formwork tradesman. On large contracts they are helped by the site engineer in locating these components with regard to position and accuracy. The various types of fixing and cast-in component can be divided into the following:

1 The socket or cast-in threaded fixing device whose configuration can be either male or female

2 The slotted or dovetail type of fixing which incorporates some cast-in component

3 A fixing or fastening which depends on the formation of some through hole or recess in the concrete

4 The steel or plastics building component, switch box, conduit power point, cooker box and similar items

5 Arrangements for handling precast concrete components.

It should be mentioned here that as new methods for boring concrete develop and resins become widely used, a considerable number of fastenings are made after the concrete has hardened. Shot-fired fixings, chemical connections and bonded treatments are used where the formwork process is carried out as for normal concrete and the fixing arrangements are made as a secondary operation.

This chapter however is concerned with the systems that require some provision to be made in the formwork for the location or insertion of a device or aperture for the immediate ex-form provision of a fixing or service component.

Fixings

The fixing should either be jigged or located in order to achieve a suitable degree of accuracy, but it should also incorporate some arrangement which accommodates any inaccuracy in the setting. It should be so detailed that the positioning does not become critical beyond the normal degree of accuracy expected from the methods used in formwork or concreting.

On many contracts endless hours are often wasted as a result of the remedial work necessary when fixings and fastenings become misplaced, either because of bad detailing, bad location or ineffective fastening arrangements. In a lot of cases these errors compound up because a designer has stipulated a location for a component which is far too accurate, and has made his fixing details too tight. Components for subsequent fixing are only drilled or provided with lugs which can be adjusted by a small amount about specific centres, so that the whole fixing process can become difficult because of cumulative errors.

It must also be mentioned that the disciplines of accuracy vary considerably between different sectors of the industry. Fixings which are embedded, for instance for plant fixing are often required to be positioned to closer tolerances than those usually needed for the concrete component or structure.

Where precast concrete is involved the location of items such as plastic plugs can cause large problems. For example, one unit may have as many as 50 or more fixings within the reveals of a cladding unit, and are intended to provide screw fixings for pre-drilled frames manufactured elsewhere.

The formwork designer during the early stages of a contract can do a great deal to expedite matters by studying the various fixings indicated by the structural engineer and rationalizing types, positions or even orientation within the structural component. There is little to be gained by using poor quality or badly designed fixings and even less by allowing those on site to cause delays and upsets to the subsequent trades as a result of inaccurate or unsatisfactory fixings.

Assuming that the labour involved for casting-in the fixings has been costed, including any resultant delay to other activities, and it is then decided to proceed with the method of embedment or transfer, the formwork designer will have to ensure that he possesses full details of the component. If it is of a proprietary nature this information will be given in various catalogues, descriptive leaflets and such like, but if manufactured by a subcontractor it will be in the form of working details and schedules. Although primarily the responsibility of the designer or structural engineer, it is worthwhile checking that there is sufficient concrete in which to embed the component, and that any additional reinforcement can be located within the

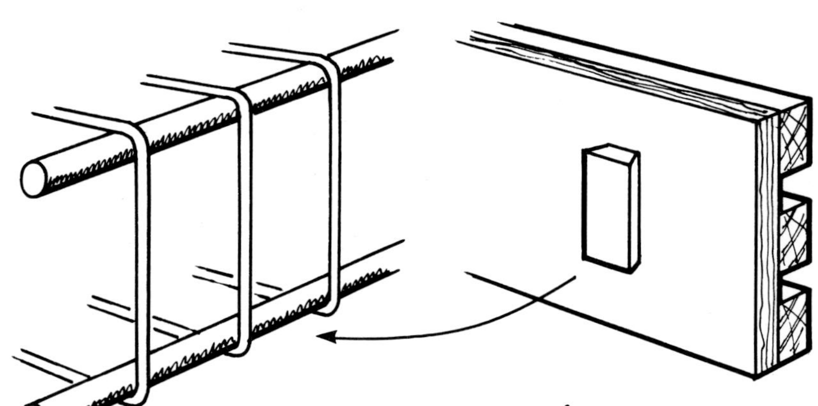

Check that cast-in components do not coincide with reinforcement.

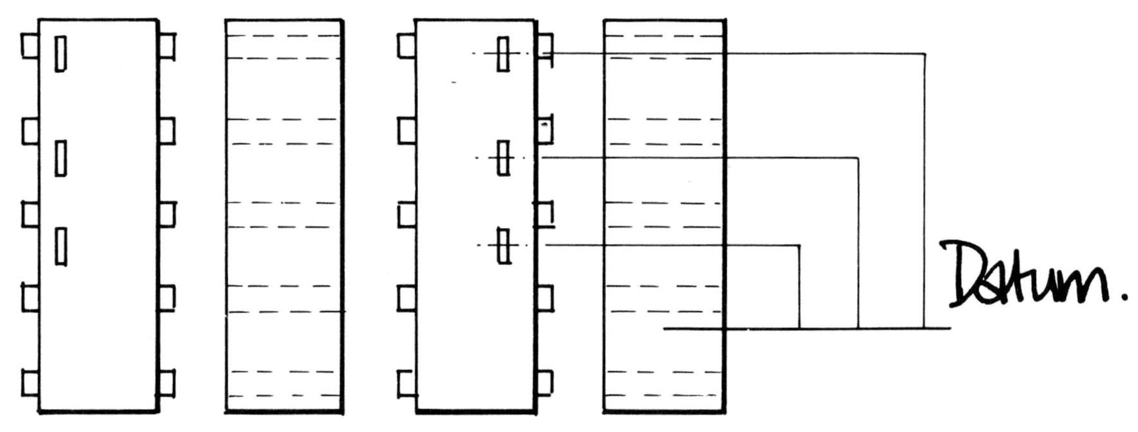

Fixing block cast in sockets etc. should be located with regard to datum marked on form.

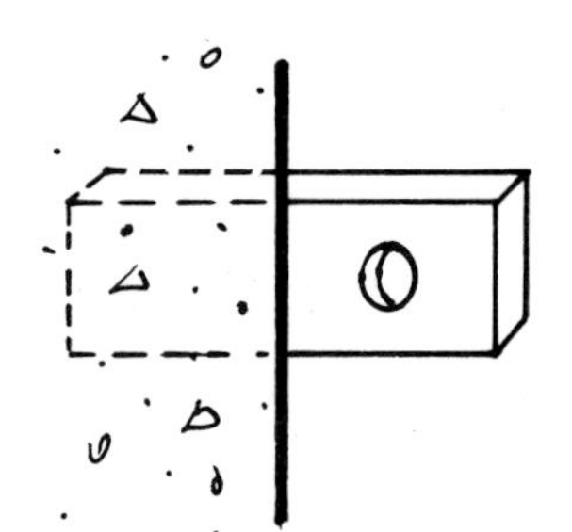

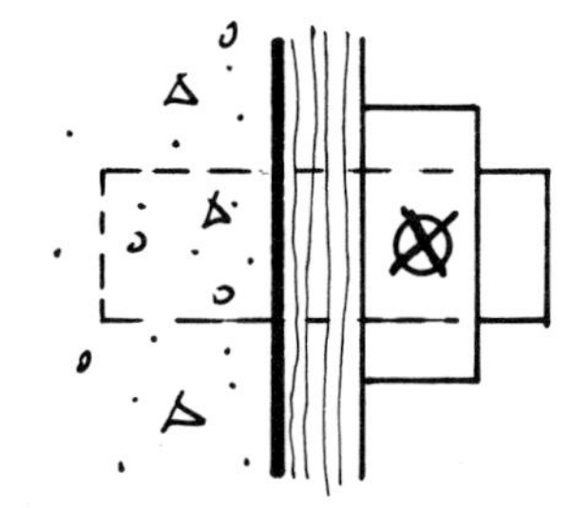

Where projecting plates are used ensure that critical dimension is observed. Bolt or pin through block to avoid displacement during concreting.

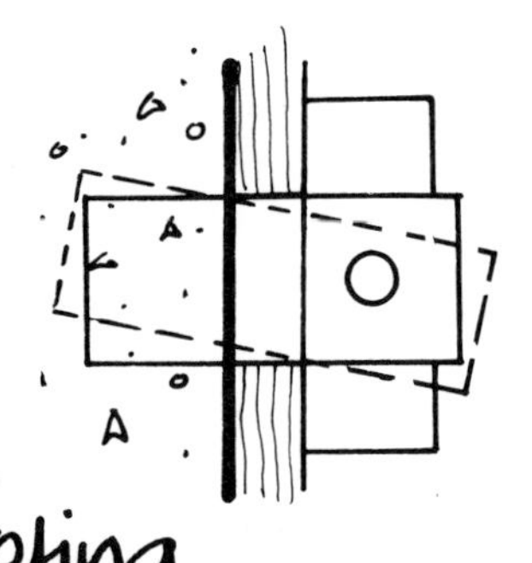

Details of cast-in fittings.

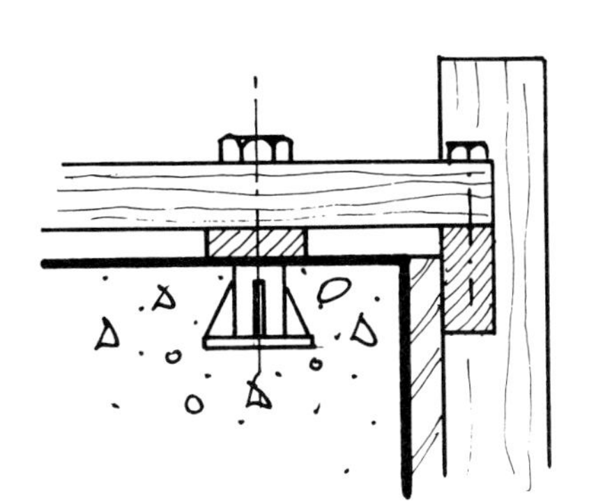

Cast in sockets should be accurately centred off.

Avoid fixings which are too close to arrisses.

Sockets wriggled into position provide poor final fixing.

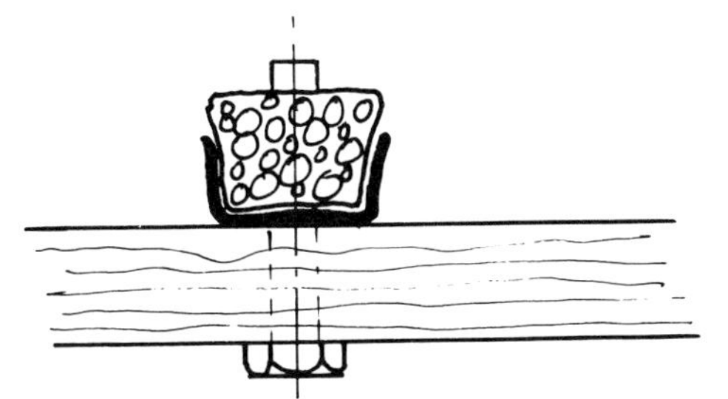

Always tape or fill masonry slot to avoid grout infiltration.

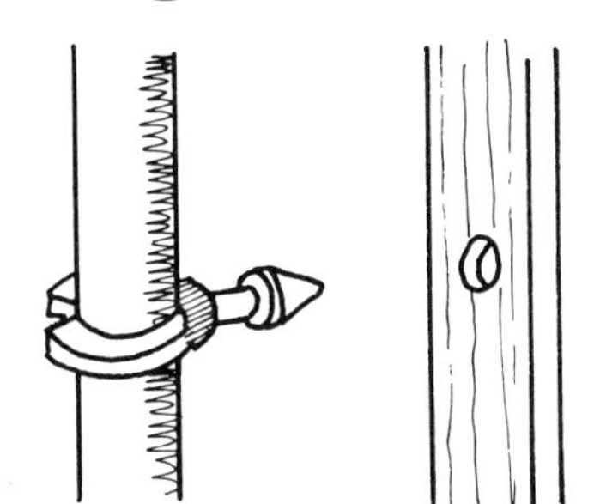

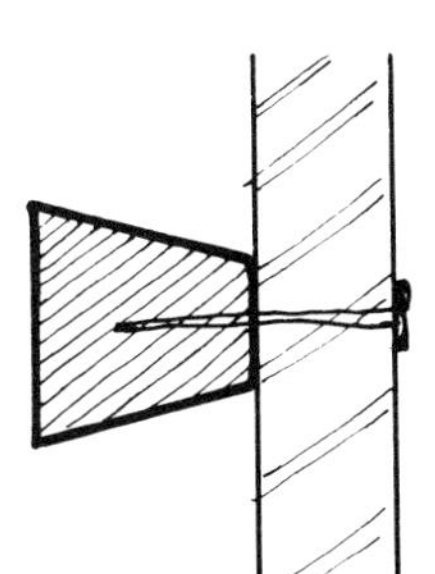

Conduit fixed by necked plastic studs.

Plastic fixing plugs push fit on pins or fixed by neck in drilled hole.

Strip adhesive used to stick plugs to steel form.

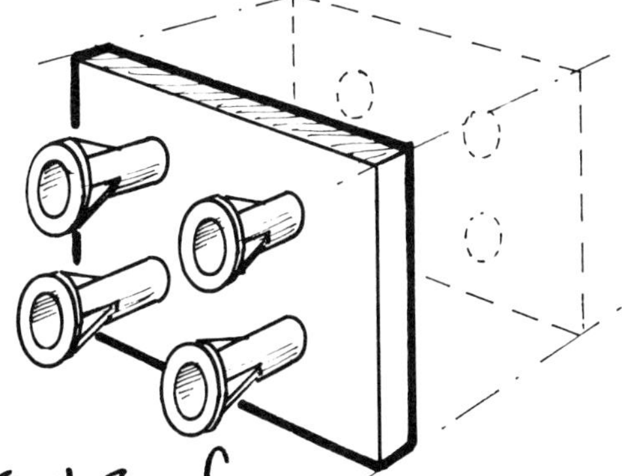

Groups of inserts set onto sub-sheathing for ease of location.

Details of cast-in fittings.

structural reinforcing cage, or that there is room between the reinforcing bars or ducts which contain prestressing tendons for the component itself. This applies to slender units, such as beams and columns where fixings adjacent to the arris of the concrete are to be provided.

Where accuracy of position is critical and to help site personnel, all cast-in fixings should be related to some datum that is easily identifiable on the form and which can thus be used in the actual process of setting out and fixing the components onto the form face. Where extreme accuracy is required, or where the repetition validates the work involved, a drawing should be prepared which indicates just the outline of the form and shows the position of every cast-in component related to a selected datum. This can then be used by the carpenter when the fixings are positioned, and reduces the time spent on interpretation of drawings and details which often have a conglomeration of confusing details not relevant to the work.

All dimensions on drawings should be shown as running dimensions to avoid any inaccuracy resulting from a series of repeated measurements, and they should relate to the datum. This datum should be a line which can be simply established on all faces of the form and which can also be related to a known line, or level on previously cast concrete.

Where the work is at all repetitious, the labour used in establishing fixing and fastening positions may be reduced by taking extra care with the initial setting out. The forms should be marked indelibly using a paint, marker or scribed line, so that those concerned with erection during subsequent stages can follow the marks and fix fittings as indicated. The marks should indicate not only the centres of the components, but also their orientation as this may be critical with regard to cast-in plugs, or indeed cast-in plates for subsequent fixing of bearing brackets, stanchions and such like.

On repetitious work the carpenter, or form fixer, often 'learns' a form, i.e. he carries a mental picture of the blocks and fixings which he has to locate on the form or mould. His memory is helped by drillings and marks from the previous operations. The supervisor, often hard pressed on a job, may not check every set up, and inadvertently, because of a blocked hole or badly smeared indentation, one or more components may become omitted. Unfortunately once a component has been omitted in one operation it is likely that the hole or indent will become completely obscured, resulting in a series of columns, beams or components being cast without fixings. The remedial work may not only damage the component structurally, but may also cause defects in its visual aspect. It is far easier to adjust a component within the form than to try to remedy an error once the concreting has been carried out. Careful attention to accuracy of location will be amply rewarded by rapid and accurate construction.

One small practical point which should be remembered is that although fixings may be installed correctly in the concrete, their locations may be masked by a film of grout or paste, with the assumption made that they are missing. Such fittings are best found by sounding the area in question with a hammer.

Masonry anchors

The various types of recess and slot formers are perhaps the most widely used fixings in concrete. They are versatile and can be easily located within a form. Several systems incorporate a fastening arrangement for attachment to the form, and can include baffles and tape which prevent grout ingress. Fastenings of the channel members may be effected by nailing, the nails being allowed transfer into the concrete during striking, by screwing and bolting using semi-rigid blocks.

Where continuous lengths of channel are specified adjacent to the edges of walls, arrises of columns and beams, concrete may be spalled away during the striking operation, and therefore bolted or screw-fixed fastenings are preferable. Tape or foam should always be used to exclude grout and thus present a clear passage for the introduction of a dovetail tongue at any point throughout the length of the channel.

Wall ties and straps

In many building situations wall ties and straps are required to allow tying in of brick panels to the concrete structure. These ties are best fastened by stapling to the form, the ties being bent to a right angle to ensure correct embedment. If a small portion of semi-rigid foam is inserted and trapped between the tie and the form this improves the removal process when the tie is straightened into its normal working position. A rudimentary recess formed in the concrete is considered helpful where mortar keys into the recess during the bricklaying and infill operation.

A safety hazard should be mentioned which is applicable to all types of fixing. The transfer method of placing ties and straps often results in nails and staples projecting from the concrete, and as these comprise a hazard, care must be taken to instruct those responsible for striking the formwork to turn over any nails and projections as the work progresses.

Switch boxes, power boxes and outlets

Increasing use is being made of cast-in facilities for services such as electricity, air and take off points, the service boxes being transfer cast into the concrete. Most of them allow some degree of tolerance on location, although many hours can be wasted where, due to inaccurate or faulty fixing, boxes become misplaced. For normal constructional work, and to withstand the rigours of placing and compacting a bolted fixing is the most suitable. The bolt may be inserted through the form, or through the box and into a captive nut, or it may indeed be inserted into a tapped hole or welded nut in the box if made of metal. It is essential to ensure that all such bolts are removed before any attempt is made to remove the forms, since failure to do so will result in a damaged form or torn box in the green or newly cast concrete.

Foam fillers or tape should be used to prevent grout ingress, but even with this precaution it is advisable to

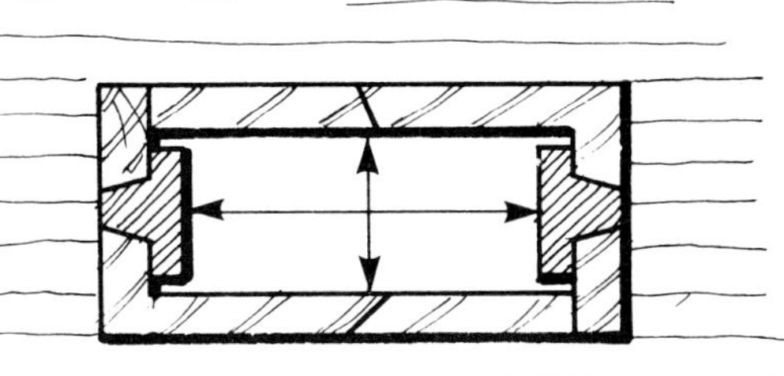

Wedges force formers against openings in form.

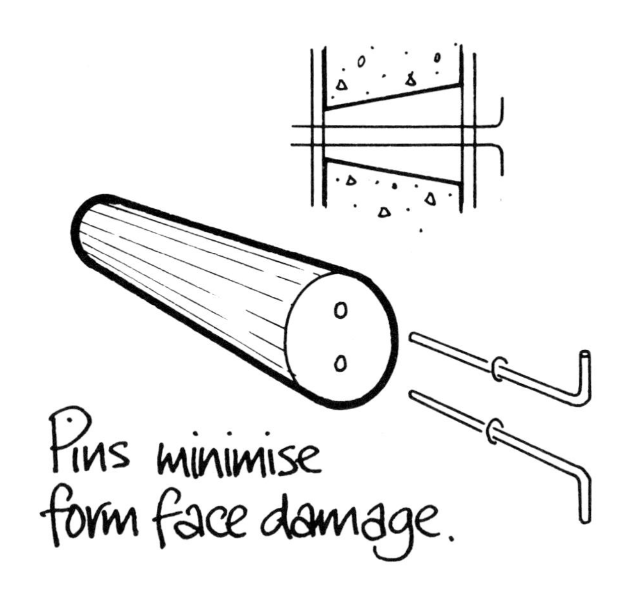

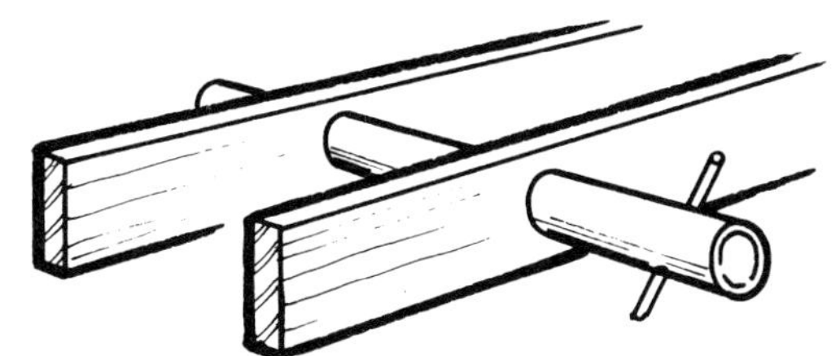

Holes can be formed by steel tubes drawn at 1–2 hours.

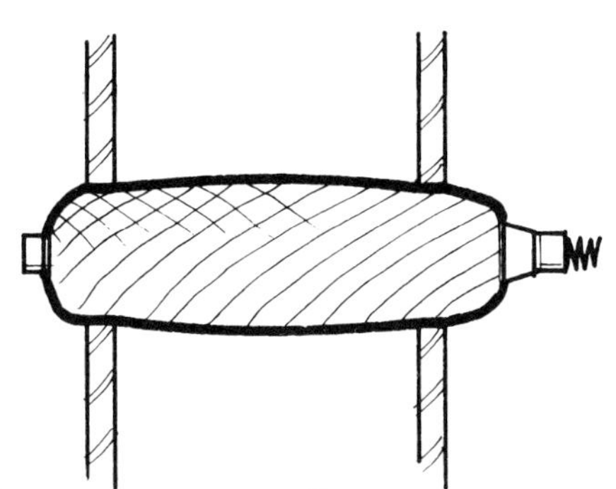

Inflatable formers make grout tight joint.

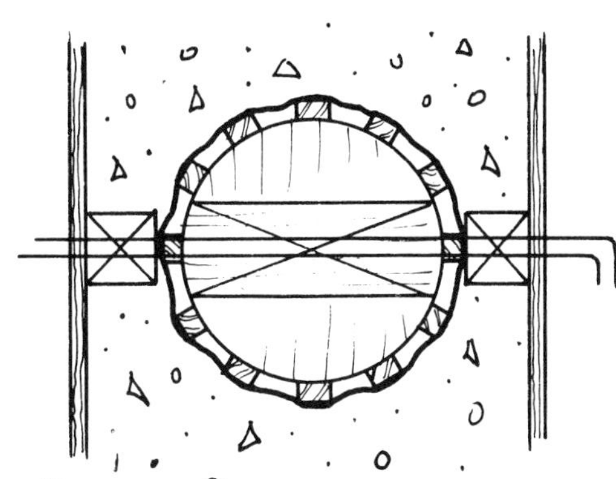

Core formed by timber diaphragms and polythene sheet.

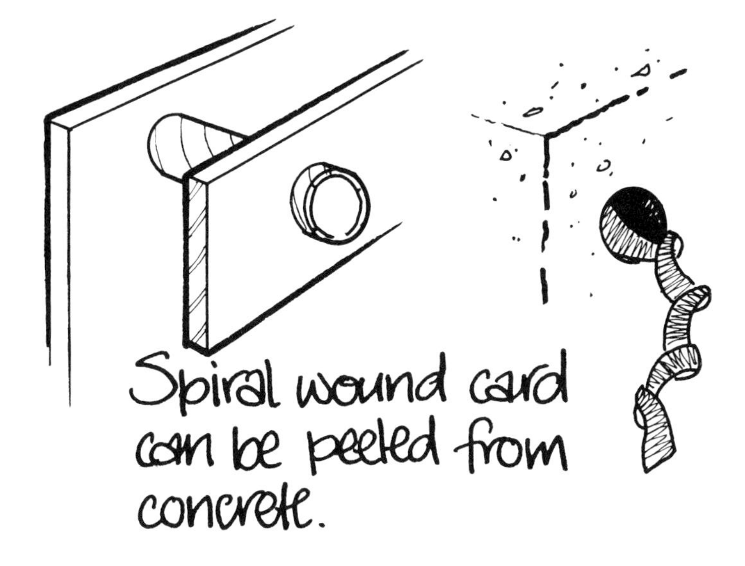

Spiral wound card can be peeled from concrete.

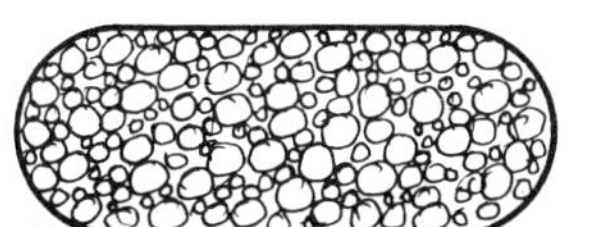

Polythene tube filled with expanded foam or granules.

Metal core former stripped by peeling rubber fillet.

Core openings and duct formation.

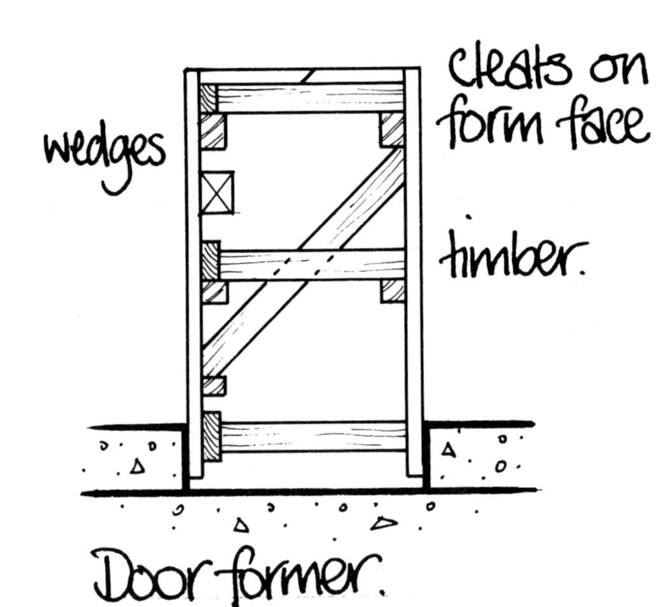

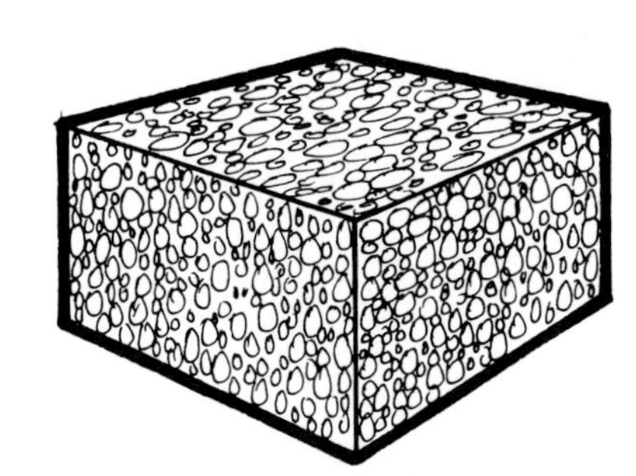

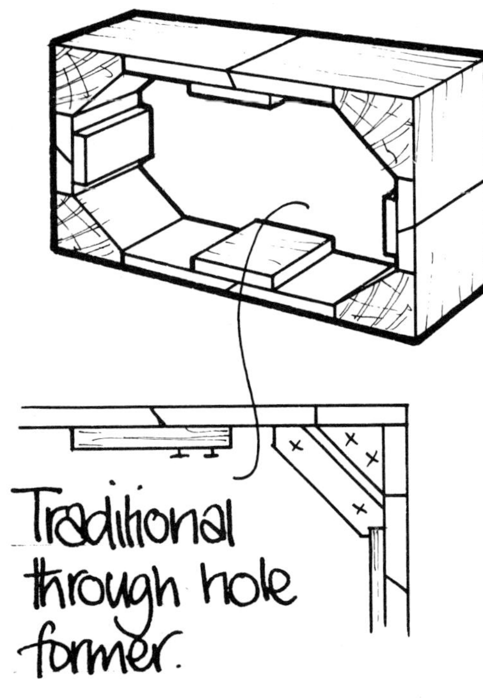

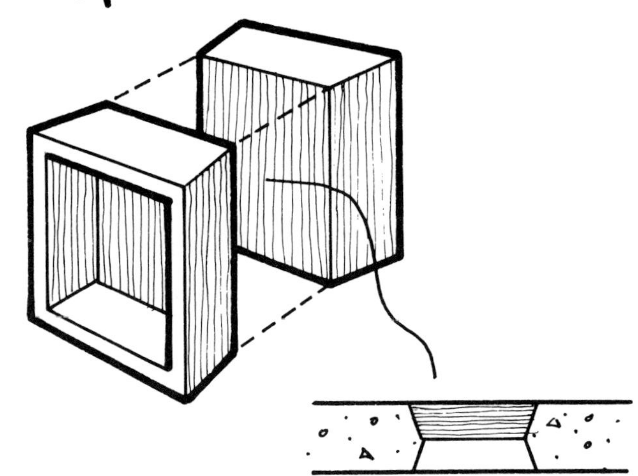

Through holes formed in halves fixed to forms.

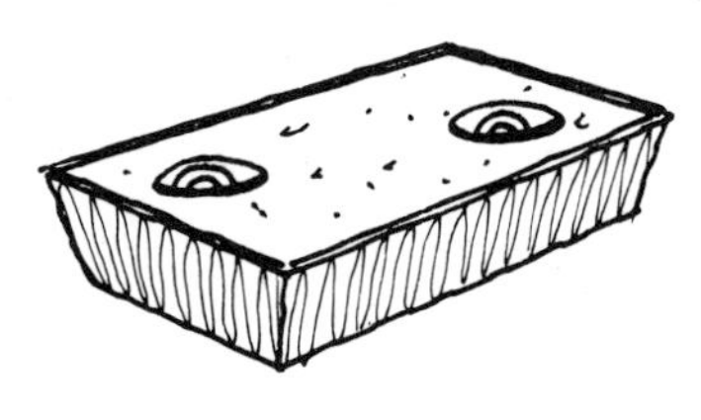

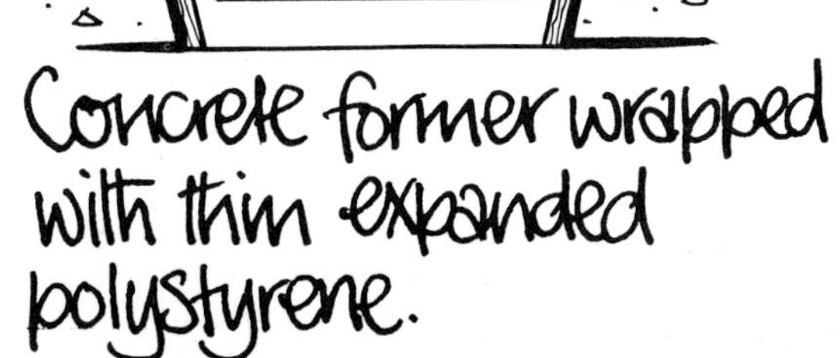

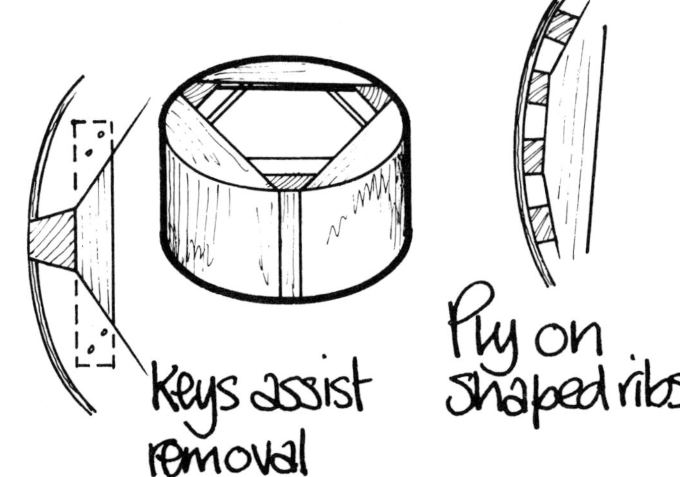

Core openings and duct formation.

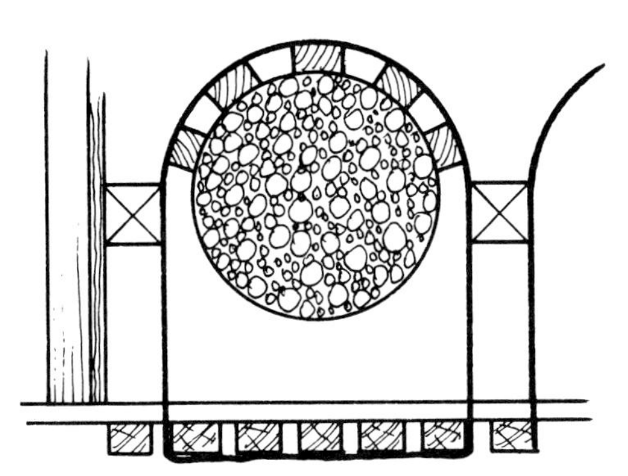

Polystyrene with timber and strapping to avoid floatation.

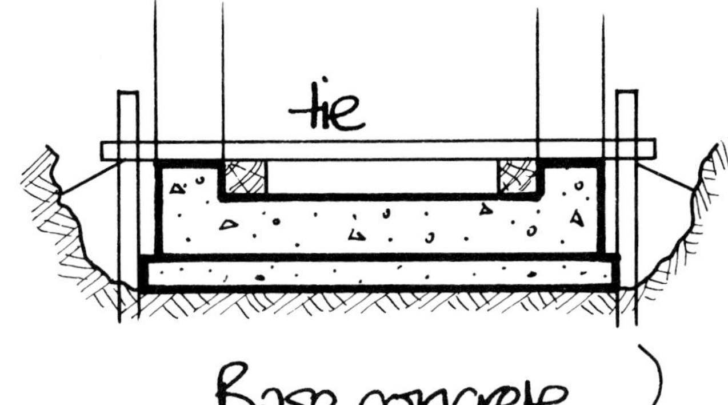

Base concrete.

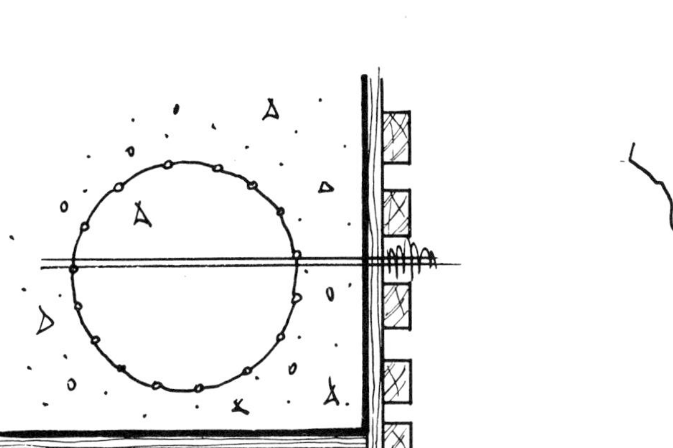

Stressing wire on formers wrapped with polythene sheet.

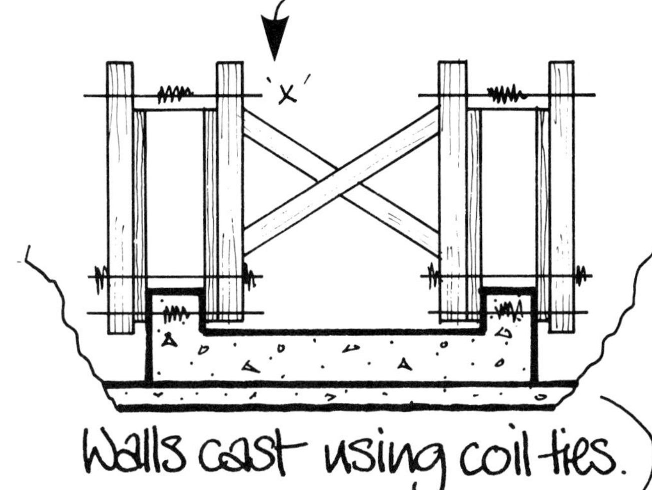

Walls cast using coil ties.

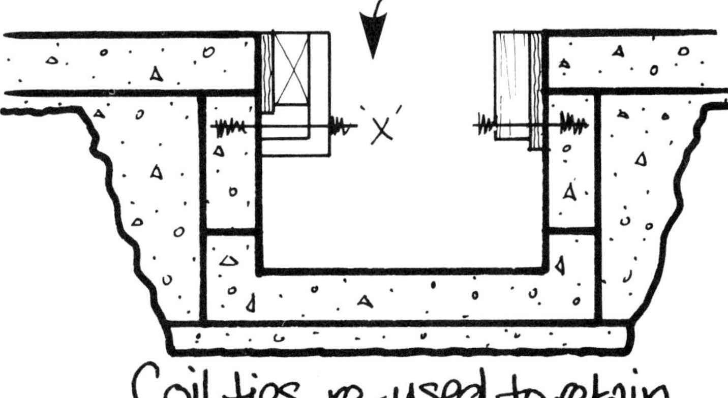

Coil ties re-used to retain slab edging.

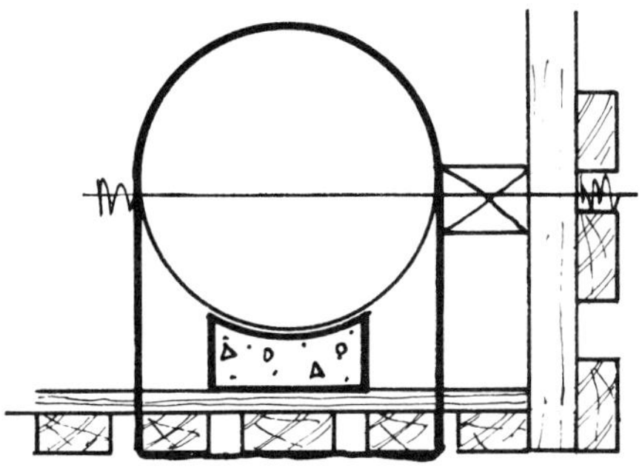

Steel, card or foil former supported on blocks.

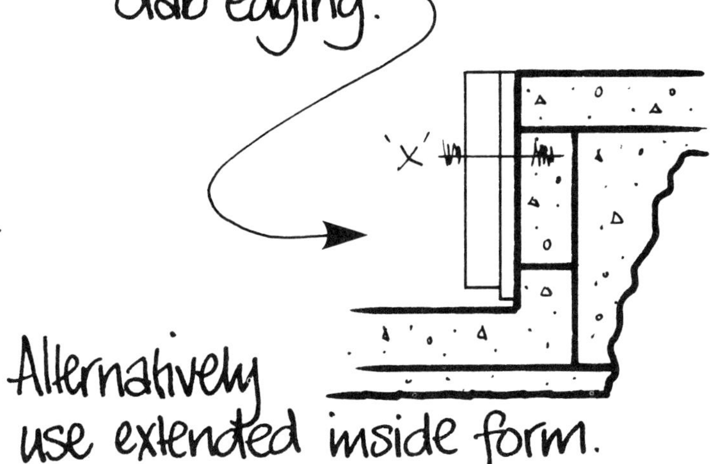

Alternatively use extended inside form.

Core openings and duct formation.

cork or plug any connecting conduit to prevent infiltration and blockage. To this end, where electrical services are being installed, it is useful for stout drawstrings to be inserted in each run, thus offering a simple means of 'proving' the run.

The techniques of conduit installation have been simplified to a large extent by the adoption of plastics which ensure that the cemented joints are completely grout-tight. Where runs of conduit are inserted, other than vertically in the form, or for long runs of vertical conduit, a length of reinforcing bar should be slipped in to stiffen the conduit during the placing and compaction processes, it being withdrawn immediately after casting.

The conduits should not be allowed to float or become displaced during the concreting operation. Plastic rings and plugs can be slipped into the holes in the form to prevent this occurring. In Europe many conduits and boxes are fixed by means of plastics pings which shear on striking leaving the component correctly embedded.

Switch boxes and similar components are often cast into walls which have been cast, battery fashion, on site or in works. In these instances care is necessary to ensure that the fixing is made in such a way that it can be removed as the outer moulds are stripped in the normal sequence of mould and unit removal working into the battery of units.

Dowel bars, projecting bolts and starter bars

Dowel bars are often used in structural concrete work, because they provide simple locations either for components with a suitably placed socket, or for components with projecting loops. Similarly, projecting bolts can be used to advantage although these could introduce restraints by the action of large washers and nuts.

Badly located dowel bars or projecting bolts can be responsible for many hours of wasted time on site. Unfortunately a designer often fails to allow for the normally attainable degree of accuracy or the variability inherent in construction, and these coupled with any movement or displacement during casting tend to require exceptional remedial efforts. When a dowel or bolt is bent or sprung to achieve a connection, local stresses are induced and the concrete may well crack or become damaged due to spalling. These connections are often located at bearing points where concrete strength is essential and if improperly handled can cause problems to everybody. Ideally, all bolts and dowels should either be firmly fixed in their exact locations by some jig or fastening, or so incorporated as to allow movement so that they can simply be adjusted to fit into a subsequent component.

Rigidly fixed bolts or dowels require considerable attention, and a formwork designer must establish clearly how critical are their positionings. He should check for tolerance within the engaging hole or recess, and whether it is reasonable for the particular application intended.

If the bolts or dowels can be made removable or if the tolerances can be increased then these should be carried out so that expensive arrangements to ensure the required accuracy can be avoided.

It should not be forgotten that the various trades will often work to different degrees of accuracy. What may be considered extreme accuracy to the carpenter who has to mould and form virtually fluid concrete, becomes meaningless when compared with the incorporation of a fitted bolt by a plant engineer. At bolting points and similar junctions the two regimes often meet with poor results. Jigging and templates are essential for achieving the degree of accuracy needed for the direct location of bolts and dowels, but only after a check has been made to ensure that some simpler alternative cannot be used.

Often, bolts used for holding down equipment or fixing steel stanchions, are plated and sleeved to allow some degree of movement once the concrete has hardened. Expanded metal sleeves can be used to form cavities in which the body of the bolt can be adjusted to suit minor inaccuracies in either bolt setting – stanchion bases or machine-base fixing holes.

After discussion with the design engineer it is often possible to omit a dowel and as an alternative provide a pocket, finally casting the dowel into its correct position once the adjacent component has been located.

For extreme accuracy, steel, or timber and ply templates should be adopted or templates made from the item to be fixed. These templates are then used when the formwork is drilled for the bolts or dowels, or for positioning them in the upper surface of the concrete, such fixings being made only after the form panel has been plumbed and lined.

Where dowel bars and projecting bolts have to remain vertical or normal to the cast surface, it is advisable to make a template in dummy form from some suitable material such as ply sheet. The dummy is then applied to the extreme ends of the projecting member. The template must be suitably restrained to maintain the bolts or dowels in the correct positions and the panel if used as stopend should be suitably braced to withstand the pressure that arises as concreting proceeds.

In any situation where a group of projecting bolts or dowels have to pass through formwork, it is necessary to ensure that a section of the sheathing can be removed. If this is not done it is quite likely that the form can become trapped by differential movement of the dowels. The removable panel, provided that an indent is acceptable, can be mounted over a clear opening in the form through which the projecting bars can pass, even though cranked or bent.

While a lot of care can be taken over lateral location of bolts and dowels the actual lengths of projection are often overlooked. If this is critical, then for a bolt a sleeve can be used with a plate and nut, or for a dowel a clamp. These will ensure that the correct amount of projection is being achieved.

Plates, bearing angles and structural connections

Where steel components have to be embedded for constructional or connection purposes, the formwork designer will need to study the details early on in the contract. It

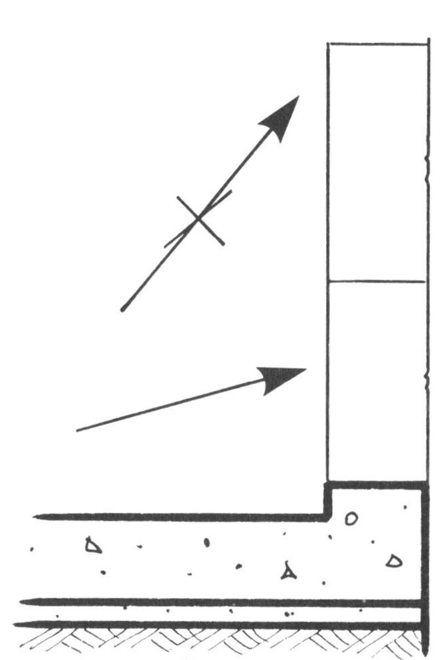

First lift simply propped. Subsequent lifts present problems.

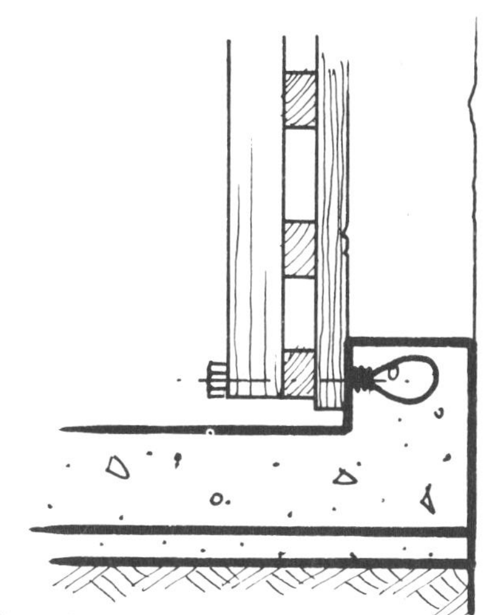

Buried anchor will assist in combatting uplift.

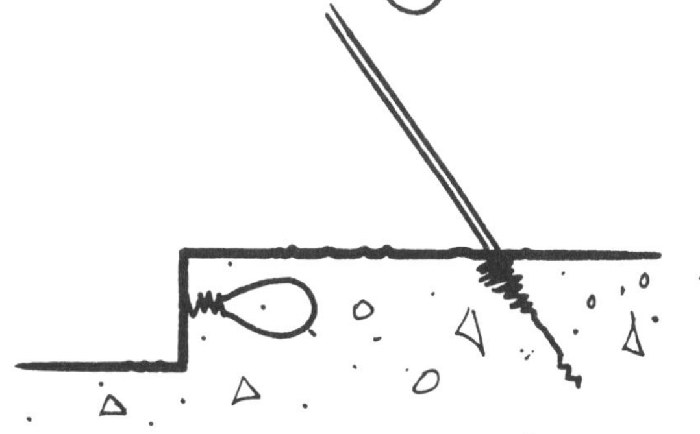

For wide lifts buried anchors replace through ties.

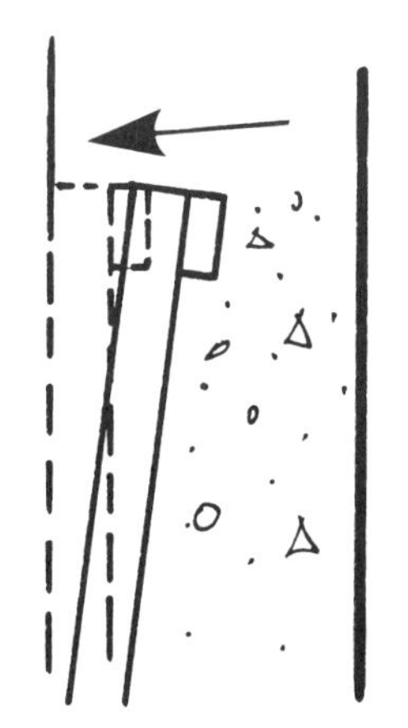

Top of form set inward to allow for deflection under load.

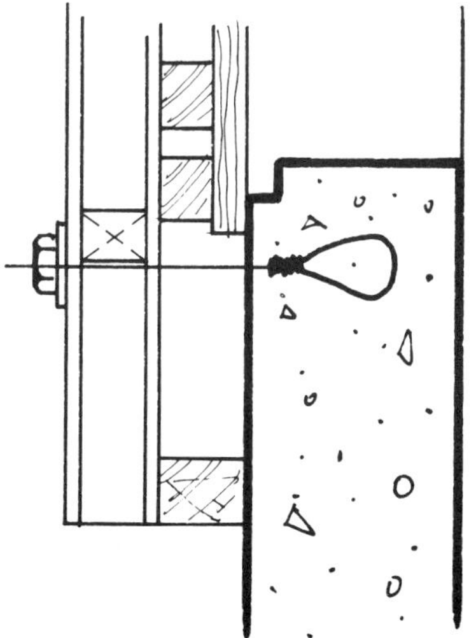

Ensure that concrete at buried anchors has gained strength to sustain load.

Cast-in fixing for single-sided walling.

should be possible for him to ask for holes, plain or tapped to be included in the cast-in component to help with locations within the formwork. This saves a lot of time and gives a greater degree of accuracy, especially where a bolted fastening can be made between a cast-in component and a form. The weight of many cast-in connections that comprise plate or angle with lengths of bars forming anchorages either makes fixing difficult, or causes rotation during the concreting operation so producing inaccuracies. It costs little to drill and tap such components or even to weld on small lugs which bear against mould lips or edges to prevent movement, little that is to say when compared with the cost of remedial work due to inaccurate work.

Cast-in sockets, lifting and fixing attachments

A check should always be made to ensure that there is sufficient concrete to allow adequate embedment and that there is enough clearance for any loops or legs of steel which are necessary to achieve the required bond between the component and the concrete. Most sockets are simply fixed by a bolt which passes through the sheathing. These bolts must be removed prior to striking of the formwork.

When commercially designed sockets are used, the formwork designer should obtain details of application and note any special instructions, particularly with regard to sealing the threads against the ingress of grout. Where sockets are to be cast into trowelled or floated faces, as distinct from formed faces, ideally, the sockets should be suspended from frames or templates and then cast into place. Often sockets are wriggled into place following the compaction process and this causes, locally, a high incidence of fine material about the socket, possibly affecting a reduction in pull-out strength.

With the increase in site-produced precast concrete, the formwork designer is now often required to design moulds. Mould design has been covered in Chapter 16. An essential part of precasting work is the handling of freshly cast concrete and the erection of matured concrete from stock. Success in both operations depend on suitable handling arrangements.

The most direct and simplest methods of handling utilize either through holes or slings, although slings can present difficulties in the initial removal of the units from the moulds. The next most direct method is to use embedded loops of reinforcement or prestressing strand known as lifting hooks. There should be a satisfactory bond between the hook or loop and the concrete. Lifting hooks are usually centered off by using straps or pipe supports across the mould – care being taken to ensure concyclic placing of the hooks where flat slabs are handled. Local recesses around the hooks allow projecting bars to be burnt off and the concrete surface made good.

Projecting bolts are used for wall panel handling, while lifting boxes are used in conjunction with either a bolt head or a nut and washer. The bolts are often used in the final fixing so that care must be taken with position and projection to achieve a satisfactory uniformity. A jig or clamp on the mould should be fixed which determines both position and projection, care being taken to avoid movement caused by vibration.

There are many patent arrangements for handling precast concrete, most of which can be embedded. The comments with respect to socket installation by casting, also apply to lifting and bracing sockets. The devices should be centred properly and then *cast* into place.

Screwed arrangements for lifting must be properly protected by grease during casting, and the correct number of threads should be engaged during handling of the precast component.

The lifting stud is becoming very popular and this is explained to some extent by the simplicity of installation, i.e. rubber split ball mounting which provides a good connection with the lifting slings. During installation the stud is firmly held in the rubber grip which also forms an accurate recess around the stud head for later insertion of the lifting hook. Once the mould has been stripped away it is possible to spring out the rubber grip which can then be re-used for positioning further studs.

A useful device for handling concrete, and for that matter a variety of other formwork and temporary works operations, is the screw anchor which provides a fixing for threaded eye or normal bolts used for fixing plates, angles or forms into position onto previously cast concrete.

On a point of safety, and particularly where loads other than those axial to the bolt for fastening may occur, some mechanical connection should be provided by means of a loop or link to the main reinforcement within the unit.

Column guards and stair nosings

Various forms of protection are transfer cast into concrete – the simplest being the column guard used for car parking and warehouse construction. These are fixed within the column using bolts passed into tapped holes or captive nuts. The bolts are removed during form stripping and the steel (or plastics) guards then remain firmly embedded in the concrete. Similarly for precast stair production the nosing member is bolted within the form, either cast on edge or inverted, and then transferred into the concrete during the demoulding operation.

When designing these or any other kind of proprietary fixings and fastenings the formwork designer must be fully conversant with the manufacturers' recommendations both for installation and use in providing the required fixing or fastening.

Some concrete help from RFA

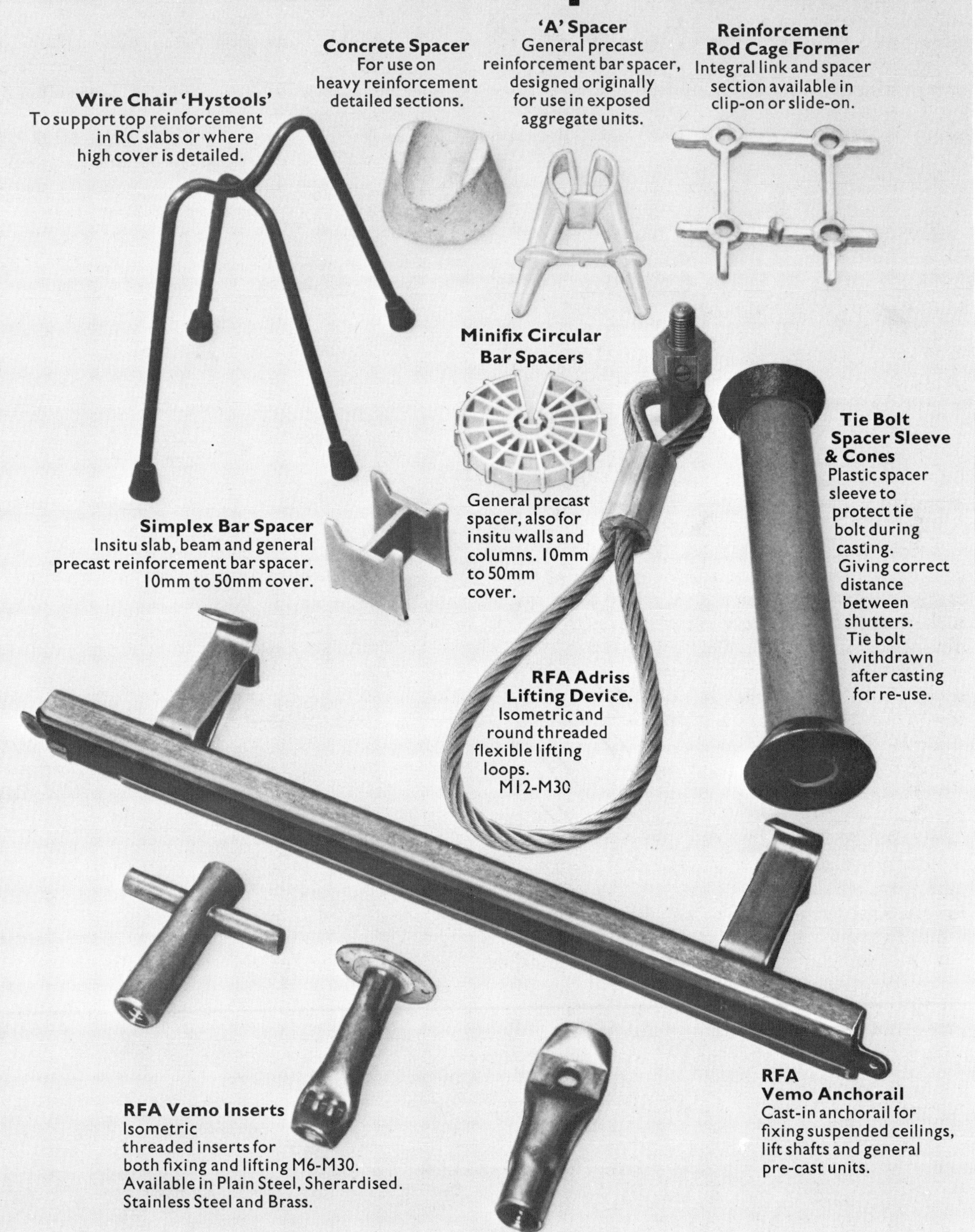

Other products include:- Tying Wire and Bag ties. Expansion Dowel Bars and Caps. Fixing Blocks. Chemical Fixings. Release Agents.

RFA (Rebar & Formwork Accessories)

Head Office: 49 High Street, Crowle, Scunthorpe, South Humberside. Tel: Crowle 543.
Southern Office: 30A Bower Hill, Epping, Essex. Tel: Epping 72989. Telex 916226. Tolberg London Ref: REBAR.

Figure 14.1 A simple, but effective instrument – the spirit level – which is adequate for most plumbing operations.

14. Setting out and manufacture – site and works prefabrication

Setting out full size

Setting out for form manufacture is best carried out at full size on some prepared surface. This can be a permanent floor in the works or yard, a few sheets of ply secured to timber bearers in some suitable situation on site, or even a previously constructed floor slab. Taking the latter, the ply sheets can either be used as form sheathing on completion or sanded clean for use in further setting out. If the floor in the works or the yard is considered as permanent facility, datum lines can then be accurately and permanently marked by sawn marks or scribed lines on steel plates let into the floor. For circular or geometrical work long radii can be marked out using a tape down the aisle or gangway, although the technique of using a shaped batten cut to the profile of the rise at half length between two known points on the circumference can be used to generate the required circle where space is at a premium. Where high standards of accuracy are involved use should be made of a theodolite and permanent stands provided for the equipment.

Initially, whatever the form material, a sheathing line, known as the critical line of form profile, must be laid down. Where a lot of geometry is involved the lines of various sections can be superimposed one on the other using coloured pencils or ball pens.

Where there is some critical relationship between the line or curves they should be set about the same station lines. When stressing ducts are to be located the various offsets should be related not only to the same stations but, ideally, also to the position of the bearers or soldier members of the completed formwork.

The full size set out provides an opportunity for checking the fairness of lines. Where transitions are involved, say for wind tunnels or profiles of tunnel intersections, the only way in which the lines can be set down full size is by bending battens around known points on the surface. For this purpose timber battens are best because they have a straight grain and are knot-free and thus avoid sudden changes in line.

Once the sheathing lines are set out longitudinally or transversly, the appropriate sheathing thicknesses can be set off normal to the face line, so that the outline of the shaped ribs or formers is determined. The actual depth of these ribs or formers can be marked and transferred onto templates. The templates can be marked by pinning the oversize template material in position on the full size set-out and marking the shape of the member by reproducing the initial setting out on the overlay. This process, while apparently a duplication of effort, means that the initial setting out is retained, and this is especially valuable where a complicated form or mould is to be subsequently built and positioned over the setting out.

The series of templates can be marked up and used for the preparation of the framing members, for rough sawing and subsequently for 'spindling' timber, or for checking rolled steel ribs or steel sheets and plating prior to fabrication.

Key details

The process of setting out can be simplified by preparing a small scale drawing which includes the critical dimensions that govern the profile, and it is helpful to those who set out and construct the actual form if all the dimensions are indicated as running dimensions from known datums. This drawing should in effect be an abstract of all the structural details. Where there is a duct, a through hole, projecting steel or some other upset to the sheathing, these should be indicated. The setter-out can then ensure that the details of cleat and bearer positions on the formwork detail are correct and that there is necessary clearance for any projections. A further detail which can be validated at this stage is whether any striking fillets or stripping keys indicated on the formwork drawing will in fact allow the removal of the form from between or around projections. In cases of doubt the clearance can be checked by moving the templates on the plan or section.

From sections and developments, as required with hoppers, eliptical and circular templates can be prepared for the bevel cuts to the edges and ends of timber sheathing, or the surfaces of the sheathing where ply or steel is being used. For ply sheathing, which is simply adjusted using hand tools, margins should be left for trimming, though where steel plating is concerned the templates must be strictly followed, in order that the profile cutter can copy the shape exactly. Once these templates have been established materials cutting lists and schedules of fixings can be prepared to outline sizes. The 'rods' or templates can be used for marking out or machining, care being taken to ensure that they are marked with the identification number of the panel or component and the number of such pieces required in the finished form.

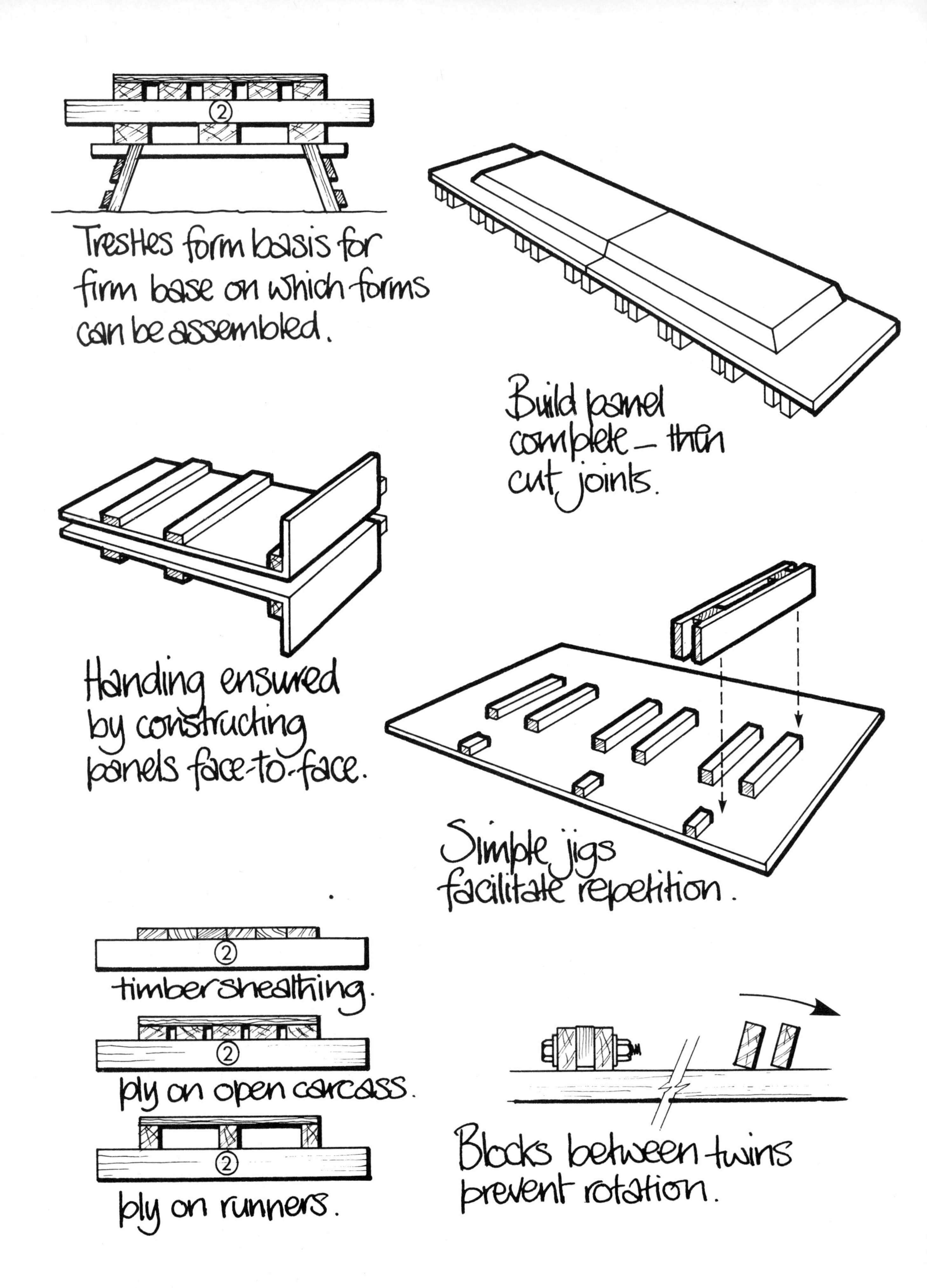

Details of setting out and manufacture.

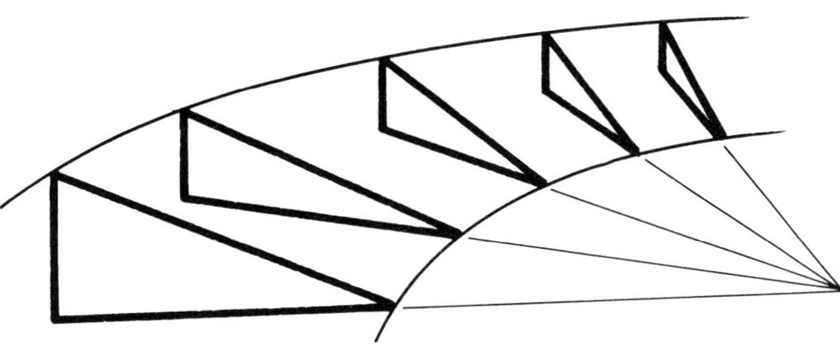

Conical form.

Construction can be simplified by locating diaphragms of known form on established lines.

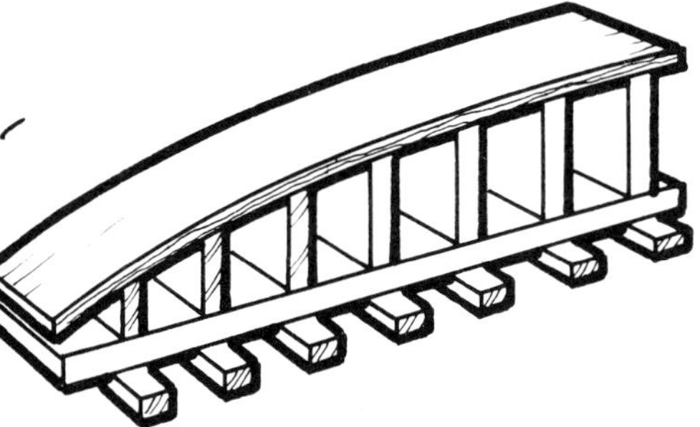

Camber generated by blocks on substantial soffit.

Ramping form generated by raising diaphragms by calculated amounts.

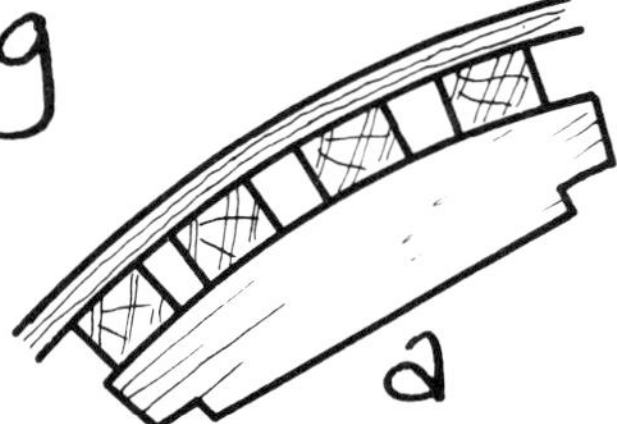

Circular forms generated by (a) open lagging and ply (b) close lagging.

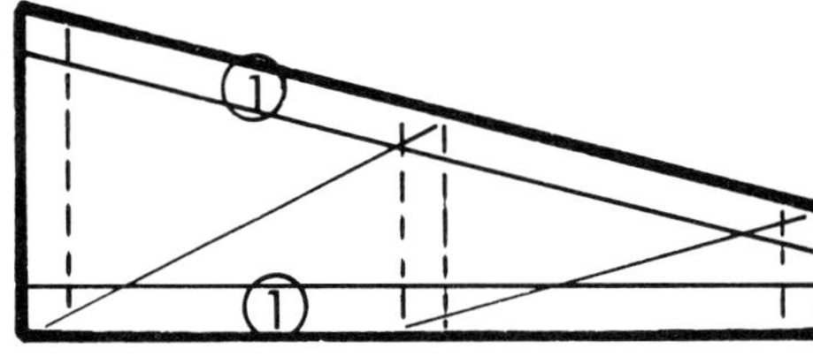

Diaphragms can be of composite construction using timber and ply.

Only face need be close jointed.

Details of setting out and manufacture.

In some precast works, when particularly high degrees of accuracy are required, the templates are made of zinc. They are not only prepared for the mould members but also for testing the finished concrete component. This degree of accuracy is rarely sought with *in situ* construction, although if the specification demands extremely accurate faces, it will be worthwhile making up templates that can be used for checking the completed form after installation and prior to the steelfixing operations. Such templates must be rigidly constructed and adequately braced.

Templates must be scribed indelibly with lines which can be related to established datums during construction. Such templates are required to correctly position the now popular ribs in striated finishes. Once the components are to hand, a start can be made on the fabrication of the form. Ideally, all framing members should be taken down to a known datum or, failing this, rough trestles can be erected to reproduce supporting points on the falsework or grillage to be erected on site.

First erection

If possible, the fabrication should be carried out over the full size set-out in which case the various points on the form can be related to the plan using large framed squares or plumb bobs. The framing members, ribs, runners or soldiers should be rigidly supported in position and this can be facilitated by working on purpose made supports or stools and the insertion of cover or splice plates. For the more critical work, the members that generate the line of the sheathing should be erected initially as continuous members, sheeted and subsequently cut when the form panels are separated for transport to site. Flat sheathing surfaces, plane curvature and flewing surfaces can readily be achieved using board sheathing, ply or steel sheet. Compound curves, conical surfaces and domed surfaces demand multi-layer construction, either as double diagonals, sheathing the final layer of boards, or as ply slips, the surface being fitted and close jointed. For relatively flat curves it may be possible to double sheet, using open jointed material, and then subsequently to sheath the face with tailored thin sheet material, either ply or hardboard. The choice of method depends on the specification and the designer's stipulation regarding permissible deviation from the fair curve. Where truly spherical surfaces are required these may be produced in light gauge metal in a panel shop. Recently, spherical corners for precast moulds have been formed by spinning, the required shape being cut from the blank.

Ideally, carcassing members should provide a regular shape from which supports can be taken to the falsework or adjacent to previously cast concrete. For complicated columns or shaped walling the carcassing should be framed to allow tie forces to be transmitted where necessary to commercial steel strongbacks or tie frames, so that distortion of the sheathing surface due to the local incidence of high stresses can be avoided. By the use of blocking pieces and diaphragms the construction can be brought to a boxed shape which simplifies tying.

As the form takes shape, a check should be made that the carcass has not been distorted by any bending or springing of the sheathing material in obtaining a fit. A good rule is that no adjustment to apparently wrong or badly fitting components should be made without reference being made to the designer and the person who produced the full size rod or template. This guards against wastage of material where the fitting or fixing sequence has not been understood or correctly followed. There are a number of similarities between form construction and boat building and the sheathing material should be applied to the framing in such a way that the stresses imposed on the carcass are equalled out as the construction proceeds.

Thin layers of sheathing material do not obviously impose great forces but the standard sheet of sheathing ply requires the application of considerable force to maintain any real degree of curvature. Where fair curves are required particular attention must be paid to the support and fixture of the sheathing in each end bay of a panel, this is necessary to avoid the tendency to flatten produced by the absence of continuity over further carcassing on framing members. It may be necessary to introduce additional dummy ribs in these areas to maintain the true form – this should be apparent from the full size set out.

Where form panel components are standardized the individual templates can be marked with bolt, screw or connection positions and these transferred by the Markerout prior to machining will considerably reduce the amount of hand work in construction. These matters are of course essentially the responsibility of the designer who may, because of difficulty in transmitting details on scale drawings prefer to indicate the position of connections on the full size rod once the true scale of the problem can be visualized. Once the form has been fabricated it is essential that all joints should be marked to ensure a satisfactory fit being obtained when the form is re-assembled. Some responsible person must check the lines and details and ideally someone connected with the site should be shown the key points regarding the re-erection and use of the form on site.

Handling points must be clearly marked and a drawing prepared of any foundation work or the location of support points which are required on site. These points should be rationalized such that to produce sloping surfaces or geometrical detail all that is necessary is that supports should be erected at certain levels in known plan positions, geometry being dictated by the formwork arrangement.

Setting out for formwork erection

As construction processes become more rapid the importance of accurate setting out becomes more apparent. The positioning of simple tubular supports can be critical particularly where multi-storey construction carried out rapidly means that supports are required for transmitting the loading of two and sometimes three storeys between slabs.

All prop centres should be accurately established, and if these are indicated on the floor slab by dots of paint, it soon becomes apparent when the centres are being exceeded. Lines of props can be checked for plumb by sighting against a known plumb feature or two end props which have already been plumbed.

The wise site engineer, or person responsible for the formwork arrangements, should see that standing supports are maintained one above the other on succeeding floors, and, again, a little preparation will ensure that the necessary degree of accuracy is achieved. With regard to the supporting systems, tie bars or lacers are sufficient for accuracy of line and plumb, and it is only necessary then to locate the lines of the main bays of such supports.

The height of a floor form, whether based on a prop supporting system or some regular table form system, can best be related to a datum which has been marked on the columns by the site engineer or supervisor. Datums must be clearly indicated and preferably dated or signed.

The final setting out for level, and hog or camber if specified, can be achieved by the use of a dumpy level as soon as a few bays of slab have been erected and stabilized by bracing, or tied to previously cast concrete. An excellent anchorage can be made by making sure that a runner is securely paged up under a beam adjacent to a column in each bay of the previously formed and concreted area.

Alternatively a raking tie can be taken from a putlog yoke secured around the base of a column. In this connection the intelligent use of a cable winch as used in steel bracing can assist in maintaining an adequate tie for erection purposes, although the continued use of wire strainers is best avoided during concreting.

For ramps, either straight or curved, datums on columns are necessary. Provided there is some agreed correlation between the datum and the eventual sheathing surface, the actual final line can be achieved by setting the ramp sub-structure, which is constructed to the agreed dimension, to the datum lines. This arrangement is particularly useful for circular ramps and where the underside of a cantilevered ramp consists of a flewing surface. For all kinds of circular ramps the engineer can assist matters by setting out a template which can be used in the construction of framed members that provide the appropriate rake to the underside of the slab. These members will, if set radially, ensure that an accurate flewing surface is achieved.

Figure 14.2 Where accuracy is of prime importance, gang casting using this commercial form simplifies column location.

Once the plan profile of such surfaces has been established and the slab edge forms erected, the engineer responsible for setting-out can help the concrete gang by providing radial marks on the top edges of the forms. These can be used as guides while the concrete is screeded to ensure uniformity of thickness of slab and a true surface. Similar indications on forms to simple flewing surfaces or to edge forms in any folded plate or hyperbolic paraboloid form of construction are also useful.

The role of the engineer

The engineer can assist those who are concerned with form erection for anything other than straightforward work by providing known reference points scribed onto mortar pads or dots of concrete at key positions on the ground. Where the construction is based on some clearly defined module, it is advantageous to set out on the foundation slab and intermediate floor levels the complete pattern of the grillage. Reference can be made to these points by anyone who is concerned with form erection or setting out, at any stage in the construction using plumb bobs or an optical autoplumb.

With new constructional techniques constantly emerging it is not unusual to find laser beams being used for setting out and providing standing optical reference lines. It is remarkable however the degree of accuracy that can be obtained by using plumb bobs and piano wires with barrels of oil to dampen movements caused by wind. There is much to be said however for the electronically directed laser beam which automatically homes onto a staff and indicates levels related to the position of the instrument within a matter of seconds.

The plumbing of formwork always makes considerable demands on the site engineer. Generally it is difficult to set up an instrument and the plumb must therefore be achieved by the builder's level and plumb bob. There will always be some movement during placing and compaction as wedges take up or pads compress, and thus it is essential that the plumbing process is maintained throughout the concreting process. Final checks being made immediately the concrete is placed. Cantilever slabs, and forms which cantilever on single sided construction, are particularly critical and the form should be pre-set such that any settlement will leave a positive camber or hog in the construction.

For repetitive uses of forms the concrete level rarely coincides with the top of the form in every case. Where the concrete level is within the form, levels should be indicated for attaching rules or joint form fillets. The setting out of such lines of ply faced forms should never be indicated by nail heads that protrude from the sheathing face, since these are particularly dangerous and can cause severe damage to operatives' hands.

At a very practical level there are a number of points which should be remembered at all times:

1 Profiles do become displaced and are often subsequently re-instated inaccurately. A check back to known datums should be constantly made.

2 It is not unknown for instruments to be moved by inquisitive people while the engineer confers with a joiner or chain man. The setting up of the instrument should be checked before a fresh sighting is taken. Heavy movements around a tripod can cause large movements.

3 Datums marked on columns or walls by other people are not necessarily accurate. One's own datums should be maintained and initialled.

4 The person on the staff or at the end of the tape should know what is expected of him, but he may not be familiar with all the signals, or how important it is to keep the tape taut and 'on the mark'.

5 When working 1 ft or 1 m 'on', this of course must be deducted when marking the distance! Running dimensions should always be used to avoid cumulative errors on successive measurements.

6 Inaccuracies must always be recorded. They may be acceptable within the requirements of the specification but if unrecorded they could cause problems, maybe months later, e.g. when services are being installed or some cladding fixed. Ideally, a dimension book should be maintained in the office for reference later on in the contract when the original 'setters-out' have moved elsewhere.

7 Time should be set aside to check back on any query whether it be a dimension or a drawing detail. The costs of remedying hardened concrete are prohibitive.

8 Key dimensions should always be checked during and after concreting, although if the kickers are correctly positioned and forms kept plumb, there should be little problem.

Checklist

Is geometry intended?

Have critical points/lines been established?

Have simple alternatives been proposed?

Has a datum been agreed:
1 For works manufacturing?
2 For site erection?

Can the substructure or carcass be brought to a general level?

How is continuity of face maintained?

How is profile to be checked:
1 In construction?
2 In the job?

Can construction templates be used on site?

How is set-out related to features of the structure?

Is provision made for standing supports?

Can continuity of support be achieved throughout structure?

What arrangements are to be made for easing supports as construction proceeds?

Where striking fillets are incorporated do they allow release from between returns?

Does the geometry trap any item of formwork?

Do cast in fittings or projecting steel trap forms?

How is steel located within forms?

Are there special problems of concrete placement?

Has uplift been considered?

Does concreting method affect the stability of the form?

Can supports be taken from previously cast concrete?

Does special nature of formwork impose loading on fresh concrete during striking?

How is form handled from inside completed component?

Who is to be responsible for erection?

Can they visit fabrication department to see assembled form?

Figure 15.1 Sculptured panels cast from models prepared by the designer contrast favourably with the board-marked concrete.

15. Exposed or visual surface finishes

General

The exposed or visual aspects of a concrete structure during its life present a commentary on the skills and abilities of those who have been concerned with its construction. Well designed and soundly constructed formwork provides a basis for successfully achieving a specified finish, although it does not necessarily guarantee a successful outcome. The concretor who has to handle, place and compact the concrete is really the one who determines the eventual aesthetic appearance. Unfortunately the techniques employed are those where there is often a lack of skill or training and instruction. As a result, well constructed and sometimes expensive formwork is wasted when the eventual results, in terms of moulded concrete, are viewed. These problems however are often solved through discussion and co-operation between the formwork designer and the site staff involved.

The formwork designer is probably the person best informed with regard to the specification requirements and he should ensure that he includes in his details all the relevant information on the established method of concreting, speed of placing, compaction equipment and such like, so that formwork operatives are kept suitably informed. During site visits and especially on those that coincide with new work, or the first uses of equipment, he should follow-up method to see that the criteria upon which he designed his formwork are appropriate and that the designed methods are being put into practice.

The placing and compacting of the concrete that is to remain exposed tests not only the skills of the concretors but also those of the form manufacturers and erectors. In

Figure 15.2 The design of this large concrete structure was carefully detailed with all the practicalities of concreting and formwork construction taken into account.

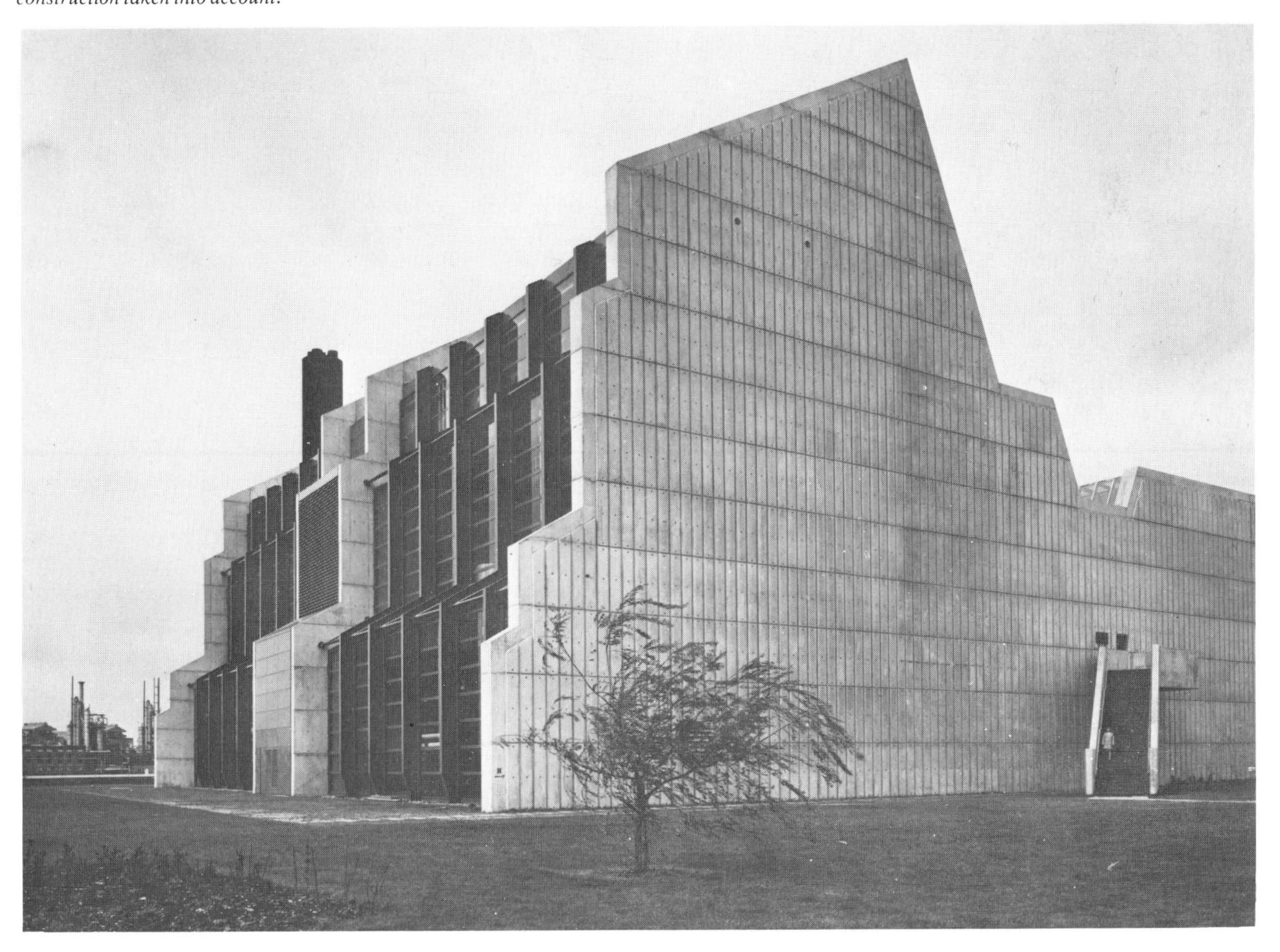

order to achieve successful concrete surface finishes of a reasonable degree of consistency, close joints and grout seals are of paramount importance. For example, not only must it be shown that the form panel joints are tight, but the actual joints between the sheathing materials within the panel construction will require special attention. Quite apart from the normal structural requirements that relate to deflection, forms for producing high quality finishes should be rigidly constructed to avoid differential vibration or flutter being set up between various areas of the form, and particularly between individual sheathing panels.

The degree of effort applied in compacting visual concrete is generally greater than that employed for the production of normal structural components. Almost automatically the supervisor gives more attention to the way in which immersion vibrators are used or to the location of external vibrators. Sometimes mechanical hammers are also applied to the form in an attempt to introduce additional vibratory effort, which will create excessive pressures, resulting in deflection and movement of the forms. To be successful, a great deal of thought must be applied to all aspects of the concreting operations. To produce good visual concrete, all the finer points of construction and concreting techniques have to be adopted; thus the particular skills of the carpenter, trades supervisor, concrete ganger, and even the semi-skilled operative who cleans and oils the moulds can have marked impacts on the final results.

It is virtually impossible to separate the various activities or allocate a particular priority because every member of the construction team has a definite part to play. A methodical approach has to be made during the preparatory stages of form manufacture after which the casting process needs to be carefully planned and executed.

Formwork considerations

Visual concrete demands good quality formwork, i.e. joints must be tight, the form face must suit the required texture and be adequately supported to withstand the compactive forces applied to the concrete. The appropriate parting agent must be selected and particular attention needs to be given to the striking and curing processes.

From the earliest stages of the form design and during the setting up of the formwork system, the designer will need to be alert to those particular areas of visual concrete which merit special treatment. It may be possible to arrange the sequence of construction such that new forms are used at the start of a particular section of the concrete structure where visual aspects are perhaps more critical. This is ideal and allows the use of special materials and construction techniques. Where this is not possible it will be necessary to reface or refurbish formwork that has been used for previous work.

For specially fabricated formwork, grout checks and ruled joints can be incorporated and closed-cell gaskets inserted into grooves formed within the form adjacent to panel joints. Special arrangements of backing members

Figure 15.3 Detail of part of a wall which has been cast from forms lined with Skulptured Timbers. The imported boards were of New Zealand Douglas Fir with a richly textured relief grain. (*Registered trade mark.)*

ensure that there is no excessive local vibration of the sheathing which could upset the concrete at the interface.

Feature formers and fillets need to be carefully designed, these being incorporated into the panel in such a way as to resist grout infiltration or scour during striking. All these are normal details of form design for structural concrete used to varying degrees by contractors and while they may to some extent be dispensible on normal commercial construction, they are essential in producing high quality surface finishes.

Edge grain of ply and end grain of timber must be suitably masked, while steel must be welded in such a way that the indents caused by the draw of the weld do not disfigure the face. Sheathing joints must be so arranged as to coincide with changes in section and, preferably, not made in the centre of otherwise plane faces. Joints should be adequately made and backed to ensure that the sheathing materials act as a continuous membrane, while backing members must be suitably spaced to avoid flutter which could upset the concrete. Deflections should be contained within the limits imposed by the specification. This avoids 'quilting' which would detract from the visual aspects of the face finish.

The carcassing must be arranged so that the continuity of face maintains a consistent reflection of light for a uniform appearance. This aspect particularly affects the joints between panels and the method of lapping form panels over, and achieving support from, previously cast concrete. Backing members must be so arranged at panel joints that the deflection of the sheathing in the end bays does not become excessive or differ to any large degree from that in the remainder of the intermediate bays. Gussets or stiffeners are required to prevent the edge member from rotating and allowing distortion. The joints in the sheathing should be tight and the backing members

INSITEX

Re-useable form-liner system

gives instant off-the-form concrete textures without secondary treatment

Equally suited to in-situ or precast work, Insitex form-liners enable concrete to achieve its full potential as a visually satisfying material of great variety.

Predictable quality, lower costs, fewer manhours and the elimination of site problems are some of the practical advantages of the Insitex System for textured concrete.

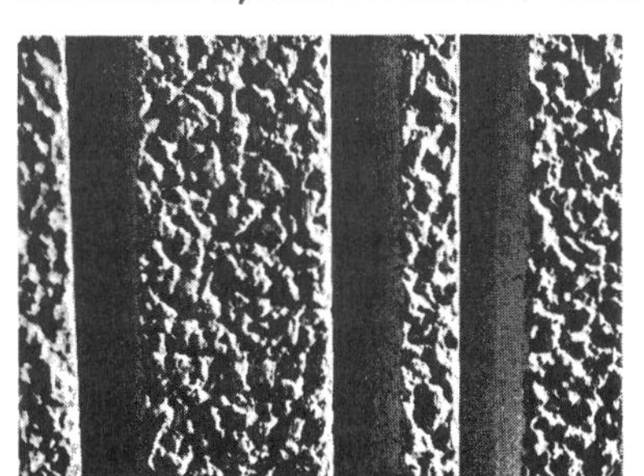

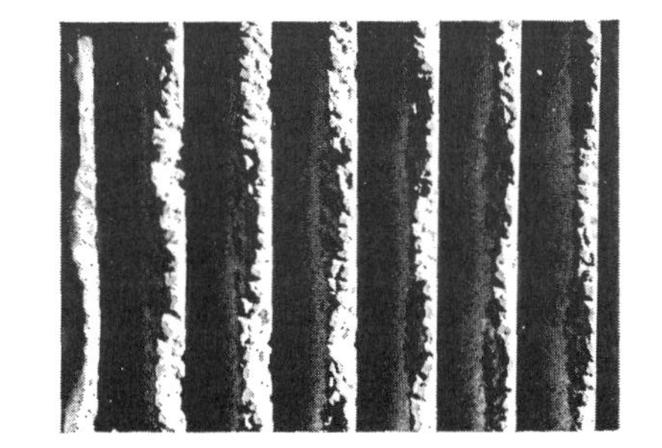

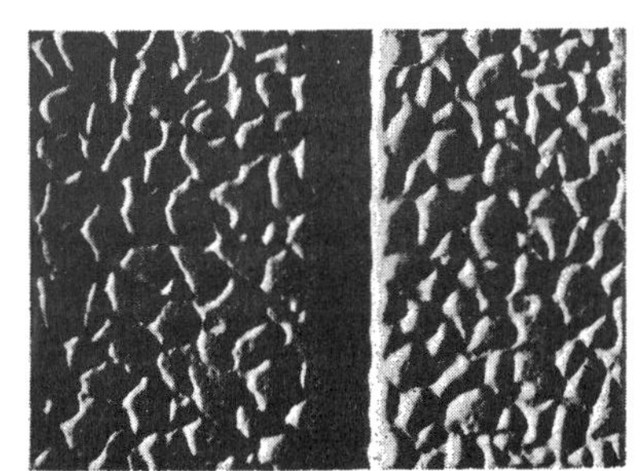

Examples of 40 mm. deep ribs (scale 1:10) from a wide range of traditional and special textures.

INSITEX PRODUCTS, 3 Stewart Street, Milngavie, Glasgow, G62 6BW
Telephone: 041-956 2681

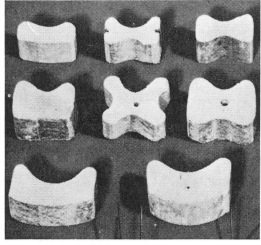

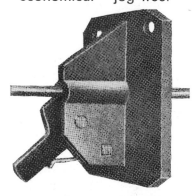

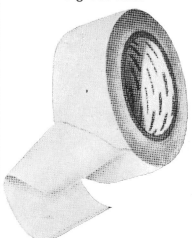

Figure 15.4 The extensive use of board-marked concrete required careful pre-planning at the design stage.

so positioned as to allow the actual sheathing edges to make contact. Where timber and ply are used this can be achieved by setting the carcassing member back from the end of the sheathing and inserting packers or washers at the bolting or wedging positions.

For joints in steel formwork, particularly with purpose made forms, gasketting should be inserted behind the plane of the sheathing. Mastics or fillers inserted within the sheathing plane should overlap in the carcassing to provide support for the jointing or filler material.

Special sheathing surfaces such as heavily grained or grooved materials, require special attention. Any facet, groove or splay should be checked to see that it does not nullify the lead or draw which is allowed for striking purposes at returns or splays. Materials that exhibit board marking require particular attention and where rigid natural board or ply is used, fins or ribs which could trap the form or cause adhesion have to be removed from the form face.

Features and fillets forming recesses and chamfers must be securely built onto or into the form in such a way as to resist any displacement caused by the stresses imposed during striking, and to avoid grout infiltration at the joint with the sheathing material. This is particularly critical because even the smallest fin of concrete, cement paste or mortar trapped within a joint can hinder the production of a high quality surface finish and provide disastrous results.

Sealers and coatings

The appearance of the concrete depends on the nature of the form face and any applied treatment, so that the selection and application of oils and parting agents requires careful consideration. Table 15.1 which has been reproduced from the *Formwork Report* of the Concrete Society and the Institution of Structural Engineers sets out the principal factors with regard to choice of materials. Pre-treatment of the form face with some sealer or coating will achieve either a completely inert non-absorptive face, or one that exhibits some degree of consistency.

Different degrees of absorption of the face affect not only the texture of the concrete, but to some extent the amount of blowholes or pinholes exhibited on the concrete surface. Colour or shading is governed by surface absorbancy, i.e. the more absorptive the face the darker will be the colour of the resulting cast concrete. This phenomenon is particularly noticeable where one board has been replaced during the sequence of casting. An examination of the series of lifts or bays seven to ten years after casting will reveal the dark mark which has resulted from the insertion after the first use of a more absorptive board as a darker patch on the concrete; this will also be easily identified on each succeeding lift throughout the job. This indicates how essential it is to achieve a degree of consistency. Bearing this in mind, especially for the more critical finishes, it is advisable that stocks of materials be retained for repairs. These should receive exactly the same coating of sealer as those used for the fabrication of the form panel.

Ideally, for most visual situations the coating provides an impervious layer between the form face and the concrete. This layer be it a polyurathene varnish, or a rubber based paint or shellac, provides the base onto which the release agent or mould oil is applied, so that scour or local damage excepted, the form should provide a consistent face and excessive colour variation is avoided.

Because an impervious face is produced, particular attention must be paid to the mix design and the placing and compacting methods to ensure that unwanted

Figure 15.5 The absorbency of one board in the form face has caused discoloration throughout succeeding lifts – now visible after some seven years. The new board was not 'tempered' to match those in the existing panel.

TABLE 15.1
Recommended release agents

Type of form face	Plain surface finishes		Special surface finishes	
	Pre-treatment	Subsequent applications	Pre-treatment	Subsequent applications
Timber – sawn	2 3 5 6	2 3 5	6 2 5	2 or 5 2 5
Timber – planed	2 3 5 6	2 3 5	6 2 5	2 or 5 2 5
Plywood – unsealed	2 3 5 6	2 3 5	6 2 5	2 or 5 2 5
Plywood – sealed	None reqd.	2 5	None reqd.	2 5
Plastics – GRP polypropylene etc. plain or textured	None reqd.	5	None reqd.	5
Steel	None reqd.	2 5	6 (5 can be used to prevent rust)	2 5
Concrete	6 (wax) 6 (wax)	2 5 6	–	–

Main types of release agent

1. (Neat oils)*
2. Neat oils with surfactant (wetting agent)
3. Mould cream emulsions (water in oil)
5. Chemical release agents
6. †Paints, lacquers, waxes and other impermeable coatings
7. Wax emulsion. (Not included above as insufficient site data available.)

* Not recommended. Not included in table.

† These are pretreatments only.

entrapped air and water are removed during concreting otherwise defective concrete faces will result.

Only the best materials should be used to obtain high quality surface finishes and while they cost much more than traditional oils or emulsions, the chemical release agents have many features which commend their use. Carefully prepared oils and emulsions undoubtedly help to achieve high quality surface finishes when used according to instructions. However certain advantages can be gained by the use of chemical agents, not the least of which is that the chemical agent remains active long after normal form oils have been absorbed or have evaporated from the form face. As the action of the agent depends on contact with the cement paste, there is ample time for the intermediate operations of steelfixing and form erection to be carried out. This, coupled with the fact that only the smallest trace of the agent is required to achieve satisfactory striking of the form, offers an extremely sound system. For the full value of the material to be achieved, however, a high degree of supervision is necessary, probably greater than that where other oils and treatments are used. For the production of visual concrete this sort of supervision is essential anyway, and the control of application may amount to not only applying the material with the recommended fine spray, but also to removing any excess or build up in corners or on features. If these build ups are neglected, local retardation of the cement paste occurs with the risk of discoloration and spalling and tearing of the concrete face.

Initially any oil or parting agent tends to gravitate towards recesses and grooves in addition to draining down the face. For this reason, care must be taken to swab away any excess at these places. For particularly critical surfaces, dust or contamination should not be allowed to collect on the moist surface of the form during its erection. Dirty hand prints can become transferred onto the face of white concrete. Many examples exist where rust from reinforcing steel has dusted onto a form face and become transferred onto the concrete surface. Pencil marks made by carpenters or steelfixers have also been known to become transferred onto the concrete face, and in fact any foreign matter that adheres to the mould coating will do likewise. Mould coatings themselves have also become transferred in situations where the basic precautions of ensuring a keyed, or more importantly, a dry surface at the time of application have been ignored.

Rubber-based paints and shellacs can become transferred in this way although more often than not the cause can be traced to bad preparation or poor methods of application.

Where polyurathene varnishes or coatings are used suitable conditions and locations for application should be chosen, especially where temperature and moisture are to some degree standardized, and where dust contamination can be minimized. The one-part air-curing coatings give excellent results in sealing form faces, possibly because curing of the coating is a continuous process. One-part materials also produce a good bond with the faces to which they are applied, so that blistering and scabbing are minimized.

Developments in the uses of surface retarders have meant that these materials can be employed for the production of exposed aggregate surface finishes. The introduction of lacquer-type retarders which solidify on the form face and are transferred to the concrete during casting has produced a reliable means of exposing concrete that is becoming widely accepted on site. When any retarder is to be used, careful field tests must be carried out with the type of concrete mix to be used in the actual casting of the structure. These tests establish the degree of retardation to be achieved under the actual conditions of moisture and temperature experienced during the actual process and are valuable for the training of site staff in the application and use of what may be to them a new material.

As with oils and parting agents, retarders require careful application. If a sealer is used it must be applied as per the maker's instructions, the actual retarder being applied systematically by roller brush or spray. Uniformity of application is essential because it is possible to discern on a concrete face those areas which have received more or less coverage. Such a defect could be evident throughout the life of a structure and while it can be corrected by a tool known as a needle gun, this is expensive. Where retarders are to be used with steel moulds the mould face should either be zinc sprayed or treated with an etchant, or even grit blasted. This provides a key for the prime coatings and improves adhesion.

Any form face which has to impart high quality surface finishes must be inspected just before erection to make sure that defects do not become transferred and it must be remembered that a concrete face will mirror the slightest imperfection. Certain imperfections are regarded by architects as functional and thus possibly acceptable. Few architects or authorities however will accept defects that arise from badly made joints or poorly constructed formwork.

The most likely causes of defective face finishes are set out at the end of this chapter. It cannot be over-emphasized however that the mixing, placing and compacting methods employed have equal effects on the suitability of the surface finish.

Concrete mixing, placing and compaction

Earlier in the chapter certain aspects of formwork which govern the colour and texture of the eventual concrete were examined, but various activities in the production, placing and compacting of concrete also vitally affect the colour and texture of the finished component.

Any situation in which the normal, standard practices of concrete production are neglected will ultimately bring about a sub-standard concrete surface. Poor aggregate storage, badly built bin dividers, ill-covered bin or stack drainage can upset the correct final mix which is essential in attaining good quality visual concrete.

Variations in the materials brought onto site, changes of the grading, the colour of the aggregates, particularly that of the sand or fine aggregate due to different sources of supply, can have major effects on the final concrete colour. Grading variations of the materials may be more critical where concrete mixes are prone to bleeding. Bad storage and transportation may even cause contamination of the aggregate, resulting in variations of colour. If only one mix is exceptional it can be sufficient to ruin an otherwise excellent area of visual concrete.

It might be thought that these matters of control are not quite so critical where the concrete is to receive some other treatment, such as grit blasting or tooling after removal of the formwork. In point of fact any variation becomes immediately more apparent once the face film of fines is disturbed. Local upset caused by vibration is characterized by the high incidence of fines and paste where there has been excessive vibration or where leakage has occurred from the form joints.

Concrete mix proportions must be rigidly maintained and every effort made to achieve consistency of these proportions and the correct water content. Workability is critical both for proper placing and for achieving adequate compaction. The water content is not difficult to control on site provided that due allowance is made during the batching process for the free water that is contained in the aggregates and sand. Standardization of free water content can be achieved by maintaining sufficiently large quantities in stock under the conditions mentioned earlier for allowing drain down and standardization. Measurements should be taken periodically to ensure that due allowance is made when the water is added to the mixer.

The adequately trained and well briefed batchman or mixer driver should be able to achieve a remarkable degree of consistency of mix, provided that the materials are stored in reasonable conditions, and that the weighgear on the plant is constantly checked. Where concrete control needs to be critical, a quality control inspector will be required to advise on adjustments for maintaining consistency. In complying with specification, the inspector should maintain checks on the materials and equipment in addition to conducting workability tests.

Mixing time is an important factor in achieving consistent mixes. Concrete mixers are often overloaded or partially loaded, and the length of time allocated for mixing may vary and thus upset the consistency of the mix. Ideally, particularly where ex-form finishes are to be expressed, the mixing time of standard quantities should be similar for each succeeding batch.

Where concrete is delivered from a ready-mixed concrete supplier, the largest possible quantities should be ordered from one source, the mixing plant being advised of certain special requirements with regard to consistency;

certain precautions over the receipt of the concrete on site must be observed. Good access should be arranged and all transfer equipment assembled and cleaned. Again, consistency of timing is important and should, where possible, be standardized.

Consistency is equally important in connection with transfer equipment, skips, dumpers and the like. Where possible, the first batches of concrete should, by agreement with the appropriate authority, be adjusted by the removal of a quantity of the large aggregate to compensate for the equipment being coated with cement paste and fines. These batches should be used in those parts of the work where the finish is not critical. Thus the concrete mix reaching the forms will be of consistent proportions and not lacking in fines and paste. Where possible the timing of the transfer operations should be standardized, and each succeeding batch of concrete given similar treatment.

Placing techniques

Once the concreting operation has been set in motion it is important that nothing should be allowed to interfere with the steady, progressive loading of the forms. Any stoppage or delay will become evident on striking of the form face, i.e. it is characterized by a dark band of fine particles, or some mark of inconsistency.

Where high quality concrete surface finishes are required concreting must be especially well organized and controlled. Access arrangements must be such that a continuous layer of concrete can be placed along the form thus maintaining a fresh concrete face of even depth throughout the length of the form. Where possible, concrete should be chuted against the less critical face of the form to avoid defects from a face deposit of grout arising as it stiffens against the form before the bulk of the concrete is placed and compacted. Chutes, tremies and elephants trunks should be used to direct the supply of concrete as near to its final position as possible.

In connection with access it may be necessary to divert reinforcement, or in some circumstances arrange with the design engineer special reinforcement layouts, to ensure adequate access for the placing of the concrete at regular locations along the line of the run. There are very few walls or deep beams that are exposed on both faces in practice, so that openings can often be left in the least critical face for chute or trunk access. Where reinforcement has been diverted for access it is essential that it should be reinstated by a suitably skilled operative under trained supervision, exactly as indicated on the drawings.

Joints for openings left in critical faces should coincide with features or panel joints, and where necessary additional features introduced to mask any possible seams. The tendency to continue placing through one of these openings until the concrete reaches the brim must be avoided, since it promotes dark joint lines. Ideally, the concreting operation should be transferred to another area at a time when the carpenters can make a good job of inserting and sealing the filler panel before the concrete reaches the lower horizontal joint.

Specifications normally stipulate that concrete be placed in layers, and that immersion vibrators be used to pass from the upper layer of concrete into the previous layer so that there is amalgamation of the layers. This is particularly necessary when manufacturing visual concrete as this is the only method which ensures that the workability fines and the unwanted entrapped air and water are drawn progressively up the form face to the top of the lift thus reducing the occurrence of blow holes and voids at the face.

Workability fines and laitance should be removed from the top of the lift and replaced by good quality concrete which amalgamates with the previous layer of concrete. This should prevent the formation of a layer of substandard concrete. Over the years many contractors, in order to avoid bad areas of visual concrete, have concreted additional depths of columns and walls and then cut away the top 300 mm or more before subsequent operations are effected – this obviously is very wasteful. The 'mucking-out' of the top of a mould is a well-known activity in precasting, particularly where finished upper surfaces are required.

At the top of a lift it is not unusual, even with the best compacted concrete, for horizontal cracks to appear, usually the result of post-vibrational settlement. This can be avoided to some extent by re-vibration of the concrete while it is still workable enough to be energized either by poker or by slicing with a thin blade. It is not generally desirable to re-vibrate with external vibrators as this may promote other defects.

When the concrete is placed in a form it should be introduced in reasonable quantities, as near to its eventual position as possible. Any use of vibratory equipment as a means of distributing concrete only induces layers or stratification of the face by layers of fines and paste so that the larger particles roll or slide along the surface of the previously placed concrete.

Compactive effort

The type of vibration applied is a major consideration and the internal vibrator is a useful tool for achieving uniformity of compaction in most situations. However immersion vibrators demand skilful application by a trained operator and often the bad results achieved reflect a lack of training with regard to the effective radius of vibration and the appropriate rates of insertion and withdrawal.

Once external vibrators have been attached and operated to a preplanned sequence of timing related to location they do not depend on human efforts to achieve results. Unfortunately locations are often badly planned and the very nature of form fixing and its attachment to previously cast concrete cancels out the vibrationary and compactive efforts. Provided that attention is paid to the correct location on each side of a lift or bay, and that there is correct orientation of the vibratory effort, external vibrators achieve excellent results. Provided the vibrators are securely fastened, it is possible to impart an excellent mode of vibration into the form, thus energizing and compacting the concrete. Possibly a judicious combination of external and poker type vibrators will achieve the best results.

Of all the methods, the technique of 'tickling' the back of the sheathing with a mechanical hammer is least likely to achieve a satisfactory long term finish. The compactive effort achieved by this process while sufficient to bring fines and paste to the surface of the form, is rarely sufficient to induce unwanted entrapped air and water to rise through the concrete, with the result that defects become masked with a thin layer of paste which later oxydizes or erodes away to display tiny blowholes and voids.

A similar defect results from badly located external vibrators which energize only a local area of sheathing rather than promote an overall uniform vibration of the form face. Here a pumping action is set up which may promote bleeding from a mix of a particular grading: alternatively the pumping action may result in the mobilization of the cement paste in such a way that aggregate transparency occurs on the form face.

Most of these comments have been related to the construction of vertical or near vertical visual concrete faces. There are special considerations which govern the preparation of horizontal or sloping faces and of course there are particular details which must be considered when the production of horizontal and vertical faces, which can be viewed jointly, are concerned.

The major problem with horizontal surfaces of visual concrete, i.e. floor and beam soffits is that at some point in the construction cycle the formwork is used as a working and access surface. It is essential that the formwork designer and planning engineer co-operate to afford the site personnel some alternative area for the storage of material, the fabrication of steel and for general access. Earlier it was noted how prone are oils and parting agents with respect to transfer of contamination and marking. This applies particularly where horizontal surfaces are concerned. Minute clippings of steel wire and layers of rust from steel or ties can settle onto a parting agent and subsequently become transferred.

Exposure to the elements may result in displacing oils so that scabbing may occur during striking. Where possible, appropriate parting agents should be selected and the steel reinforcement tied into cages other than on the critical soffit areas. When finally laid in position, the reinforcement should be supported on plastics stools or chairs which can be clipped into position before the cage laying takes place. Where ties have to be made in situ the adoption of proprietary clips avoids the problems incurred by traditional tying and clipping using steelfixer's nips.

The concreting process must be carefully controlled and, ideally, smaller batches than may be the case in straightforward structural concreting operations should be placed. Again, the concrete needs to be deposited in small quantities as near to its final location as is practicable. This avoids 'tide marks' where larger batches cause a scouring action along the form face thus displacing oils and parting agents and generating ring marks which are impossible to remove. All sheathing joints and laps with previously cast concrete must be adequately supported where necessary, and additional chocking pieces inserted along the line of the joint to prevent grout infiltration and curtains caused by secondary deflection of the sheathing face.

Wherever the exposed finish surface comprises soffit and vertical face, the concreting arrangement should be such that continuity is maintained. If this is not possible, for example where a complete form is erected for a wall or spandril springing from a slab, then a feature or fillet must be incorporated to allow the timing of the operation such that some stiffening takes place in the slab, before the vertical upstand above the slab is concreted. The timing of this operation is particularly critical if any amount of surge below the inner form is to be avoided. Similarly, if too long a time is taken then secondary deflection will cause the formation of curtains as the deflection of the outer form face allows grout to percolate outside the face of the concrete to the slab. On this basis alone it is generally preferable to arrange for the upstand walls, spandrills and parapets to be cast as secondary operations onto kickers with a suitable feature employed to mask the joint.

Liners and textures

A great deal of exposed concrete is cast against liners or linings of plastics, timber, rubber, card and similar materials. While in the majority of these cases the work is successful, occasional failures do occur. Apart from new

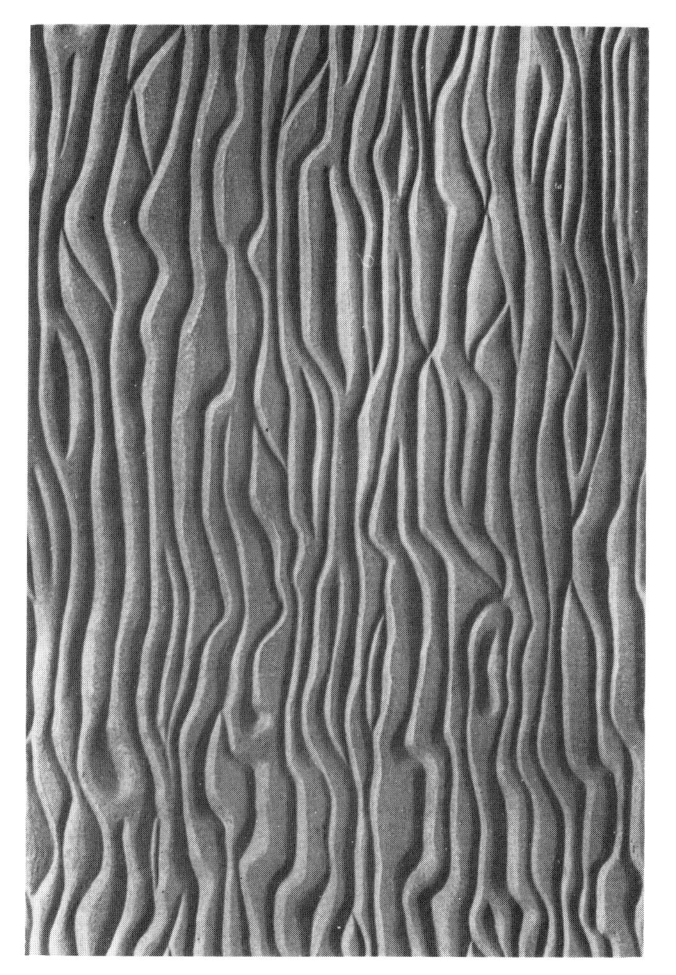

Figure 15.6 Detail of the exotic surface seen on the panel in Figure 15.7.

Figure 15.7 The manufacturer produced dense polystyrene moulds from which concrete masters were used to breed glass reinforced plastics moulds for unit casting.

and improved materials, even though suppliers carry out exhaustive research on their products, the failures are more often than not due to mis-application. This usually arises from a lack of understanding of the material's properties – both chemical and physical.

Chemical failures are often the result of bad storage, poor mixing, wrong application, or due to an ommission in the reading of instructions on labels or in the suppliers' literature. Physical failures are caused by incorrect application or where a material has been overstressed or subjected to wear for which it was not intended.

A criticism often levelled at the construction industry is that it is slow to innovate. This is not difficult to appreciate when the constant flow of new materials and materials applications is visualized. However sometimes a company adopts a new material or technique, uses it haphazardly often in considerable quantities and perhaps is ill-prepared to introduce it in the most suitable way.

Many cases have occurred where a particular site has employed materials successfully, and reported good results to head office. As a result the material may be sent onto another site where the requirements and circumstances are totally different – failure would then be inevitable. Once the decision has been made to use a particular material, one person (often the formwork designer) should make an intensive study of the attributes of the material, i.e. suitable applications, makers' recommendations and such like, and then follow through all aspects of purchase and application. By doing this an objective assessment can be made and useful data accummulated for later circulation to all interested parties.

Figure 15.8 Rope-faced concrete finish – an interesting application of a traditional method of providing 'keyed' surfaces on concrete.

As an example a case can be cited involving a new material where the structural designer had specified a very deep profiled soffit to beam members. The contractor elected to have prepared plastics liners for insertion in the form. In practice the heavy detail tore out of the liner and became embedded in the concrete. There was little doubt that the fault lay in the lack of substance in the liner proper, and once the thickness was increased the work proved to be extremely successful. In the early stages the

Figure 15.9 Producing rope-faced concrete. The rope is torn from the concrete to provide a feature in addition to exposing aggregate at rib peaks.

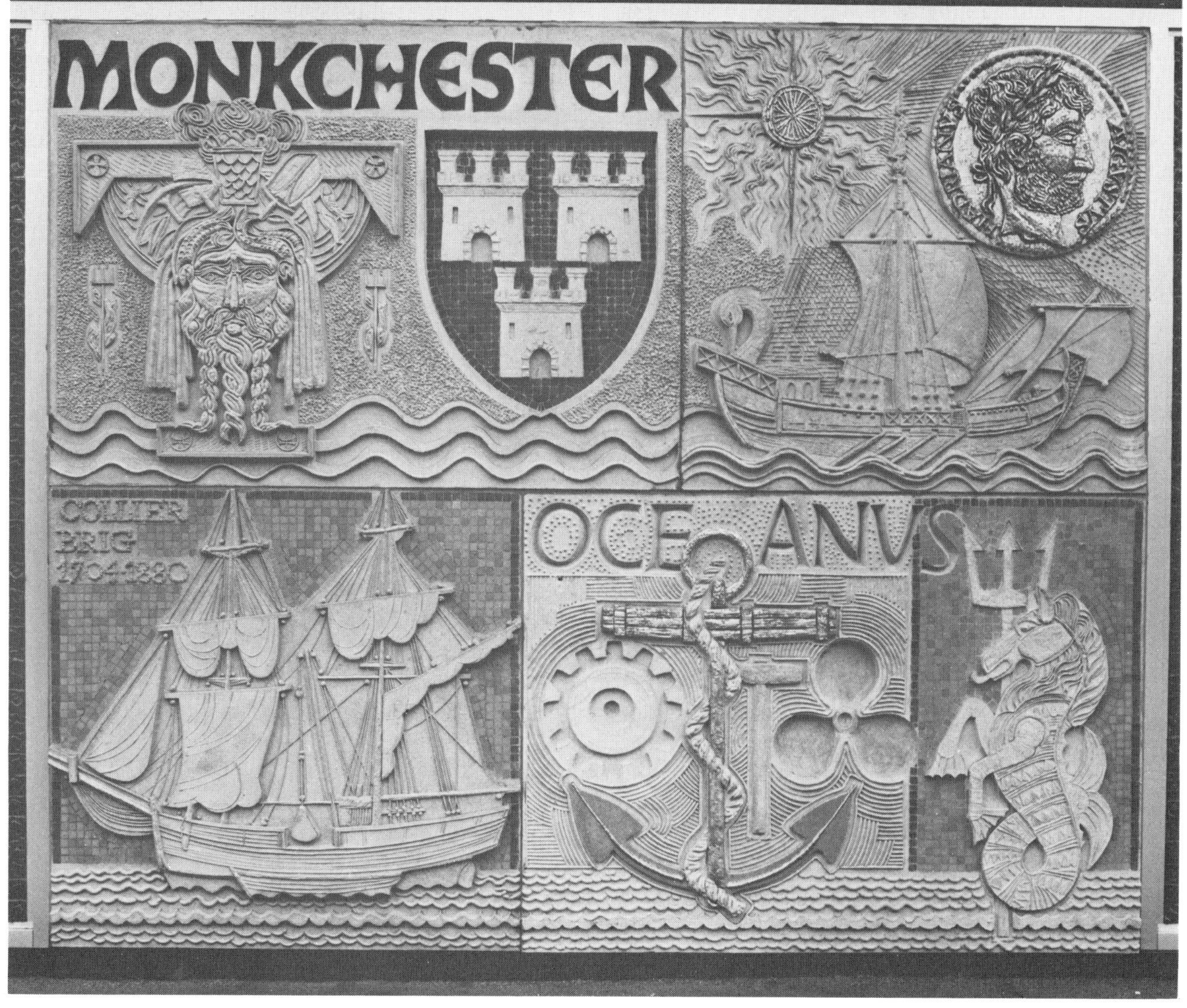

Figure 15.10 Sculptured panels cast from an artist's master moulds which were carved from expanded polystyrene. These were cast by a specialist contractor and subsequently decorated.

supplier had continually emphasized the need for an adequate thickness of material with some reinforcement included – this was disregarded.

This sort of thing proves the need for careful co-operation between supplier and contractor during the early stages of promotion of some new material.

Where board marked concrete is being produced any attempt to lap a form over concrete to exhibit the profile formed by a board in the previous lift should be avoided. It is obvious that, as timber is a natural material that exhibits considerable variation in grain pattern, no board can satisfactorily seat onto a profile generated by another. Some allowance should always be made for recessed joints or features between lifts and bays, so that smooth forms can be seated back to smooth concrete and thus provide a grout-tight seal.

Associated activities

In addition to the activities of the form designer, the concretor and the formwork carpenter, a number of other trades become involved in producing exposed or visual concrete finishes.

Possibly the person with the greatest impact on the appearance after those just mentioned is the steelfixer. He handles rusted steel across or onto forms and substantial pieces of bar or complicated cages of steel, so it is not unusual for the critical faces of moulds and forms to be badly damaged by impact, or marred by transference of dust. The steelfixer either works immediately before or after the formwork tradesman and should receive instruction, particularly where a finish is critical necessitating special care to be taken over the preservation of the form face.

Reinforcement and in particular indented or deformed bars which are dragged along form faces will destroy precious grain markings or possibly tear the corners of feature formers. Even the marks of a steelfixer's boots or gloves can mar the form face and become permanent defects on the concrete face.

The most common type of defect occurs where tie wire clippings have become trapped within a form, or where rust has accumulated on reinforcement which has not been turned back by the steelfixer. Careful attention must therefore be applied to cover control and to avoiding clippings becoming lodged within a form. These sort of defects have occurred even on sites where the utmost attention has been applied to the cover for reinforcement.

The handling of forms during the striking operation and to and from storage, and in re-erection has been discussed elsewhere in this book, but it needs to be emphasized however that the actual handling of the forms prior to the concreting process can be critical with respect to the eventual finish. Form panels may weigh many tonnes and when being swung across a site or hoisted into position it is inevitable that they will rotate on the crane hook, because of wind or stress due to the bond. Panels which rotate in this way are difficult to control and are a danger to all concerned in the construction operation. Tail ropes should therefore be used to control panels and the operative in charge must be capable of communicating with the crane driver by using the approved code of hand signals. Good control can be thus maintained and the likelihood of panels becoming scarred by contact with starter bars or of corners being damaged and rendered unsuitable for the necessary close fit, is considerably reduced.

Possible causes of defective surface finish

Normal structural concrete

Problem	*Probable cause*
Colour variation	Inconsistency of mixing and placing; of curing and of striking times. Variations in compactive effort, and differing conditions of form face.
Stained faces	Incorrect treatment of face; badly applied oil or parting agent.
Mottled or patchy faces	Excessive local vibration due to flex in panels or poor compactive technique.
Retardation of face	Excessive quantity of parting agent.
Spalling or tearing of the face	Insufficient amount of oil or parting agent; early form striking. Adherance of fines to faces during early part of placing operation.
Excessive corner damage	Early form striking; poor compaction.
Leakage	Omission of seals; bad fit against previous lift. Wrongly located form ties. Secondary deflection of form face or carcass; local deflection at kicker. Non-continuity of form carcass.
Cracking at corners of openings	Rigid formers restraining concrete from shrinking. Excessive force required in striking.
Blowholes	Mix badly graded. Poor compaction. Impervious form face. Dumping of concrete and entrapped air.
Water runs	Mix badly graded. Thin form face. Pumping or flexing of face.
Cracks at face	Loss of cover. Settlement after compaction. Bad reinforcement position preventing compaction.
Curtains or snots	Poor contact between form face and previously placed concrete. Secondary deflection of sheathing.
Holes or voids	Congested reinforcement. Choked formwork. Bad placing techniques. Insufficient workability.
Undulations on face	Local deflections due to excessive rate of fill. Lack of carcassing members.
Crazing and dusting	High intensity of fines at face. Rich mix. Smooth impermeable form face. Lack of curing.
Darkening of face	Delayed striking. Water movement. Absorbent form face. Inconsistency of coating.

Visual and decorative surfaces

Problem	*Probable cause*
Inconsistency of exposure	Bad concrete placing techniques. Retarder coating inconsistent; wrong type of retarder. Incorrect timing of exposure.
Bad striking of features	Form construction inappropriate. Insufficient lead or draw. Incorrect timing of striking.
Inconsistency of colour	Variations in fines content. Colour variation in fines. Pitch or slope of face. Inconsistent placing. Variations in depth of lift.
Sloping lines or stratification	Bad placing techniques. Breaks in concreting operations.
Leakages, honeycombing at joints	Omission of seals. Wrong joint detail. Excessive form overlap. Lack of sheathing support at joint.
Rust marks at face	Incorrect cover tie wire. Nails and cuttings in forms. Snap ties of incorrect length. Pyrites. External contamination from plant, scaffold and such like.

Figure 16.1 The hyperboloid lattice frame and the panels for this water tower are of precast concrete.

16. Moulds for precast concrete

General

Generally, precasting can be adopted to advantage when the geometry of a particular component is critical, or when there are sufficient components of a particular size to warrant establishing a precast yard on site, or where special requirements demand the sub-letting of complicated units to a specialist supplier of precast concrete. Much of site precast concrete production is carried out in order to simplify the progress of the in situ concreting operations, and a lot depends on there being repetition in the contract.

Items which can be precast on site include:

Piles

Beams, columns and simple structural components

Sections of retaining wall

Spandril panels

Crosswall panels

Floor panels and ancillary units.

Situations often arise where the skilled use of precasting techniques can reduce the problems of those concerned with in situ concrete construction, especially where impossible sections are detailed, or where congested reinforcement needs to be incorporated or, particularly, when exotic finishes are required. To overcome these problems site-produced permanent formwork or precast sections can be introduced, these being supported by some form of grillage, or bolted into the formwork arrangement for transfer into the final locations.

Where isolated items are to be precast such as stair flights, special beams and corbels, for later inclusion in the in situ structure it may be that a simple mould can be established either adjacent to the batching plant, or at a ready-mixed concrete delivery point. If filled when deliveries of an appropriate concrete mix are being taken this type of mould provides a buffer activity for those who are involved with the general steel-fixing and concreting activities.

Mould details

Moulds for precast work usually, though not always, cost a great deal more than forms for in situ concrete although the higher costs are offset by the large number of uses which can be obtained from them. The degree of accuracy, provision for preparing the surfaces, and mechanization are often greater than for normal in situ formwork although perhaps not more than with purpose-made or special formwork.

Moulds are usually over-designed mechanically to ensure foolproof casting, and indeed are often provided with renewable faces or liners to lengthen their casting life. They should be sufficiently robust to remain straight while in use, but it is always advisable to fix bases to a casting deck or grillage. Long units can be maintained straight by arranging for at least one side to be supported by, or attached to, members cast into the casting deck. Alternatively, units may be cast each side of a wall especially cast for the purpose which can include pockets for some bolting or tie arrangement. This ensures not only the straightness of the units, but maintains squareness of section, which is always a critical aspect of all precast production.

On precast work the designer has the option of selecting which way up the units should be cast. This depends on the required surface finish, the geometry and configuration, the amount of projecting steel and the location of connections.

When large quantities of units are to be cast, it is economical in the long term to provide an extra number of bases or pallets, so that side moulds can be used twice or perhaps three times in a shift, the castings remaining undisturbed on the 'spare' bases. Extra feature formers need to be provided where early removal of the former would otherwise damage the 'green' or freshly placed concrete. This provision of additional base moulds and feature formers is particularly valuable when due to the requirements of the specification the units have to be left on pallets for up to three days, since it allows a daily casting cycle.

Mould materials

Particular gains can be achieved by the careful selection of mould materials. For example, steel should be used for highly stressed parts of a mould or section where wear or scour is likely to occur, while timber profiles can be used to form special features particularly where considerable repetition is likely. Cast plastics or extruded alloy sections provide economical ways to meet the structural demands and resist scour or tearing.

For small units, timber and ply moulds can provide 30 or more uses, and if re-lined even more. Glass reinforced plastics and casting polyurathanes can be used to produce

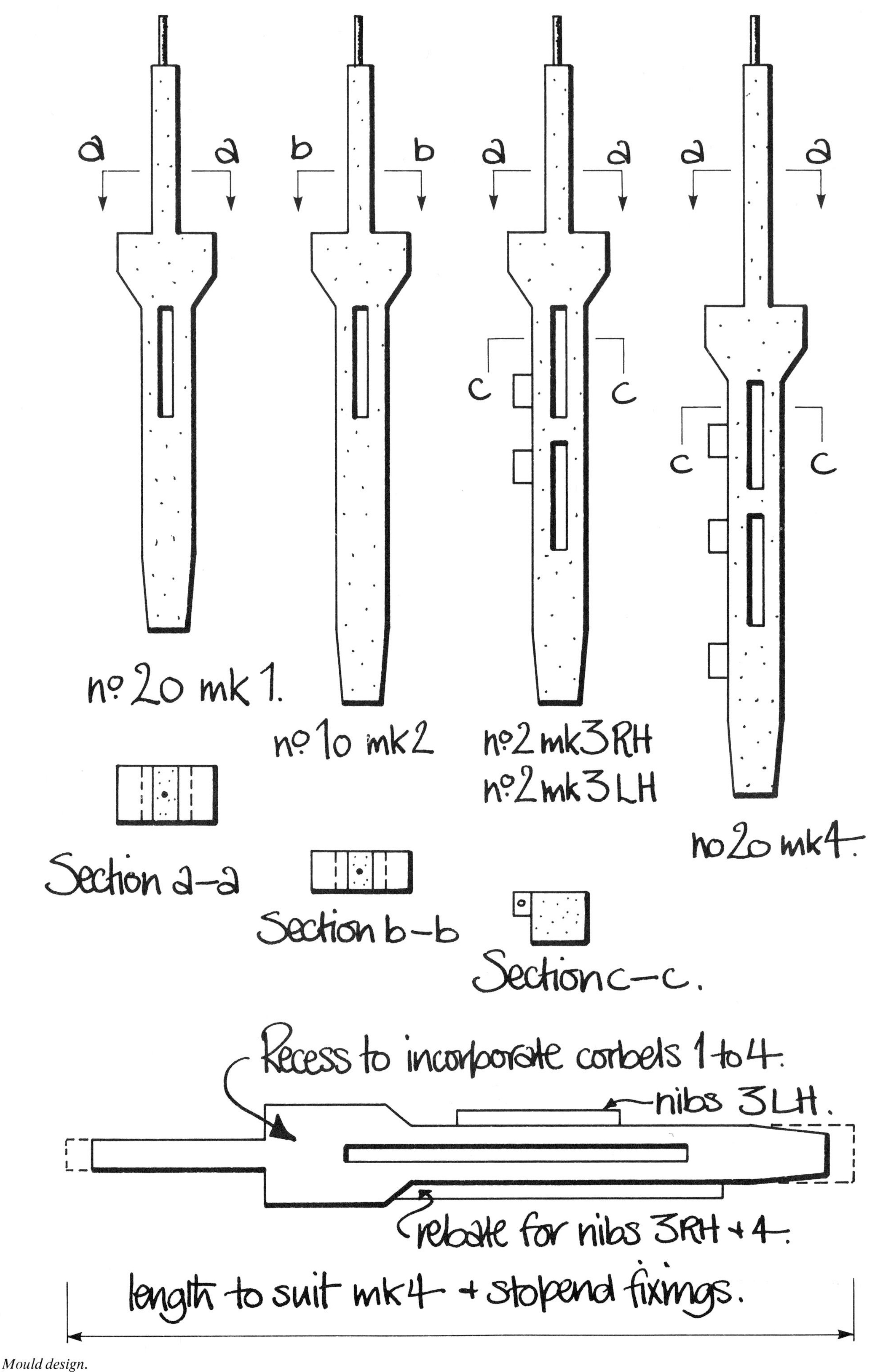

Mould design.

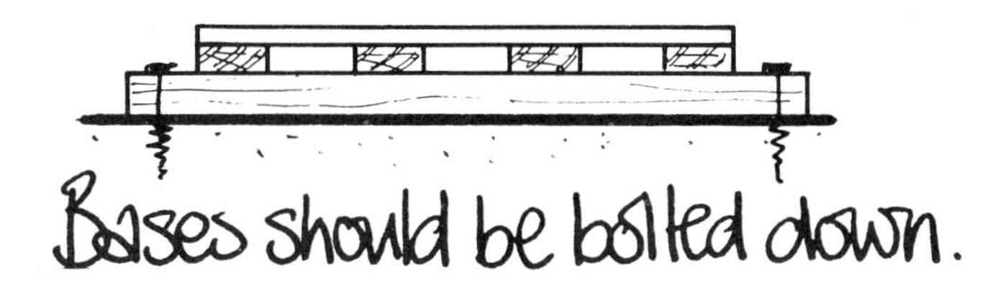

Bases should be bolted down.

Concrete bases can be tied with steel and clad with timber.

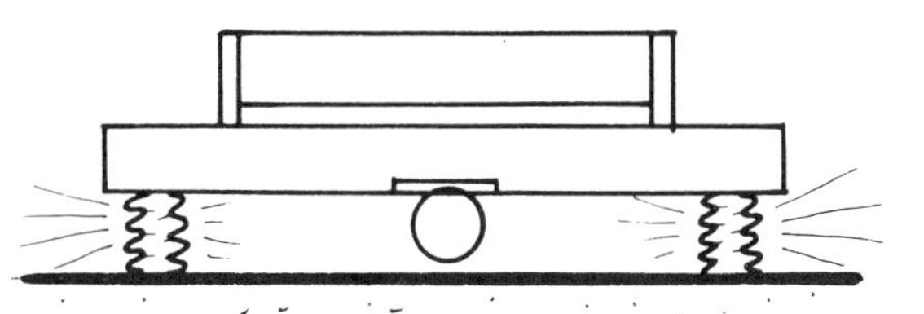

Moulds for facing panels require flexible mounting to promote even vibration.

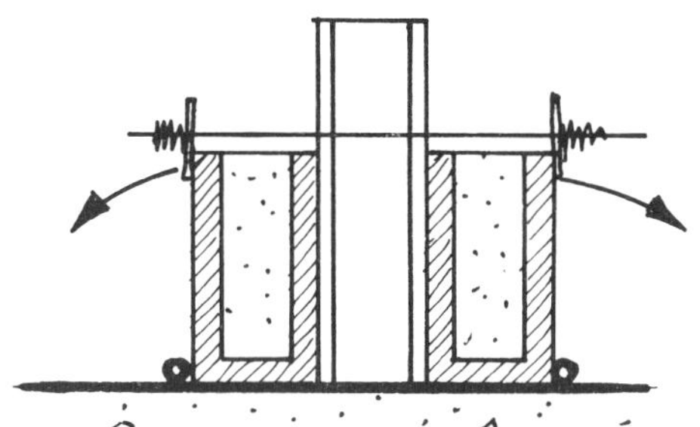

Moulds may be set to r.s.j.'s to maintain line.

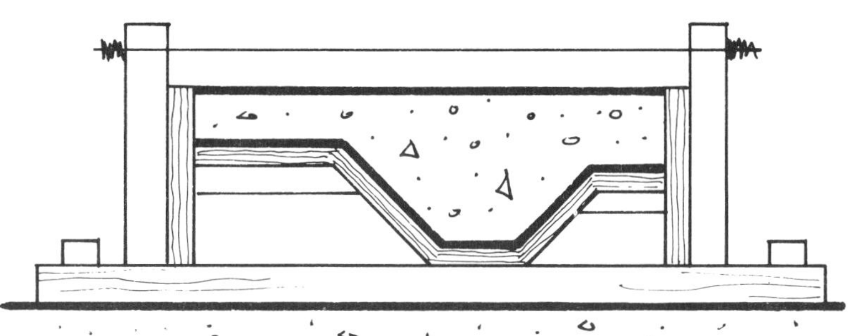

Generally, profiles can be incorporated into basic 'box' mould...

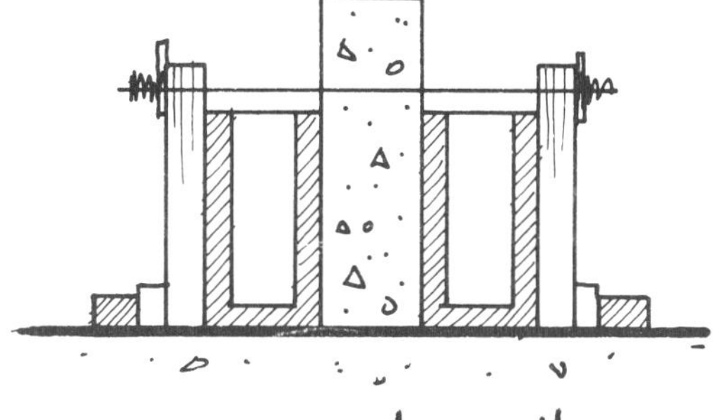

...or concrete walls.

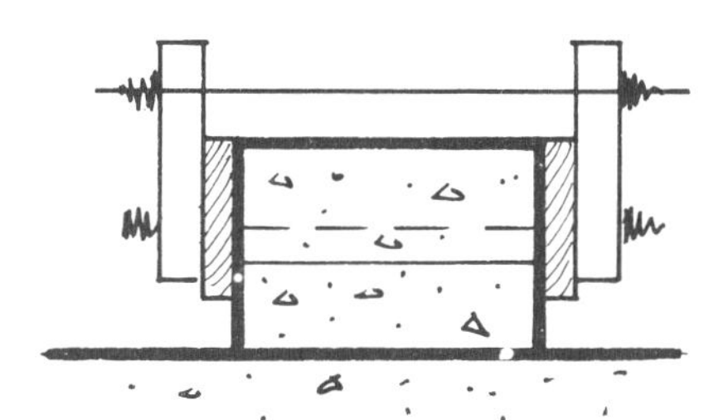

Stack-casting saves space and moulds.

Space must be allowed for various activities. Steel stock and stacking depend on casting and delivery rates.

Mould design.

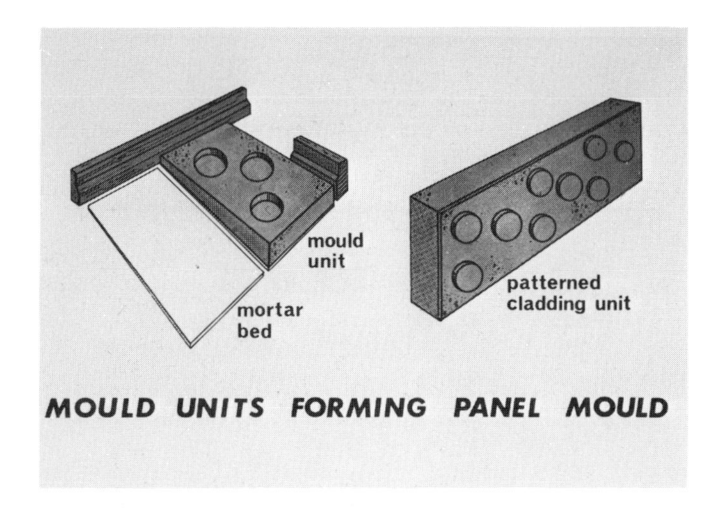

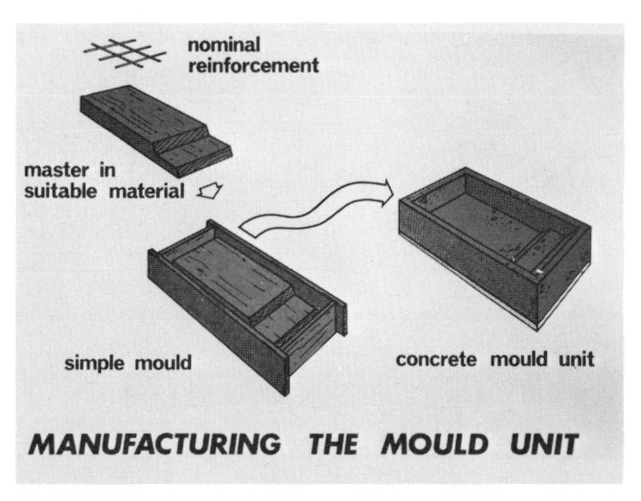

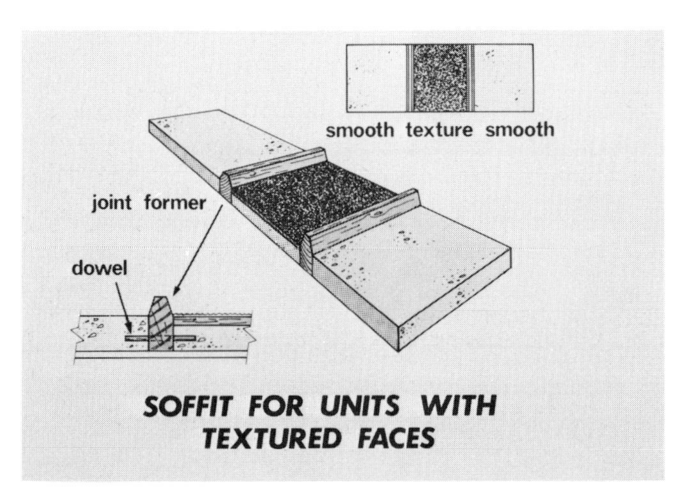

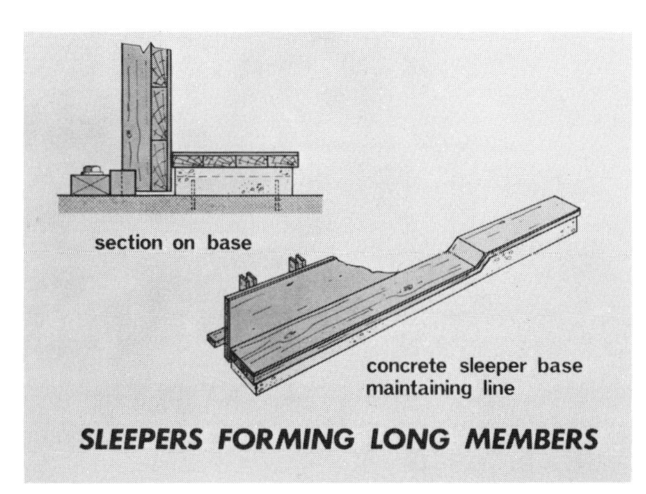

Figure 16.2 Four applications which demonstrate how concrete can be used in mould construction.

several hundred casts while steel moulds can achieve up to 1000 uses. Obviously, the number of castings obtained will reflect the amount of money spent on the mould, the skill with which they are designed, the care with which the moulds are used and the amount of repair and refurbishment expended during the casting programme. Much precast work is carried out on the basis of accelerating the curing process, and this is important when the mould materials are selected.

Generally, precast moulds are subjected to intensive vibration using either external or poker-type vibrators, so that the moulds must be robustly constructed to withstand the stresses induced by the vibratory efforts.

Concrete itself is of great value to the site precaster because it can be used for the substructure, the sheathing and base construction and also for long-line type moulds for piles and structural units where it combines the functions of sheathing carcass and stressing bed. The illustrations throughout this chapter show the basic methods of mould construction, all of which can be undertaken by the normal site carpenter and trades supervisor.

Ideally the forms, or moulds as they are called in the context of precasting, should be constructed under cover and the necessary basic machines for woodworking, cutting and welding should be provided.

Of course the precast mould can be ordered from a specialist supplier, ready made. If this is done points of detail and preferably some notes or drawings which indicate the site requirements should be prepared by a suitably skilled person.

The whole process depends on a skilled approach to the mould design, close supervision of the casting process and in particular a careful and detailed approach to the mechanical handling problems posed by special details of the precast components. The mould designer must therefore use all his skills to ensure that the stripping and removal processes are made easy and that he obtains the best value from the expensive materials that are used in the mould construction.

Moulds used in works production are often mechanized and with the advent of cheap and reliable hydraulic and pneumatic rams and equipment site-produced moulds (and forms) should incorporate arrangements whereby hinged sides and stopends and self-ejecting feature formers are air- or oil-operated for speed and efficiency. Many hours can be spent in extracting formers, and lifting and moving mould sides and as such these activities demand a good deal of cranage. Any uncoupling that can be effected by the use of simple travellers, winches or hinged mechanisms will undoubtedly provide early returns for the money invested. Rams and pumps can be used with a variety of form and mould components. The speed of operation can reduce labour costs and in turn may, for precasting, allow the achievement of more than one casting cycle per shift. There are times where the value, in

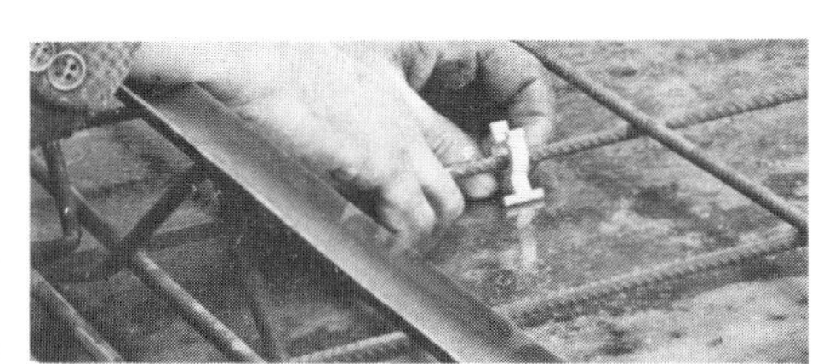

terms of increased uses achieved by the early release of the main items of lifting equipment, are such that the investment in other equipment such as mechanical nut runners, electric or air-powered winches, and small gantries over gang or battery moulds are soon repaid.

Operating costs can be further reduced by adopting construction techniques in which moulds are produced as one-piece items, so reducing the number of pieces to be handled and thus improving the accuracy of the mould. Induced inaccuracies can result from the distortion of parts as a result of frequent striking and re-assembly, poor maintenance and the build-up of paste and particles in joints.

Mould designers need to pay particular attention to the requirements of the engineer and architect with regard to the location of steel reinforcement, location of ducts and covers. Ideally, the design of the reinforcing cage should be integrated with the mould design. This applies particularly where steel cages have to be positioned without spacers being apparent on the face. For current cladding production or cross-wall construction, the steel reinforcement location is critical with respect to both appearance and durability of the component. The steel cage must be securely fastened to the inner mould or form in such a way as to preserve its location, and be supported against the vigorous compactive effort applied to the concrete.

Figure 16.3 Some of the formwork used for the lattice beams shown in Figure 16.1.

For the actual physical design of the mould, the form designer must separate the range of components into families or groups of similar units which can be cast from a reasonable number of moulds in the programmed time. The grouping should be such that excessive modifications are not required between castings, and such that the profiles can be achieved from sound form or mould arrangements. This is achieved by designing the mould as a container which can be used to cast the largest cross-sections of the largest component in a group, the eventual profile being generated by pads, packers and stoolings inserted into the complete form or container.

It is obvious that a good balance must be achieved between the number of moulds used, and the degree of alteration work carried out to achieve the profile of the units. Most adjustments are accomplished within the length of the mould by the re-location of stopends, the changes to sections being obtained by stools and such like. The designer should satisfy himself that any change to the mould does not detract from the accuracy of line or maintaining continuity of the mould.

When corbels and nibs are incorporated in the structural component, the carcassing of the mould should be arranged to fly past such projections for some distance on each side of the protrusion. This may appear to be somewhat expensive on materials, but the resulting accuracy of line and profile should provide good returns in terms of performance and speed of erection.

One factor which is critical in precast production is the lifting of a component which often has to take place at an early age. The mould design must incorporate some means of accurately locating lifting devices, lifting loops, screws, studs and sockets, plus numerous plugs, connections and cast-in components. These must be cast in, rather than installed, after the concreting has been finished, so that full load-bearing capacities are obtained in terms of the local concrete strength.

To bring about speedy and safe working there must be a clearly established programme of work that is fully understood by the operatives involved. The working area must be kept clean and free of obstruction. For example, racks for mould retention between castings reduce the number of hazards caused by moulds that fall or tip over. Where tilting frames are used, operatives will need to be instructed on how they are used and warned of the dangers of venturing under frames that are suspended from lifting equipment. Wherever possible, battery moulds should include inbuilt gantries for mould panel handling. This ensures that assembly work can proceed without delays caused by crane availability.

Stacking and mould storage areas should always be carefully kept in such a condition that workers have safe access at all times, and that the possibilities of progressive collapse of stacks are avoided.

Just as with normal in situ formwork practice, it is desirable that there should be drawings for each mould assembly together with notes and sketches of pads, stoolings and arrangements for modifications.

Every site, and indeed every person involved with precasting, has preference for particular types of construction and particular ways of tying and bracing moulds. While there are obviously some techniques which ensure a sound, grout-tight mould assembly capable of producing accurate components, the main requirements are those involving simplicity, speed of dismantling and assembly, and in achieving the required number of economical uses.

The techniques shown in the various sketches have all been tried in practice, although before adopting them the designer should discuss their applications with the site supervisors and site management to establish any existing preferences.

When considering a site casting yard the designer must allow sufficient space for the following:

1 Aggregate storage (special)

Figure 16.4 Tunnel segments being cast from concrete moulds in gang format. Moulds progress through site casting shop to filling, curing and stripping stations. (Edmund Nu Hall.)

2 Concrete batching and mixing

3 Storage of steel

4 Steel fabrication

5 Cage storage

6 Casting

7 Finished stock storage.

The layout may require that steam or electric supplies have to be laid on for accelerated or normal curing, quite apart from services that include power, lighting, water and air. All these are in addition to the normal provisions for lifting and handling concrete and finished components. A sound base slab is essential, this should be levelled and concreted and during the concreting operation suitable benches, sleepers or socket grillages should be installed on which to secure the moulds.

The majority of profiles encountered can be moulded using pads and stooling arrangements on a basic carcass or within a basic mould arrangement.

Ideally, construction should be started from the layer of bearers that support the outermost unit face formed by the soffit. Ply diaphragms, stiffened by softwood members or steel frames of welded construction based on the basic bearers, can be used to support succeeding layers of sheathing. Side members may be hinged to, or secured by, the basic bearers and their accurate location should be ensured by inserting stopends and such spacers as are required by the length of the unit.

The way of casting should be selected such that most of the critical features are cast from mould faces and that the geometry of the unit is generated by the mould. Concreting thus is simplified and the finishing operation is reduced to one which involves only screeding and handling of plain units.

When mould construction is detailed, projecting steel should not be allowed to coincide with the basic carcassing members. Cast-in components must be properly mounted onto the sheathing face.

Just as precasting is adopted to simplify the construction of in situ profiles, so also can it be used to simplify the casting of precast elements. Any critical corbels, nibs, or surface finishes can themselves be precast in readiness for insertion into the moulds that are to be employed for casting components proper; this avoids unnecessary and complicated mould arrangements.

Accuracy and surface finish are prime considerations of precasting, and in turn require detailed attention with respect to the installation of gaskets, for reducing grout loss, and templates, for obtaining correct profiles.

One basic factor which should feature in all considerations with regard to precasting is that of concrete maturity and whether it can be suitably achieved in the factory. Generally, the greatest stresses are imposed on the concrete when it is being demoulded in its green state. Concrete which has attained a strength of 5 to 10 N/mm^2 can usually be safely demoulded, due care and attention should be paid to the use of spreaders, tilting tables and similar means of support. The characteristic 28-day strength should be achieved before the units are structurally installed. This strength requirement varies from contract to contract and necessarily determines the amount of stacking space which must be allocated for the curing and storage of the precast components.

The skills required for placing and compacting precast concrete vary little from those employed for in situ construction, but special demands are necessary where particular finishes, slender sections or massive components are produced.

The whole subject of precast concrete production has been covered by the author in *Precast Concrete Production* (published by the Cement and Concrete Association in 1973).

Figure 17.1 The location of ducts within a structure is a critical factor in form design.

17. Formwork for prestressed concrete structures and components

General

Prestressed concrete offers many advantages when compared with ordinary reinforced concrete. The introduction of a prestressing force can effect a saving of 20% in the volume of concrete required and 80% in reinforcement. Although higher strength concretes are required and better quality steel, under favourable circumstances, the cost reduction can be about 40 to 50%. On some structures a saving of 20% has been recorded.

Prestress can be introduced into concrete by two main processes known as pretensioning and post-tensioning.

In the pretensioning operation concrete is cast around tendons which are forcibly extended between abutments, the prestressing force being transferred into the concrete by the transfer or release of the tendons after sufficient concrete strength has been achieved to ensure adequate bond between the tendon and the concrete, continuously along its length.

Post-tensioning involves the manufacture of concrete components which contain some reinforcement and the introduction of prestressing forces by tendons that pass through or around the concrete once sufficient strength has been achieved. In this process the prestress is generally transferred into the concrete by some form of anchorage, either at designed positions along the component or at the ends. For circular tanks and containers the forces are uniformly applied by the tendon around the periphery of the structure.

Post-tensioning techniques

Post-tensioning techniques are often employed on site for large span components, bridges, tanks and containers and of course special structures such as oil rig platforms. The details which are critical to the formwork (and falsework) are covered by the checklist at the end of this chapter; by studying these questions it will be apparent that the designer is mainly concerned with the position of the ducts in the concrete and the formation of the correct profile and location of the anchorage arrangements. Any degree of inaccuracy with regard to these will result in failure of the prestressing process, or at the least, extreme difficulty in carrying out the stressing operations.

The structural drawings should indicate the profiles of the components or segments to be post-tensioned and the positions of the prestressing ducts relative to these profiles. The catenary or profile of the ducts are usually indicated by offsets at given stations along the line of the components. Reference to plan details and key sections will identify the horizontal location of the ducts. Generally, ducts change their locations three-dimensionally along the line of the components, so that it is important to interpret all the details correctly, and to relate both the plan and elevation of duct positions.

Any sketches of duct profiles and positions should be made separately so that this vital information is isolated from all the other dimensions on the drawing. By doing this the details can be rapidly transferred to the forms when the reinforcement is being fixed and the ducts located.

For cambered soffits or profiled units a table of stations and offsets which relate to some clearly defined datum line should be established. This datum line can be marked on one face of the formwork and easy reference can be achieved by measuring from the line to establish points which generate the line of the duct. The catenary profile is best set up by running dimensions laid down on a detail for site use.

With regard to the fixity of ducts, this is best achieved by using steel grillages welded to the vertical stirrups. Positive location can be further helped by inserting dowel pins at critical positions through the form within the run of the duct. For repetitious work the formwork sheathing should be reinforced with steel or timber, the locations marked so that during succeeding operations the locating pins are re-inserted in each set up. All too frequently repetitive work is carried out from memory of previous operations and there have been occasions where omitted vital restraints have allowed excessive displacing of the ducts or formers.

All arrangements must be such that they resist the tendency for the ducts to be displaced sideways during the concreting operations or for them to float when the compactive effort is applied. During concreting of deep slender sections the pins should restrain the ducts and the tendency for the reinforcing cages to become displaced.

To prevent local displacement or wobble between the location points, the tendons should be inserted into the duct before the concreting operation takes place. Failing this and where short transverse or vertical prestressing ducts are involved, lengths of stiff reinforcing bars should be inserted to help stiffen the ducts between supports. All ducts must be effectively sealed against the ingress of cement paste.

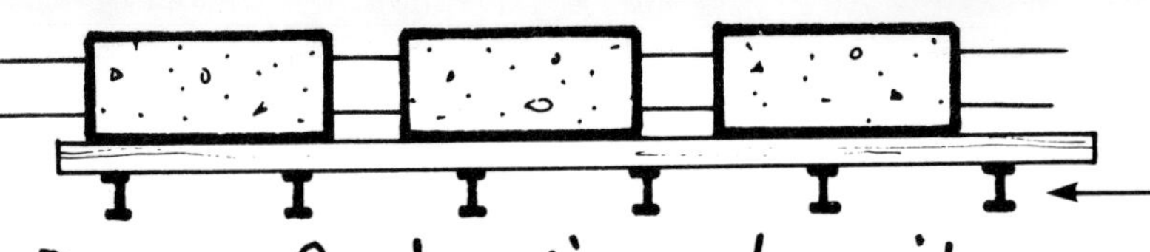

Forms for longline p/t units must allow for movements towards fixed anchorage at time of transfer.

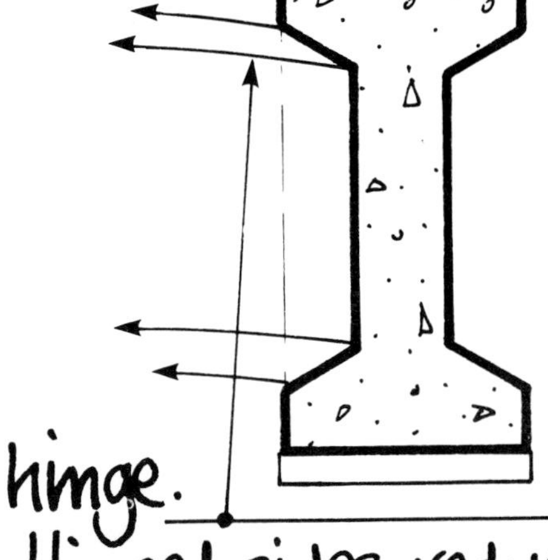

Hinged sides reduce handling. Take care with geometry.

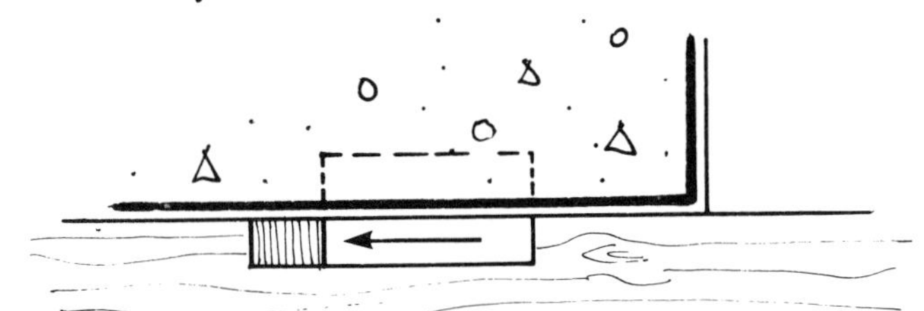

Bearing plates must be free to move within pockets in base.

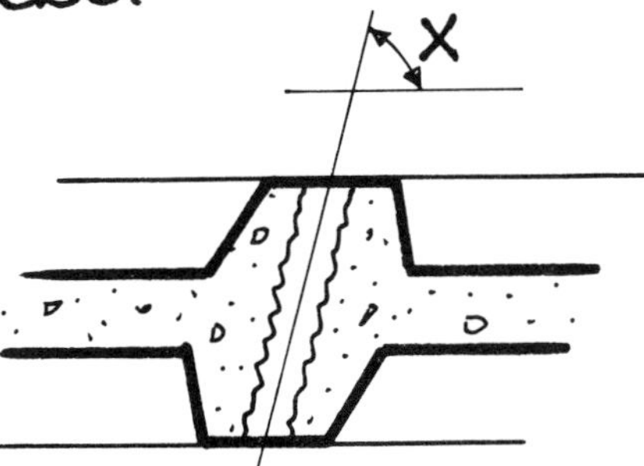

Absolute care is required to achieve correct skew (and handling!)

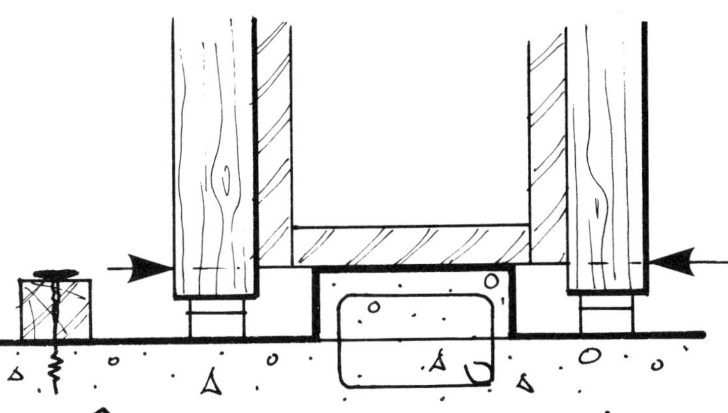

Concrete bases maintain line and provide stable foundation for soffit sheathing.

Sides must be accurately located to ensure square ends and accurate profile.

Cambers are simply formed by screeding concrete and subsequently lining with timber or ply.

Formwork for prestressed concrete – details.

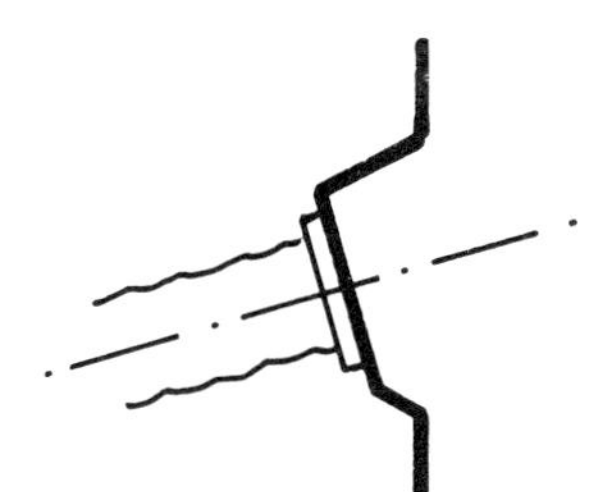

The line of ducting must be normal to the bearing plate.

Duct joints must be grout-tight-tape or use shrink-fit plastic.

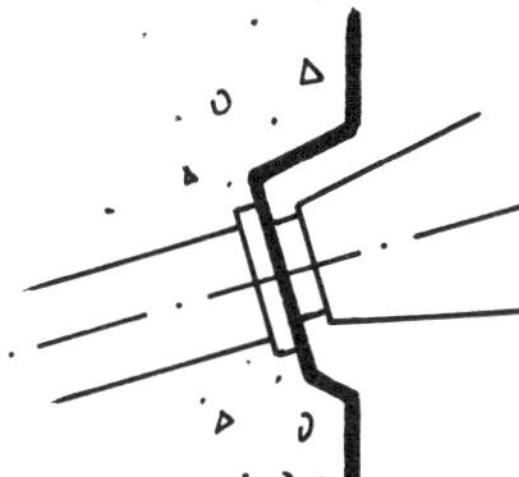

Recesses must admit the nose of the jack.

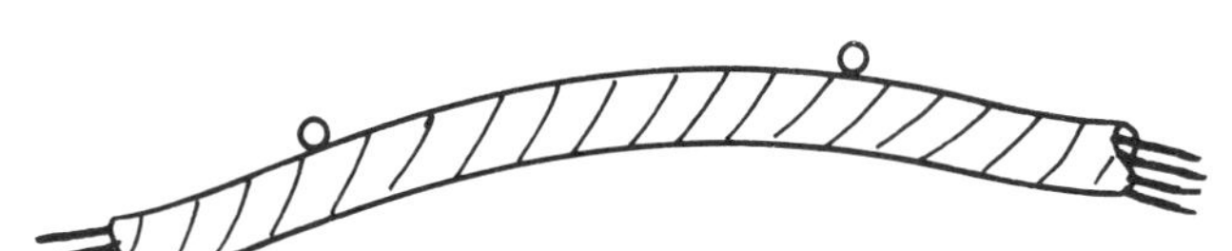

Insertion of tendons into ducts prior to concreting will avoid displacement and friction.

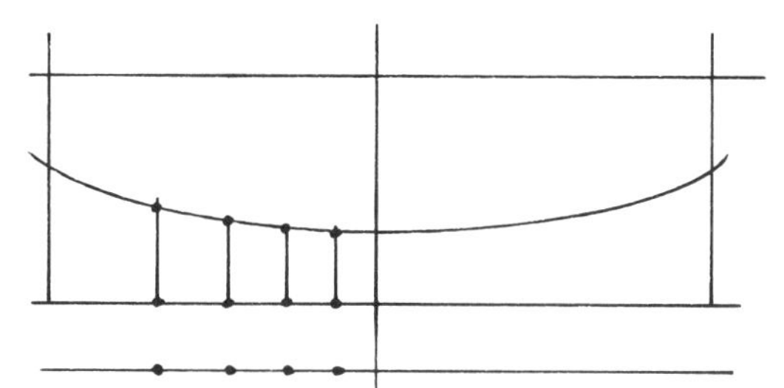

The duct catenery must be exact.

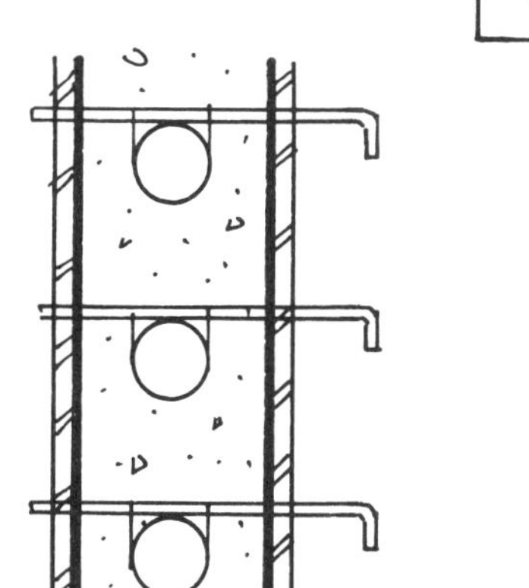

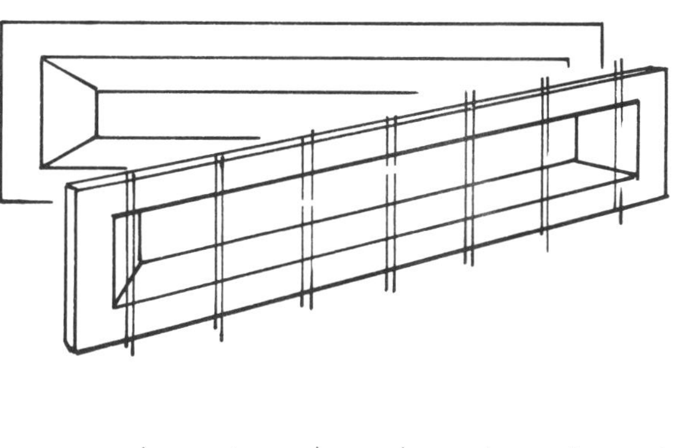

Ducts may be positioned by bars welded to stirrups or by pins through form.

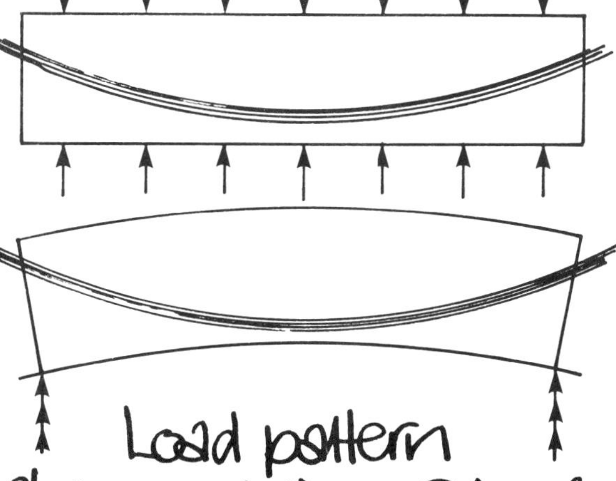

Load pattern Changes at time of transfer.

Duct details.

To avoid the need for too fine an adjustment of the formwork, weep holes or grouting tubes are best inserted into holes that have been bored during the actual form erection. Clips or cleats which locate the side forms longitudinally to register with the soffit can be attached for achieving accuracy.

It must be remembered that there will be some degree of movement of bearing plates and projections such as corbels and features during the post-tensioning operation, and arrangements must be made to ensure that featured side forms are eased from the concrete before the prestressing operation is begun. Due clearance must also be allowed for any projecting nib or steel connection so that some degree of longitudinal movement can take place during transfer.

One particularly critical aspect of prestressing is the change in the load pattern due to the introduction of the prestressing force. A concrete beam, slab component or shell in its freshly cast and unstressed state imposes a uniformly distributed load on the soffit formwork and thence into the falsework. Similarly, precast segments impose a uniform predictable pattern of loading when they are initially landed into position on the falsework. As the prestressing operations proceed and the component becomes self supporting over the designed span a change occurs in this load pattern which gradually becomes in effect point loading at the supports. This can be better understood when the hog or camber resulting from the application of the prestressing force is considered. This aspect of prestressing is critical with regard to the disposition of the support members beneath the soffit of a relatively large beam or slab because the increased load can be such that it may overload the scaffold or prop supports adjacent to the bearings.

Transverse ducts can cause particular problems, especially where components are set on the skew as encountered with some types of bridge structure. Repetitious use of beam side members helps to achieve a fair line across the bridge.

During erection, the correct location of opposing beam side members, and thus the correct angle of the duct need careful consideration. Where possible, steel bars should be made to pass through the ducts in the cast components to check the position of the holes for the formers in the panels of adjacent beams. These bars stiffen and line the ducts and are particularly useful where transverse diaphragms are being cast between previously cast main beams. While inflatable formers prevent grout ingress they do not provide a check on line.

The actual formation of the ducts is discussed in Chapter 13. However, it must be emphasized that any former used for post-tension duct formation must be completely grout-tight along its length and in addition it should be rendered grout-tight where it meets with form sides, stopends or anchor blocks.

The main quantities of reinforcement are located at bearing positions and adjacent to the scarfs or joints of post-tensioned components. The form panels at these positions should be made such that they can be removed with the minimum of disturbance to the main soffit and cheek forms; this then gives adequate access for inserting and checking reinforcement, the critical bearing plates and anchorages at the stopend position. It is usual to incorporate within these panels the often complicated formers that terminate an I-section, and provide the squared or rectangular concrete section which terminates the component proper.

Because the profiles of stopends and the blocks which form the recesses for the anchorages are always critical, manufacturers' details regarding the post-tensioning equipment must be followed carefully. Bearing plates or anchor blocks are generally bolted to the face of the blocks which form the recesses, so that the line of the duct and the aperture in the anchor, whatever the form, must be continuous and fair. The recess block also forms the recess into which the nose of the prestressing jack is inserted when the cables are being tensioned. It must be of suitable size to accommodate the jack and have sufficient clearance for movement of the jack during application of the force. The formwork designer must be in possession of the full details of the particular prestressing system to be used, so that he knows that the recess formers are in accordance with the maker's specification. Provision must be made in the fastening for the removal of the pocket former such that it does not disturb the anchor bearing.

The provision of the necessary clearance at the ends of ducts for the insertion of the jacking equipment over, or between, the free ends of the wire or tendon, and between the previously cast concrete and the component being stressed, is a critical detail which must be discussed with the planning engineer.

For precast segments of bridge or elevated roadway construction, the formwork designer should study the implications of the various aspects of casting. By some arrangement of the way of casting it may be possible to simplify duct location, and either render the fixing of the ducts more rigid or avoid their displacement by the concreting process. This choice of casting is discussed in the previous Chapter.

The design of forms or moulds for segmental construction and some forms of normal post-tensioned work should include some arrangement for casting the joints between the units. Where the joints are to be of vibrated site-cast concrete, sockets or anchors can be incorporated in the units adjacent to the joint and the forms bolted back into these. The inclusion of a fillet or recess former improves the appearance and conceals the actual construction joint.

Forms for prestressed concrete

The most general site application of the pretensioning process is that in which prestressed piles are produced on a long-line basis. This is often used on large civil engineering contracts where the cost of transporting precast piles otherwise proves prohibitive. In such situations concrete is often used as a mould material, the segments of a gang mould being cast around master units made from timber. The mould segments are laid on a prepared foundation and linked together, possibly by post-tensioning. This process has recently been used in the production of massive double-tee units for civil engineering applications. Arrangements are made to line

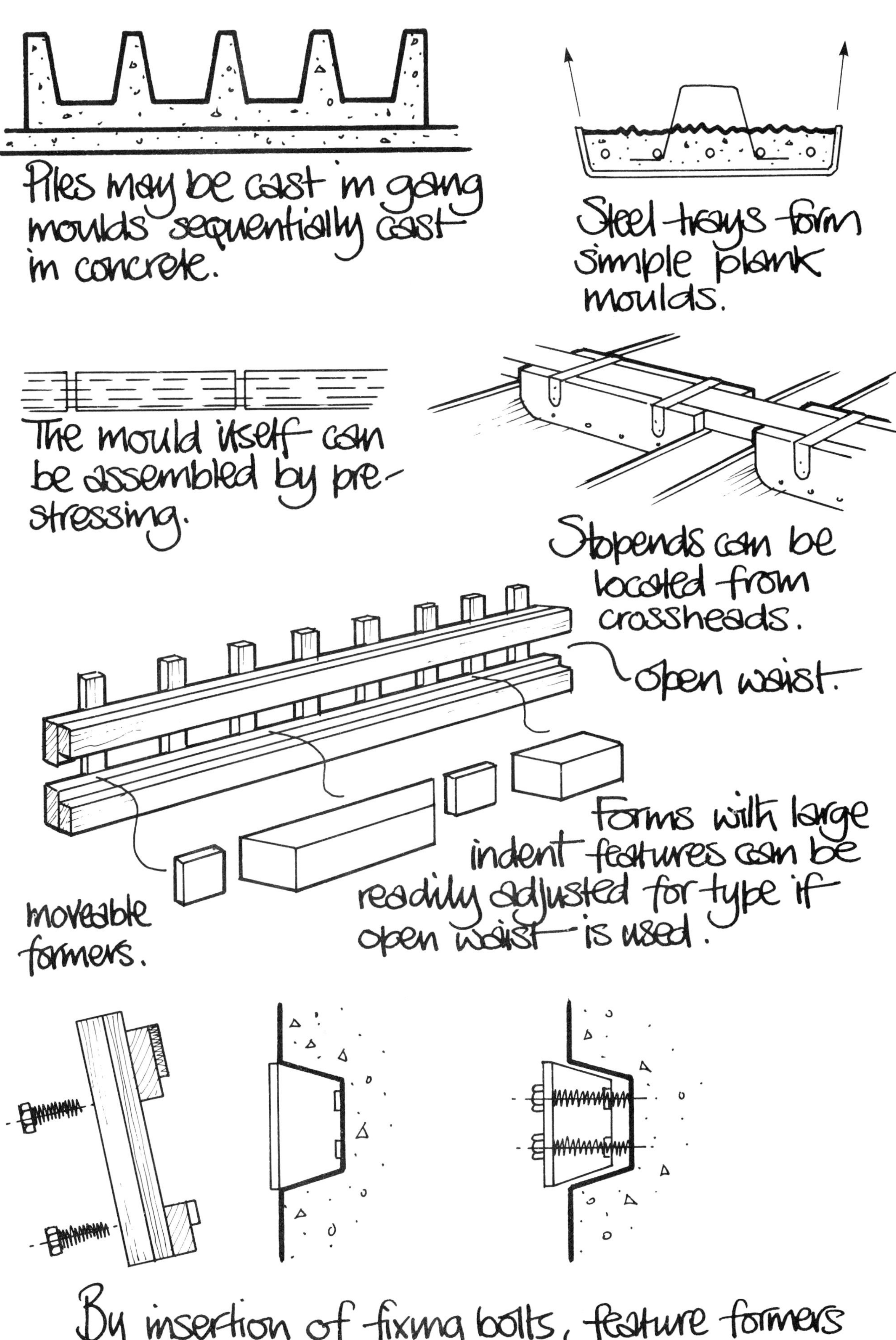

Moulds and forms for prestressed concrete.

the units accurately, while the resulting moulds are used for the repetitious casting of piles or similar units. The amount of lead or draw required to enable the withdrawal of the freshly cast units after transfer of the prestress needs to be carefully considered. A lead of 1 in 12 is sufficient, but in some circumstances this can be reduced. In fact the movement of the units within the mould at the time of transfer of prestress to some extent helps with the demoulding process, because it breaks the bond between freshly cast concrete and the formwork. This advantage is not realized however on long beds where the friction between the concrete unit and the concrete mould, or form is sufficient to overcome the force that results from the recovery in the free wire or strand between the units. For long beds and large units, such as piles, some arrangement is necessary to break the bond between the unit and the mould face by local jacking. When units cast in long lines are prestressed, the contraction of the free wire or strand causes them to slide along the bed towards the abutment to which the fixed ends of strand are secured. Because this movement scours the form face, some allowance should be made for the movement of any projecting reinforcement or connections. Mould side members must be withdrawn before transfer in order to avoid their becoming trapped or pinched during movement. The early removal of mould sides from featured or I-section beams avoids any tendency for the web or narrow sections of the beams to be cracked as contraction takes place in the concrete component. However, curing must not be neglected during early demoulding techniques.

Casting and demoulding sequences need careful consideration when forms are designed for use on prestressing beds. It is necessary to determine the direction of work so that there is a correct interlock of adjacent mould panels and parts. The stopends should allow ease of thread up of wire or strand, but also positively position these tendons during the casting activities. Where solid cores are used for the production of hollow units it may be necessary to split stopends horizontally to allow core positioning and clamping. Stopends must be accurately positioned yet readily free to allow movement along the bed during transfer. When static moulds or trays are used, the most practical methods of positioning stopends depend on the use of some crosshead or transverse member clamped to the mould proper in such a way that it may be simply released before the de-tensioning process. The crosshead can incorporate projecting brackets which maintain the plumb of the stopend.

Where prestressing is carried out on the long-line basis, the stopends are generally positioned either by stopend cleats, or bolted connections to the sides which are removed when the sides are stripped and thus do not restrict movement during the transfer of prestress. Where sides are used to locate stopends and the square of the unit thus depends on cleat or bolt location, some arrangement is necessary to accurately position the sides relative to each other and to the base. This maintains squareness of the ends, and positions bearing plates relative to side features. This is particularly critical when beams are cast for skew bridge construction and where transverse and vertical prestressing ducts are included in the cast component. The form designer must make a careful check on 'skew' from a study of the drawings to ensure that the beams are cast to the correct hand to suit site location.

Cheklist

Formwork for prestressed concrete

What method of prestressing is employed?

How are ducts to be formed?

How are formers to be located?

Can tendons be inserted prior to concreting?

Are there special anchorage arrangements?

Do grouting tubes clear backing members?

What is the size of the prestressing jack?

Is there space for nose insertion?

Is there space between units for jack insertion?

Is the catenary clearly indicated?

Will the application of prestress change the loading pattern on the soffit or pallet?

Are projecting members free to move at time of transfer/prestress?

Are there feature formers or projections on the form which may become trapped?

Is an anti-floatation arrangement incorporated?

Are tell-tales incorporated?

Will camber or hog cause loosening of intermediate supports?

Are there special points of lift or support?

Is there some provision for fixing formwork for capping concrete?

Is there provision for fixing formwork at joints?

If tendon location is by pins through side forms, is there some arrangement to prevent side forms lifting?

Are side forms located longitudinally relative to base?

Has the 'skew' dimension been interpreted correctly?

Figure 18.1 Checking for cleanliness and accurate location of reinforcement.

18. The preconcreting check

General

Before any concreting is carried out a suitably qualified person should check that the formwork is safe, that it complies with the specification in all respects and that the practical details have been correctly interpreted. A recommendation given in the interim report of the advisory committee on Falsework by the Department of Employment and the Department of the Environment, is that there should be one person nominated at each site who should be given responsibility for all aspects of a particular *falsework* set up from inception to removal. This person is referred to as the 'temporary works co-ordinator'. He would be given the authority to stop the work should the falsework not comply with stated criteria. The 'permit to load' which he issues is conditional and would not be valid for longer than 24 hours. It is suggested a copy of the permit should be lodged with the engineer, or clerk of works.

A similar – rather less formalized arrangement – would prove worthwhile for formwork and may well prevent many accidents and avoid the wasteful processes of removing 'condemned' components and the need for considerable quantities of remedial work to sub-standard costs.

The *Formwork Report* prepared by the Joint Committee of the Institution of Structural Engineers and The Concrete Society particularly mentions the need for careful supervision at all stages of formwork construction plus a thorough inspection which should be carried out by the supervisor once the work is finished and before the concrete is placed. The report explains that the nature of the inspection depends on the type of job in hand and lists certain key points.

Checks and inspection would be subject to the quality and accuracy standards established for the job as discussed earlier. The person making the checks should be fully experienced to be able to make a suitably prognostic

*Figure 18.2 For this 'small works' application, much attention has been given to plumb, lining and bracing. (*Royal Borough of Bognor.*)*

examination, and be able to foresee the likely results of particular methods of support or strutting. He should also be capable of appreciating the loads and pressures that are encountered.

The scale of the work, the intended degree of accuracy and standard of surface finish would govern the selection of the person concerned with the preconcreting inspection. Although in the general range of building and civil engineering works the trades supervisor responsible for the formwork would be considered capable to make the final check, special cases which would require inspection by some skilled engineer can be envisaged. For example, where a formwork arrangement is of a particularly critical nature with regard to accuracy, or where the formwork passes over a public access way.

The checklist at the end of this chapter is by no means exhaustive, but it should prove useful to the person who is responsible for a preconcreting check. It should serve as the basis for a more specialized list for a particular type of structure, for example post-tensioning work or some complex architectural requirement.

It includes items which do not necessarily relate to formwork, since it is impossible to divorce details of reinforcement, access and cleanliness from the main topic. All these are critical to a successful concreting operation.

In addition to the preconcreting check, the value of a standby tradesman and continued engineering surveillance during the casting operation cannot be over-emphasized, the tradesman to deal with simple adjustments and the post-concreting levelling and lining, while the engineer is concerned with checking movements against datums and providing final lines and levels.

Checklist

Access

Can concretors gain unobstructed access to fill form?

Can skips or pipelines be brought to form openings?

Is there a safe scaffold platform for the operatives?

Is ladder access to working platform secure?

Anchorages and bearing plates

Have correct fittings been installed?

Are duct formers securely attached?

Is position and line correct?

Are joints grout tight?

Ancillaries

Have brick ties been inserted?

Is masonry slot in position?

Any other fixings and sockets?

Are through hole formers installed?

Has weather bar been fixed?

Bracing

Are all bracings securely connected?

Do braces go to sound ground or matured concrete?

Are braces angled to avoid uplift?

Are braces suitably tightened?

Carcassing

Is there any evidence of wracking or distortion?

Are all connecting bolts suitably tightened?

Are the main continuity members installed?

Cleanliness

Have forms been blown out and/or debris removed?

Has extraneous material been removed from adjacent scaffold?

Concrete

Has surface of previous lift been prepared?

Are embedded ties in mature concrete?

Has rate rise of concrete been established?

Are screeds and tamps available?

Connections

Are brackets, sockets, plates for cladding correctly positioned?

Are through holes or projecting loops located as per drawing details?

Cores

Are cores correct size?

Are they positioned correctly?

Has arrangement been made to combat uplift?

Have cores been lubricated for removal?

Curing

Is curing membrane available?

Can cover be supported clear of concrete surface?

Is insulation available?

Ducts

Are ducts correctly positioned?

Are joints grout tight?

Is there positive location against uplift?

Are tendons inserted to maintain line?

Are tell-tales attached to monitor movement?

Formers

Are formers correct size?

Are they correctly located?

Are sockets and slot formers secured to formers?

Are formers braced?

Does fixing resist side thrust and uplift?

Inserts

Are inserts correctly located and securely fixed?

Are conduits supported at correct distance from form face?

Are power boxes and switch boxes taped to exclude grout?

Level

Is level of form correct within tolerance?

Lighting

Are lights available for work after dark.

Are torches available for inspecting deep lifts?

Line

Is line of form correct?

Has camber been included?

Are continuity members installed?

Oils and coatings

Have the correct materials been used?

Are edges of form oiled to prevent build up?

Power

Is power available for vibrators and equipment?

Is safe low voltage current in use?

Reinforcement

Is steel correctly positioned?

Are there positive arrangements to maintain cover?

Are projecting steels suitably supported and jigged?

Is any reinforcement to be inserted as work proceeds?

Is it available?

Safety

Does scaffold have sound foundation?

Is scaffold adequately tied and braced?

Is scaffold complete with toe boards and guard rails?

Are helmets, goggles and ear muffs available?

Sheathing

Are materials appropriate to required finish?

Are joints tight?

Is there sufficient lap over previous cast?

Have seals been inserted?

Is sheathing supported against previously cast concrete?

Has sheathing been oiled or coated with release agent?

Has excess material been removed?

Are features securely fixed?

Has grout check been installed?

Have all gaps and holes been filled?

Stopends

Are stopends correctly positioned?

Are they securely fixed?

Have stopends been oiled?

Has excess oil or parting agent been removed?

Have holes and gaps been effectively sealed?

Has grout check been installed?

Striking arrangements

Have stripping fillets been inserted?

Is there sufficient clearance for tie removal?

Is there sufficient clearance for form removal?

Have screw pads or jacks been adjusted and lubricated?

Is there sufficient adjustment available on rams and screws?

Supports

Are supports based on suitable foundation?

Are supports correctly positioned?

Are supports plumb?

Are supports tight and restrained against movement?

Have lacers been installed?

Is external bracing in position?

Are standing supports correctly positioned?

Are forkheads wedged to ensure axial loading of props or tubes?

Have correct pins been used at adjuster?

Are secondary heads suitably tight?

Tableforms

Is infill correctly positioned and supported?

Are legs jacked or castors locked?

Has bracing been properly installed?

Are runners correctly seated in forkheads?

Ties and anchors

Have correct types been used?

Are centres correct?

Are ties suitably tight?

Have correct washer plates been used?

Are correct spacers installed?

Have correct studs or cones been used?

Washer plates

Are correct sizes installed?

Are washers square on members?

Have washers been securely wedged in tapered work?

Perhaps the most important checklist pertaining to construction, particularly bearing in mind the implications relating to safety, is the one included in the recommendations of the Final Report of the Advisory Committee on Falsework, under the chairmanship of S L Bragg, which deals with the particular duties of the temporary works co-ordinator (principally concerned with falsework but essentially also concerned with formwork).

Among his duties the temporary works co-ordinator must pay particular attention to the following:

Is the design brief adequate and does it accord with actual conditions on site?

Has each element of the design been checked by a competent person and has the falsework been considered as an integrated whole and approved by a competent person?

Has the design been passed to the engineer and have his comments been acted upon?

Are the actual loads encountered on site, particularly the live loads, no greater than those assumed by the designers?

Is there a realistic programme for the delivery of materials to site?

Have there been any changes in materials or construction? Are these significant? If so, have they been referred to the designer and his approval obtained?

Has each element of the falsework and the whole assembly been inspected and the faults rectified or alterations to design approved?

Does the loading programme agreed on site accord with the designer's assumptions and intentions?

Figure 19.1 The effort applied in providing standing supports is wasted if the removal of the supports is not properly organized.

19. The striking of formwork

General

A critical activity in the sequence of the formwork operations is that of striking formwork from concrete. Although the formwork designer and supervisor are usually well aware of the problems involved when the design and method statement are prepared, unfortunately, in many instances, the site supervisor and operatives fail to appreciate the importance of skilled and organized striking techniques. This often results in formwork becoming damaged and panels being rendered unfit for re-use and subsequent destruction of the surface finishes. In the worst cases structural components are badly damaged. Rather than rely on the site operatives possessing sufficient knowledge and experience to carry out formwork striking satisfactorily, an agreed safe method of working should be designed, detailed and so co-ordinated that the process is carried out:

1 At the correct time in relation to concrete maturity

2 In time to release the forms for subsequent operations

3 Without damage to the concrete structure or formwork arrangement

4 With the minimum of site labour for the actual dismantling of the forms from the concrete face.

Timing of operations

Just when formwork should be struck has been a contentious topic for many years, the arguments generally being resolved at site level on the basis that as the results of the formwork operations remain the contractor's responsibility he should take all the risks in the removal of formwork. Many attempts have been made to specify a hard and fast rule on removal time, but these usually end up with engineers and contractors finding themselves at loggerheads. The engineer is always anxious that the concrete shall have achieved sufficient strength to fulfil its structural role and withstand the effect of temperature extremes, whereas a contractor will maintain that he has invested money in the formwork and must be given the opportunity to achieve a return on investment by frequent and early re-uses of the equipment.

Recent research* has provided criteria that can be used as the basis of a programme of striking times. This has been aimed at establishing the concrete maturity which results from a combination of circumstances. Maturity at a particular time is derived by multiplying the time (in hours) from when the concrete is placed and the temperature (°C) above a given datum during that time – the result being expressed in °C/hours.

* See CIRIA Report 36.

Times are published which are necessary in the case of two types of cement and a range of cement contents for the concrete to achieve particular levels of strength, allowance being made for the effect of various sheathing materials and degrees of exposure. The temperature of concrete at the time of placing and the ambient temperatures were taken into consideration in the preparation of the figures.

Tables have been computed which set out striking times in hours after placing. Adherence to the times given should ensure that the concrete has achieved sufficient strength to sustain a given amount of dead and imposed loading and resist mechanical and frost damage.

The figures quoted have been subject to some contention on the grounds of conservatism. Research continues, however, and future work may well lay less stress on the strength required to resist mechanical damage. Whatever else is said, it must be stated that the report represents a vast improvement on the age-old arrangements where striking times were stipulated in a four-column three-line table with just a few moderating factors outlined in several brief sentences regarding temperature considerations.

Table 19.1 Periods before striking concrete using ordinary Portland cement (from CP 110 – *The structural use of concrete*)

Type of formwork	Minimum period before striking	
	Surface temperatures of concrete	
	16°C	7°C
Vertical formwork to columns, large beams	9 h	12 h
Soffit formwork to slabs	4 days	7 days
Props to slabs	11 days	14 days
Soffit formwork to beams	8 days	14 days
Props to beams	15 days	21 days

CP 110: Part 1: 1972 states that the periods noted in this table are not intended to apply where accelerated curing or slip forms are used. Where it is not practicable to ascertain the surface temperature of concrete, air temperatures may be used, although these are less precise. The Code further states that in cold weather the period should be increased according to the reduced maturity.

As an example, for soffit formwork it would be appropriate to increase the value for 7°C by half a day for each day that the concrete temperature was generally between 2 and 7°C, and a whole day for each day that the concrete temperature was below 2°C.

In the light of earlier comments on the timing of striking related to the casting of concrete, Table 19.1 is given as a guide to students. Professional designers and supervisors work on the basis of maturity and concrete strength. Whatever the striking period, the provision of suitable curing methods must immediately follow the removal of the formwork and the concrete must be protected from low or high temperatures by some means of suitable insulation.

The practice of casting cube specimens for compression testing to provide some assessment of the development of strength in the concrete is thoroughly recommended. The specimens should be stored under similar conditions as the concrete in the structure. These cubes can be used as a guide to the concrete strength. A general rule is that the formwork can be struck from structural *in situ* concrete in flexure when the cube strength is 10 N/mm^2 or twice the stress to which it will be subjected at the time of striking.

During the past few years a major contractor has been carrying out research on formwork striking times with the intention of evolving constructional economy through the rapid re-use of forms. By using this system, it is possible to reproduce in cubes the conditions which prevail in the concrete in a structure, thus achieving a more realistic specimen for compression testing.

Early removal of formwork

The cost of formwork materials are now such that major returns can be achieved for time and money spent in devising ways to obtain *earlier re-use* of the formwork, while avoiding damage to the structure by early loading of components, mechanical or frost damage.

Proprietary suppliers of formwork have made a great impact with the introduction of the 'quickstrip' or double-headed prop. For many years those involved with formwork have achieved a quickstrip effect by introducing striking fillets into traditional decking or soffit arrangements. Pads were supported by props during and after striking and also by additional props which were allowed to stand and support the concrete structure when the soffit sheathing was removed. Attempts were constantly made to release formwork early, and this sometimes meant that re-shoring had to be carried out. This re-shoring process involved the removal and subsequent replacement, often by some unskilled person, of props which were capable of applying excessive point loadings to the underside of the slab or onto sections of the structure which possessed insufficient strength to resist the loading. The process was often combined with that of crash striking where large areas of sheathing were suddenly released to fall onto the floor beneath.

Recently, because of available information, tighter specifications, improved equipment and a direct result of the increased costs of materials, formwork systems have been evolved which incorporate standing supports and regulate the position of the standing supports in such a way that the stresses in the components are less than those to which it will be subjected in its designed role.

Prop positioning and standing supports

Mention has been made elsewhere of the desirability of careful regulation of prop positions in normal formwork arrangements. This is more critical where props have to provide support for the formwork and act as standing supports to the concrete. The lacer bars or tie bars, which are parts of the proprietary system, act as an 'assurance' in terms of correct spacing and plumbing of the equipment.

In the planning or formwork design process, a plan of each floor should be prepared to ensure that the standing supports are correctly positioned on succeeding floors, and that the loads are transferred, as near as can be judged, directly from one storey to the next. The positioning of the standing supports should be discussed with the design engineer and a carefully detailed grillage, which relates to the grid lines of the construction, prepared for site.

Standing supports within a structure transmit dead loading and a proportion of the working load from floor to floor, or from beam to beam. It should be remembered that until the structural components have achieved their design strength, and then deflected under load, they do not contribute to supporting either the dead load or the live load from the structure or constructional work above. Failure to realize this will result in the lower unpropped components within the structure becoming overloaded; alternatively excessive stresses may be generated within the lower flight of standing supports.

The deflections to be expected within the structural components are due to the designed load being applied to components in which the concrete has achieved the specified characteristic strength. Strength can be calculated on the basis of maturity and where the loads being transmitted are exceptional, the form supports should be gradually eased to avoid structural damage or damage to equipment. However, it is more than likely that any easing of the props would be carried out under the supervision of an engineer depending on the prevailing site conditions.

It should be mentioned that the possibility of overloading in props arises in the context of post-tensioned concrete work where, as a structural element is being prestressed, the loading is transferred successively along the beam to the ends as the camber or hog develops in the unit. This transfer continues until the total load is transmitted to the points of reaction.

Loosening of the formwork supports may facilitate striking, although this has caused accidents where items of packing have fallen or toppled onto people working below. The sudden release of supports can damage the structural component, and all stripping operations should be so arranged as to be gradual, application of force being applied carefully where required.

Once air has been introduced at the interface between the concrete and the form the formwork can be released for removal. All supports must be so arranged that free movement is restricted to avoid any impact of the moving form on the concrete.

Panel removal and appearance

Many panels can be struck by by a gentle peeling action in which one corner is eased from the concrete and

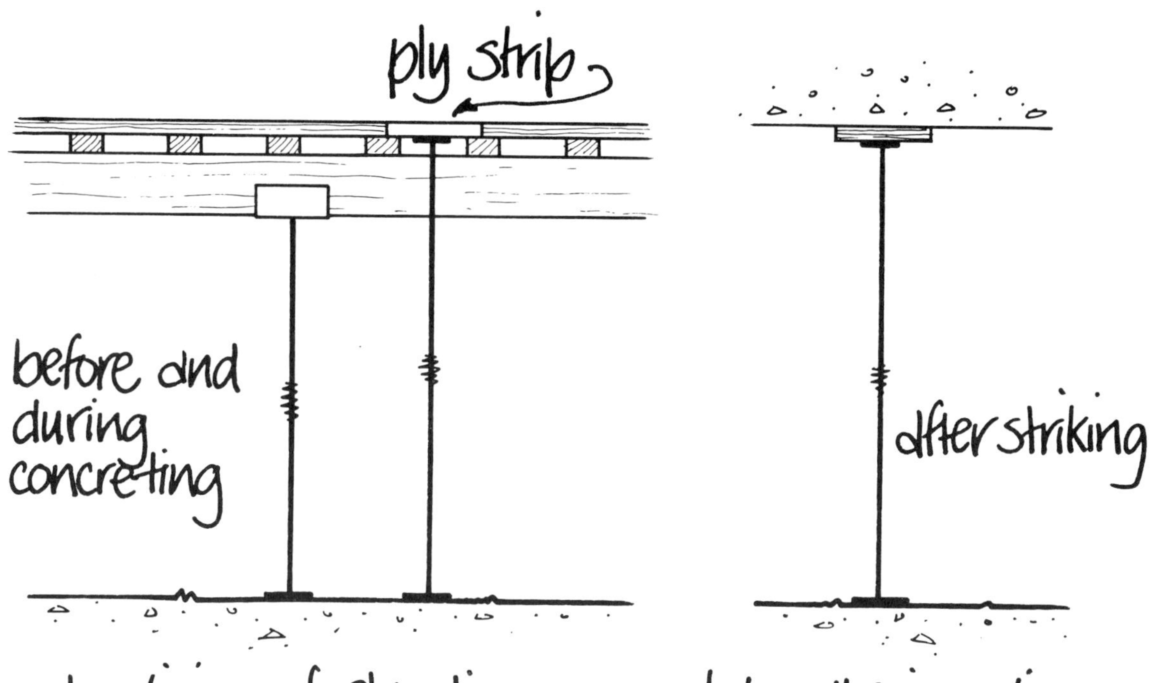

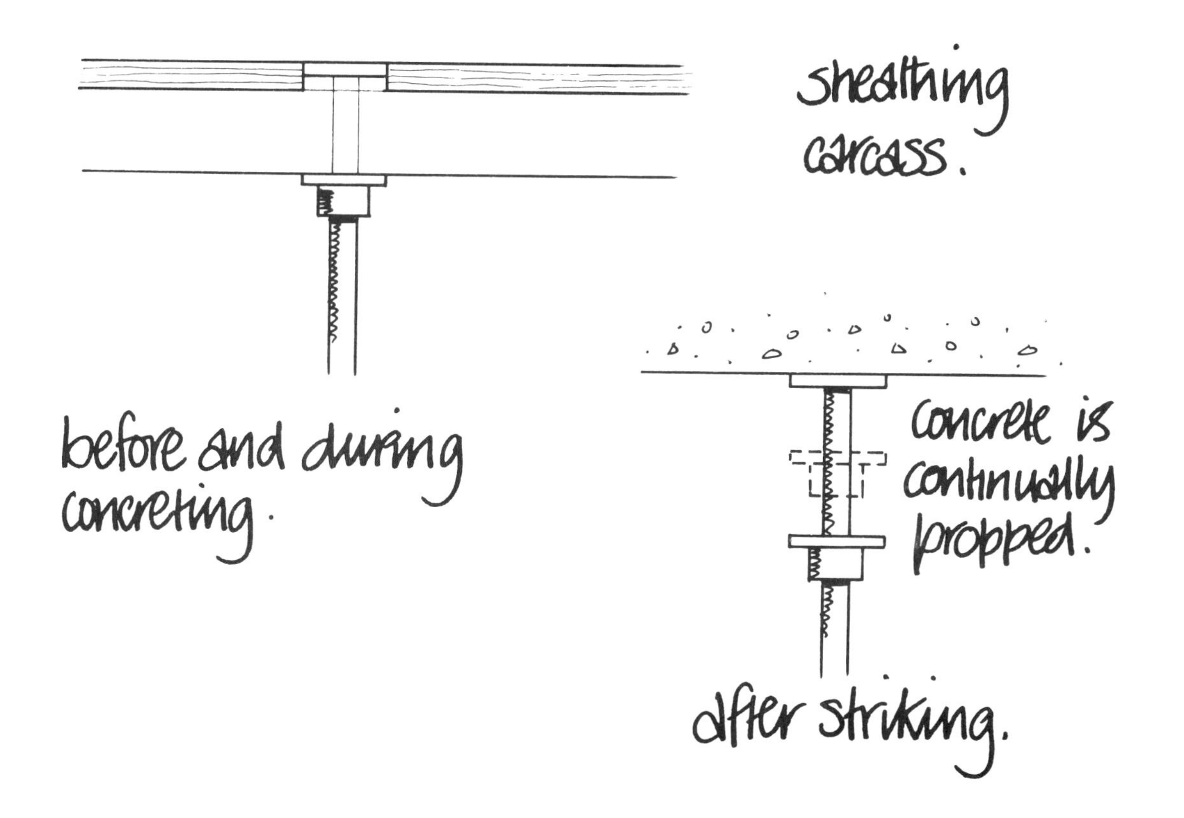

Use of secondary headed prop or quickstrip facility to free sheathing for re-use.

then continuing progressively across the face of the unit. Where necessary, thin wedges should be used to prise the sheathing away from the face. Once this initial partition is achieved panel removal becomes quite simple. With massive steel forms the self weight of the panel helps with the striking operation. Screw pads bearing on the concrete face, pneumatic rams and hydraulic jacks can be used and are in fact adopted for some of the mechanized systems of formwork described in Chapter 7. Also, compressed air can be introduced as a means of breaking the vacuum at the interface.

The effectiveness of the parting agent and the selection of an appropriate solution should be considered in formwork design. This will be based on the form face material, concrete temperatures, ambient temperature and means of curing, since these provide a more satisfactory selection than the more usual one often based on purchase price and common use.

The timing of the striking operation has a marked effect on the appearance of the final concrete surface. For consistency of finish, consistency in striking times is necessary. When appearance is not important it is generally preferable for the formwork to be removed at the earliest opportunity, taking into account resistance to mechanical damage and frost attack. This provides the opportunity for early re-use and also takes advantage of the texture of the fresh concrete at the form face, fresh concrete having something of a soapy texture. Recent techniques which have involved the use of one-piece forms for column construction and the removal of formwork without dismantling, have depended for their success on the careful timing of the operation so that the form parts from the concrete by a shearing action thus reducing adhesion and plucking. An interesting commercial technique employs a fine rubber diaphragm which is lightly stretched over the form face by vacuum during casting, after which the diaphragm is used to force the form from the face by the introduction of compressed air between the back of the diaphragm and the form.

Figure 19.2 A problem in the making! Although the formwork arrangement is simple to erect into position, care must be taken to devise a simple way of extracting from beneath the newly-cast structural slab.

Formwork handling

Formwork must be designed in such a way that it can withstand the normal rigorous handling received during the striking operation. All connection points for cranes or plant mush be securely fastened, bolted or welded to forms. The most satisfactory connections are those which are bolted and combined with plate washers behind timber members. This should always be fixed to the main carcass. For steel forms, bolted plates with washers or heavy welds are used. Lifting equipment must be covered by an appropriate certificate and registered in accordance with the *Construction (Lifting Operations) Regulations 1961*.

Where cranes are used during the actual striking operation a number of precautions need to be observed:

1. The operation should be carried out on the instructions of one person. The crane driver should receive all instructions either by radio or telephone, or by some acknowledged system of signals.
2. The crane should only be used for lifting, lowering and supporting the formwork and no attempt should be made to use the force of the machine to remove the form. (This is often attempted by operatives who rock the crane hook and the bond when attempting to free the form.) Extreme strain can be applied to a crane which virtually amounts to a test to destruction each time the crane is used to wrench the form from the concrete.
3. No formwork panel, table or item of equipment should be lifted until there is a suitable place for it to be landed, ready for installation in its next working situation.
4. All panels and batches of units should have a guide line attached which can be used to steer the load and prevent accidental impact with previously erected forms or the concrete structure itself.
5. Situations which call for the relocation of lifting tackle during the striking operation should be carefully planned to avoid the form panel or associated equipment being temporarily lodged against recently cast concrete. This particularly applies to the horizontal removal of table forms from within a concrete structure.
6. At no time must an operative or supervisor position himself over or directly adjacent to loading. The sudden loosening of the form can cause violent movement. (The source of a number of accidents during handling.)
7. A responsible person must check that all removable ties and connections into or through freshly cast concrete have been loosened or dismantled. Even a large number of nailed fixings for cast-in blocks or formers can trap the form panel and render the striking operation dangerous.
8. All loose fittings, tie rods, nails, screws and similar materials together with any grout, flash or dust should be cleared from the formwork to avoid accidents being caused by their falling during striking.

Striking method

There must be a clearly defined method during the striking operation. For simple formwork arrangements it is sufficient if the leading hand is shown the main components, and the sequence of removal. Any special stripping arrangements must be indicated and copies of relevant drawings issued for guidance. With regard to the timing of the striking arrangements, all instructions should include a checklist, prepared by a responsible engineer, to ensure that formwork is not removed too soon after casting.

Formwork arrangements, complicated either because of the scale of the structural component or the geometry, should be provided with written instructions, preferably in an illustrated manual which clearly indicates the sequence of operations, the fixings and fastenings to be released and the means of providing standing support to the structure.

With special formwork, in particular formwork which has been fabricated by a specialist supplier, one of the supervisors responsible for the use of the equipment on site should visit the supplier's works to see the assembled formwork and receive instruction in the methods of removal and handling.

One aspect of form handling which often causes problems, actually relates to the casting of the concrete. Handling systems are considered during the design process but what is overlooked is the fact that the placing of the concrete completely alters the handling situation. While it is quite easy to lower a form panel or even a complete form unit, comprising sheaths, carcassing and supports, into position ready for installation of the reinforcement and concrete, it is not so easy to remove it after casting when the fresh concrete curtails the headroom and blocks access for lifting and handling. This can be simply illustrated in the case of slab formwork where considerable attention is paid to lifting hooks and means of transferring forms from one bay to another.

Of course there are cases where, due to the lack of suitable sized openings within the structure, a decision is made to keep the formwork permanently in place. In such situations it is general to use cheap, disposable and often much used materials, to form voids beneath foundation rafts or within the heavier sections of the structure. The cost of even the cheapest disposable material may tend to promote the use of permanent formwork such as a concrete plank or slab.

Special cases

Apart from the normal care which must be employed in formwork striking, some special types of construction do demand exceptional care, to the extent that in some cases the striking arrangements may be stipulated in the specification. The sort of situation where this might arise is that in which the mass of the formwork is greater than the live loading, e.g. a shell or barrel roof where a tendency for pinching occurs, or a situation where formwork, momentarily trapped by the structure, could result in damage to the new concrete. There should be an established preferred striking sequence aimed at ensuring that a complicated structure can assume the designed deflected shape on striking.

Partial striking, which is often carried out to recover useful equipment for immediate re-use, must not be allowed to induce deformations and it should be noted that projecting cantilevers are likely to be damaged by the wrong sequence of striking. Here the act of removing the soffit sheathing while leaving the extreme end of the cantilever supported, has the same effect as incorrectly placed reinforcement in a simply supported beam. Projecting bars and cast-in connections can cause difficulty during striking and the form designer should make sure that sufficient clearance has been effected by the provision of pads and removable sheathing pads adjacent to or around such projections.

Sometimes reinforcing steel can be bent along the face and secured until the form is removed when it can be straightened. For substantial casts, large areas of formwork give problems of adhesion and although care can be applied in the selection of parting agents, hang-ups may occur which demand the application of some sort of force.

Many purpose-made forms are designed to incorporate some type of jacking arrangement or screw which operates on the concrete face by a bearing pad. Where there are indent formers or features these can be utilized as pressure pads and the jacking arrangement allowed to bear upon them. This arrangement is a logical development of designs where projecting features are left in, or on, the concrete face during striking and thus reduce fraction which would otherwise arise from return or re-entrant faces.

When all the direct mechanical attachments have been released the jacking effort can be applied. The backing members or carcass to which the jacking force is applied should be designed to sustain the massive local forces that arise. Usually the weight of the form is utilized to complete the striking, but where this is not effective, arrangements must be made to insert jack packs or toe jacks either between the carcass members of the two opposing forces, or under the soldier members, between the carcass and the base slab.

If difficulties occur in achieving a positive bearing for the jacks to provide either horizontal or vertical support, it may be possible to bolt back substantial brackets (blocks of concrete may prove useful) using inserts cast into previous lifts or bays, or to use the tie arrangements from previous formwork operations. The timing of the striking operations for deep features or indents is important. The mould coating and parting agent together with the controlled application of the required force, combine to ensure easy removal of the forms.

Another factor, particularly critical, is that of flexibility of the form face. A form may work adequately in forming concrete which is within the specification with regard to deflection (generally related to span, typical figures being $\frac{1}{270}$ or $\frac{1}{360}$ deflection–maximum related to a particular span). For coffers or deep formers this deflection which takes place in the returns or cheeks of the opening can be sufficient to negate the positive lead or draw allowed in the design. This means that the form may be wrecked during withdrawal.

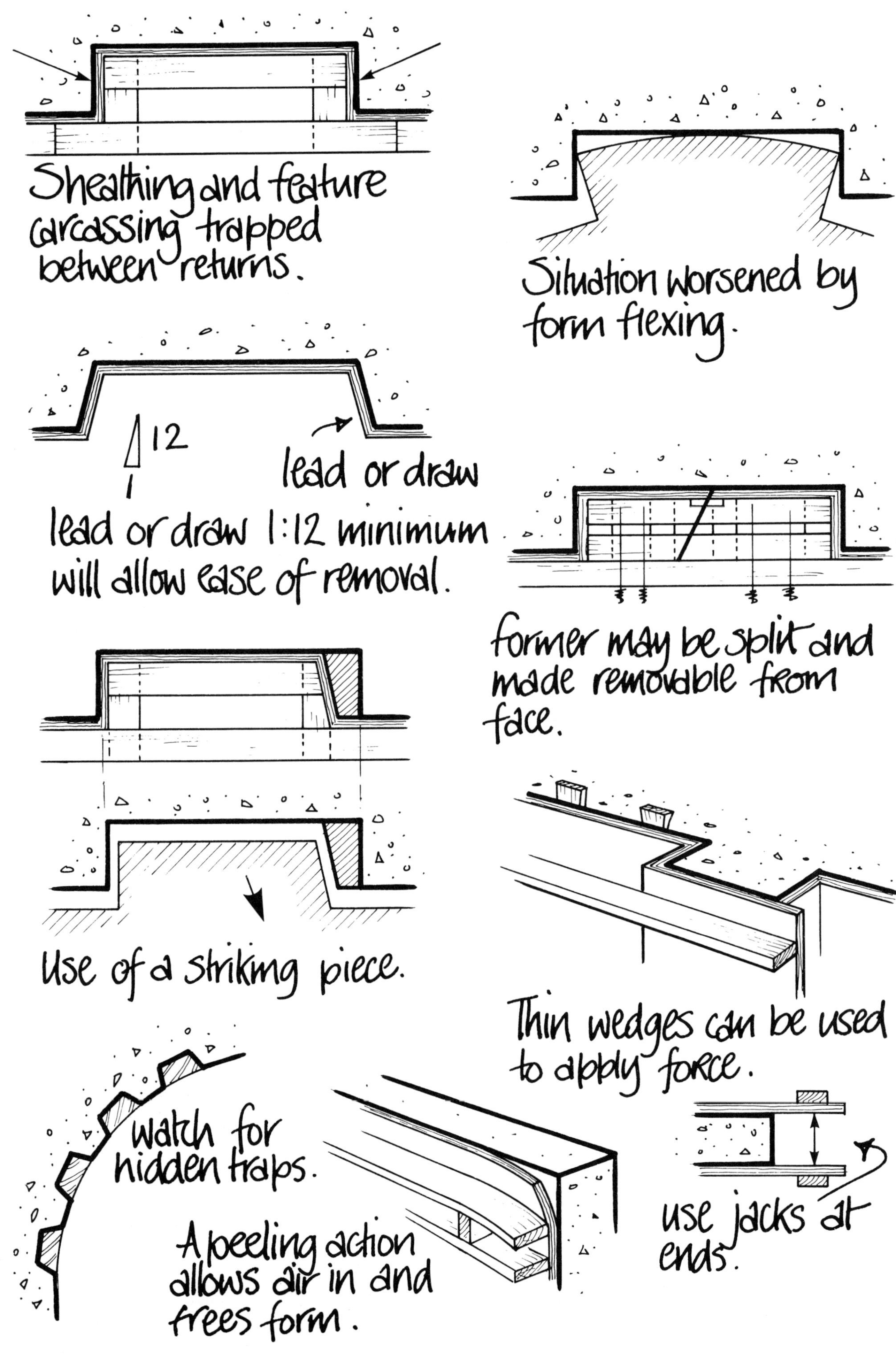

Striking of formwork – details.

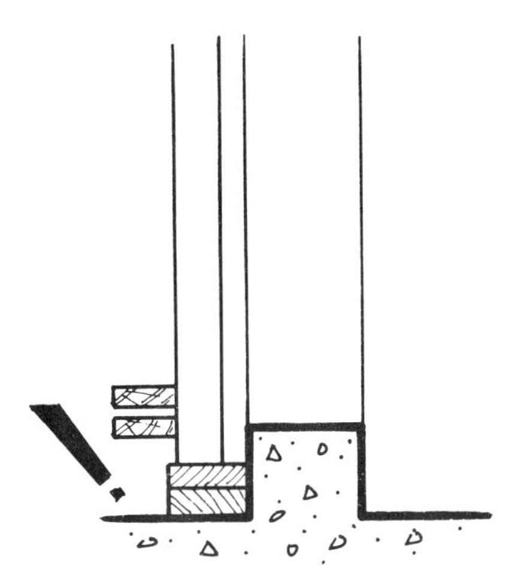

remember to remove blocks and wedges.

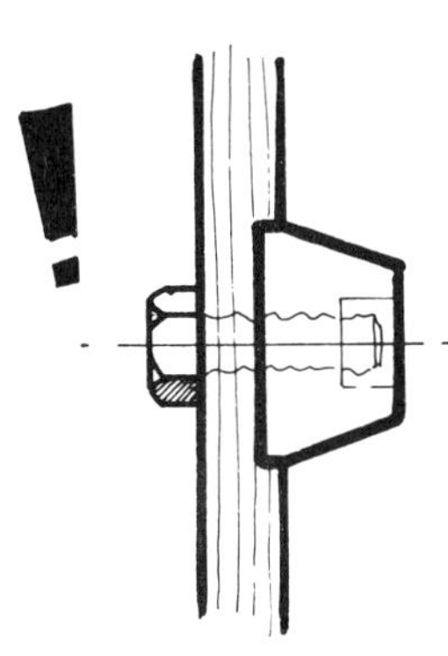

Bolts securing features must be freed.

timber or [section

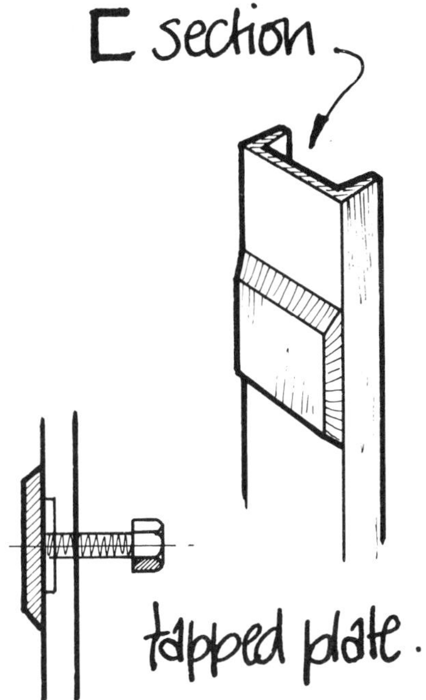

shear key formers provide jack points.

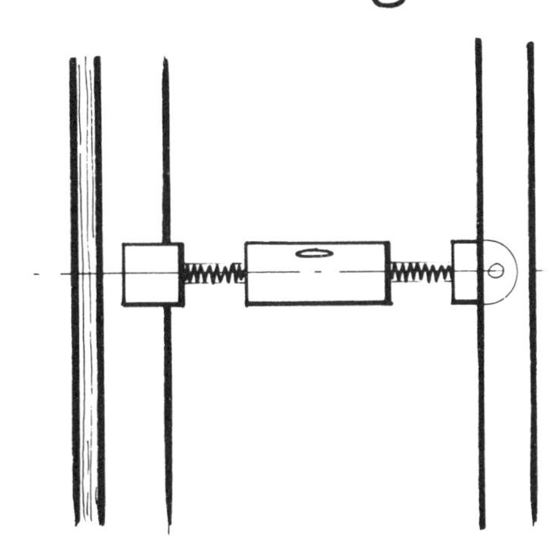

Turnbuckles from traveller control form movement.

as last resort drill concrete at strong point in form and drive form free.

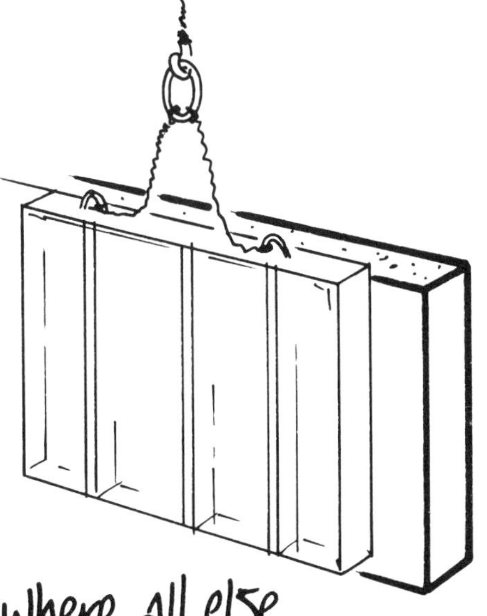

where all else fails, leave form suspended on slack chains when it will gradually free itself as air infiltrates.

Striking of formwork – details.

All fastenings to features and formers should be taken through the face nearest the concrete, so that the stresses generated during striking are adequately dispersed within the material or structure of the former. When coffered floors are being laid, an air inlet should be provided on the faces of the formers. The introduction of the air breaks the suction between the former and the face.

Projecting bolts or steel angles for mechanical engineering purposes require the use of jigs to ensure that the projecting components remain normal to the form face and allow the removal of it without entrapment. The steelfixer should be instructed to straighten reinforcing bars for direct removal of the form, although it is better that a former be inserted to allow clearance and prevent pinching.

Stripping fillets and other methods of striking formwork

The incorporation of various fillets and keys within the sheathing and first layers of carcassing ensure that any configuration of formwork can be easily removed. The striking operation must be carried out in such a way that the formwork does not become damaged so that it can be re-used with the minimum of work in setting it up again. Shocks or extremes of force should be avoided.

Various mechanical means of striking are available such as those employing rams, jacks and screws. The simpler systems usually depend on the traditional skills of the carpenter in that splay cuts and bevelled fillets avoid pinching or trapping between returns, the faces of adjacent beams, or under ribs and projections. Splay cuts in the sheathing face avoid pinching or trapping although the direction of the pitch and the amount of lead or draw must be sufficient to allow the adjacent panel components to be freed from between returns.

Where it is possible to allow a small lead or a draw at panel ends, the panel can be withdrawn without an intermediate joint. Often the structural designer does not allow any such lead, in which case should the formwork system comprise a large panel construction, to avoid panel joints and thus upset the system, a stripping fillet is inserted within the sheathing adjacent to the return face. This may be arranged either to remain in contact with the concrete following removal of the main form panel or, alternatively, the fillet may be first removed to allow the main panel to be taken away from the return.

Methods which depend on the removal of the main panel and the subsequent removal of the striking fillet are the most reliable. They are consistent with the panels being suspended or supported, in such a way, that they cannot rotate about their horizontal axis and thus bind between the remaining concrete return and the striking fillet that remains in contact with the return.

Former removal

Where small panels or feature formers are used on three or even five faces with square corners, as for a strip chase, the former will require splitting in two directions, each joint being splay cut to allow the withdrawal of the individual components. It is helpful if the feature or former is set into the sheathing face to avoid the components becoming trapped by a fin of concrete which would otherwise infiltrate behind the feature members. A very small fin is enough to trap a former and the combination of the fin with atmospheric pressure can even trap a form panel of one tonne or more and thus cause expensive damage since a heavy mechanical effort is required to remove the former.

Ideally, formers should be removed quite early on, i.e. before the normal contraction of the concrete causes any binding. Shrinkage and binding can render timber members irremovable especially where the end grain provides a key for the concrete. When formers are manufactured, end grains should be masked to avoid binding.

Careful consideration should be given to the way in which individual boards or ply sheets are lapped at joints so the formation of traps caused by grout infiltration is avoided. When formers are to remain within thin concrete components for some time, a gasket must be inserted to permit contraction of the concrete without damage to the former.

Where lead and draw is kept to say 1 in 12, as compared with the more normal lead of 1 in 4 for smooth surfaces, it is necessary to allow for some mechanical assistance in withdrawal. The former which is fixed by a jacking bolt remains in position once the main sheathing is removed, and re-insertion of the jacking bolt then forces the former out of the concrete as the bolt end impinges on the concrete. Alternatively it impinges onto a plate which presses against the concrete. Small pneumatic rams can be employed in the same way to ease the withdrawal of the former.

Formers should be quite smooth, similar to waffle formers. This prevents binding at the return edges. The flow of air is thus not impeded between the former and the concrete during striking. The technique adopted for waffle construction, e.g. blowing air through a valve helps to remove stubborn formers because the potential vacuum between the form face and concrete is easily broken. Smoothness of face, masking of end grain, early removal and the maintenance of lead draw or fillet angle are the keys to simple removal of formers and features. The exclusion of grout, which would otherwise form fins and trap the formers, is another decisive factor. A good soaking in mould oil prevents swelling of timber or entrapment problems.

Where a former is trapped, rather than demolish it or risk wrecking a form panel, it is advisable to drill from the opposite side of a wall so that the former can be punched through. If the ends of a former are accessible it is often possible to start off removal by the insertion of very thin, tapered wedges between the concrete and the former. If steel formers are trapped it is a simple matter just to weld on a lug or bolt on a clamp so that the former can be bridged and screwed away by the application of bolts inserted through the bridge.

Very stubborn formers can be removed by using a clamped or welded steel or timber bridge piece in conjunction with flat hydraulic jacks or toe jacks, which are inserted behind a bridging piece.

Checklist

What does the specification say?

Has the resident engineer agreed to the proposed striking time?

Have the cubes been crushed to indicate concrete strength?

What size are the component parts of the system?

Is there is special sequence to be observed?

Are the relevant formwork details available?

Are the standing supports indicated?

Has a special time to be observed to ensure consistency of surface finish?

Is special equipment required for handling?

Will the crane be available?

Are the slings and chains available?

Is there safe access to the forms?

Are men working below?

Is there danger to the public?

Are there any cast-in fixings likely to be displaced during striking?

Is the concrete to be protected once formwork is struck?

Are curing and insulating materials available?

Who is to be responsible for the operation?

To whom does he report in case of difficulty?

Is there a manual available?

Are the operatives experienced in the techniques?

What is to happen to the panels/materials?

Is some stacking space available?

Can all connections between the concrete and the form be released from behind the form panel?

Figure 20.1 Ad-hoc form design and erection often leaves much to be desired. Note angle and length of bracing props – and their location!

20. Formwork failure

Definition

Failure does not necessary involve dramatic breakdown or collapse of the formwork system, although the results of failure can be equally drastic in terms of loss or damage. Failure in the context of this chapter embraces all situations where, due to induced or inherent factors, formwork does not provide the intended concrete surface, degree of accuracy, number of re-uses, economic return for investment or a satisfactory reward for the efforts applied by operatives. Most formwork failures occur as a result of some faulty design, or clumsy construction and dismantling. Some are caused by inherent defects in materials which at first sight appear to be satisfactory. Every failure in performance should be examined to establish how it happened and how best any similar failure can be prevented on future work.

One of the aims of this book is to provide information for the designer, the tradesman and the student on all matters which relate to formwork, and therefore because of failures that have occured it is necessary to include a chapter on this important subject.

Specification

The most critical aspect of formwork failure is that which results in an inaccurate concrete profile or finish. Inaccuracies can be caused by difficulties experienced in setting-out, from poorly selected clauses with regard to tolerances or by the problems of handling large components that need fine adjustments.

Often when original drawings are compared with a completed structure, it is evident that very few of the dimensions are indeed critical as to the function of a component, and thus the constructor could, had he been so advised, have concentrated on these points and the effort required to achieve the required degree of accuracy would be very much reduced.

The Concrete Society/Institution of Structural Engineers *Formwork Report* provides useful guidelines for those specifying limits of accuracy and surface finishes to concrete. Set down in the Report are recommendations which cover dimensional accuracy and materials selection; these extend to the selection and use of oils and and parting agents in the achievement of required standards. The Report indicates that certain specified finishes are not only difficult to achieve but also that the cost of achieving them can be prohibitive. An example of such finishes is where concrete is specified as being smooth ex-mould quality where through ties are not permitted. Smooth concrete is particularly prone to exhibiting crazing at the surface, even the smallest inaccuracy is accentuated by the light reflected from the surface and the exclusion of through ties demands massive backing or carcassing to avoid deflection and discrepancy.

Deflection and inaccuracy

The deflection of the sheathing face is a factor which, while initially being regarded as a matter of accuracy, may well result in a failure in terms of final appearance where the resulting quilting detracts from the overall aspect of the work. Under normal working conditions it is possible for the deflections in a sheathing to vary on succeeding lifts or bays. In dealing with problems of deflection the designer should not only examine all calculations with regard to the stiffness of the sheathing but also the other factors such as rate of fill and rate of stiffening which can contribute to increased pressures on the form face. Water-resistant mould coatings may influence deflection just as much as the grain direction at the face and the conditions of moisture and heat which result from fresh concrete. Other factors which govern the deflection will be the grade and composition of the ply and the rigidity of the fastenings that locate the sheathing on the backing. For steel forms, and in particular those made of thin steel sheet, face deflections of the sheeting may exhibit quilting on the finished concrete, and where there are problems with the grading of the fine aggregate, flexing because of the vibratory effort may well cause pumping, with water runs up the face of the finished surface.

All these could be said to be failures in formwork terms, and considerable study through a succession of operations may be needed to isolate the particular mode of failure. Space restricts any attempt at a comprehensive coverage of form failures although the items noted are those most often met with in normal commercial concrete construction and in particular concrete work involving visual aspects.

Inaccuracies stem from incorrect initial setting out, mis-interpretation of drawings, difficulty of adjustment, particularly where large panels are being used, and from movements which occur during the concreting operations. Failures due to inaccuracies are particularly expensive because apart from the remedies of removal, or recasting, attempts to 'master' an inaccuracy by adjustment to finishes on subsequent installations can prove to be very costly quite apart from the problems of delay. Quite often nothing appears to be wrong when the formwork is first struck and it takes the appearance of several bays or lifts to realize that inaccurate work has been carried out. For multi-storey work problems may not become apparent until the time when services have to be installed. The precasting and concreting check can be used effectively to control inaccuracy and obviously the checking and correction of errors before any concrete is placed is the proper economic approach. Corrective measures can be very costly when worked out on a unit basis when materials, access, power and labour supervision are taken into account.

A major formwork failure results when there is a discrepancy in the profile but it can be controlled to a large extent provided that continuous members within the carcass are incorporated into the system to maintain continuity of line. Heavy walers and strongbacks, even to the extent of planned overdesign, will help to maintain accuracy. The provision of kentledge, push-pull props and braces which allow fine adjustment help to reduce an inaccuracy that is inherent in the system, while Jumbos and soldier members which bear back against previously cast faces provide vertical alignment. Attention must be given to the clips and overlaps which govern the formation of nibs or upsets to line. Whenever possible, landing rings and continuous corner members should be employed as these ensure continuity of line.

Primary and secondary deflections occur at points within the sheathing and support system where casts are interrupted, and where the casting operations are such that a form face continues over several lifts, the casting being carried out daily or independently. These positions need careful attention. Cast-in pigtails or sockets at the top of each section limit the secondary deflections and prevent nibs and runs from forming.

Local defects

A common formwork failure occurs where an opening former for a doorway, window or check-out becomes displaced during the casting process. Formers for openings are usually supported by buttons or cleats that are fixed to the sheathing surface. Either considerable impact or uneven side pressure can cause the former to rotate or become twisted. Although it is difficult to accurately position cleats or buttons on both faces of the form the problem can be overcome by providing substantial timber or steel ledging on blocks which are secured to the back of the formers so that the formers are kept square to the face.

Where large internal formers have to remain in position for two or three days after casting, a simple gasket of rubber or felt should be incorporated to accommodate shrinkage within the concrete. This avoids cracking at the corners of the opening due to the restraint exercised by the head or joint member.

Sometimes problems arise in achieving a satisfactory fit below solid cill members in opening formers. These can be overcome by leaving the cill of the opening unformed, and positioning a cill member once the concrete has reached the appropriate level at the checks of the opening.

Superplasticizers, now available as admixtures, provide flowing concrete which can easily be introduced into the least accessible parts of the form, i.e. around congested reinforcing steel or into difficult features. Care must be taken in the design process, where superplasticizers are used, to cater for the resulting increased concrete pressures.

To avoid failure induced by either excessive application of compactive effort, which may result in displaced or damaged formwork or from insufficient application, access doors or traps are required in deep lifts to prevent disruption of the face; these must be carefully installed and efficiently sealed. In the more critical geometrical forms, it may be necessary for men to enter the forms, or for traps to be provided for the insertion of vibrators and the introduction of concrete into the form. These should be carefully manufactured as part of the main form so that as they are inserted while concreting is carried out they are placed in without force and do not become distorted.

Figure 20.2 Ambient conditions, rate of fill and the use of admixtures contribute to problems of deflection.

Heavily damaged areas must be cut out and timber joints set into the face. Where bolt hole positions alter on successive lifts, grout leakage, honeycombing or staining can mar the concrete surface as a result of water movement. Steel cover plates or plastics studs can be inserted on timber forms or ply faces, to cover or seal holes neatly to prevent further failures.

For visual concrete, the formation of ribs, nibs and corbels can effect a failure. The profile of the concrete upsets the framing arrangement by distortion or displacement during the casting operations. A tendency to overfill will result in material trapping and bring about damage during striking. Provided stripping fillets are carefully inserted this kind of failure can be avoided whatever the type of construction used. The backing and support members must be so arranged as to give continuity without distortion of the form.

Stair forms which are used to give a finished tread often promote failure where, due to surge of the concrete, an unintended indent is formed adjacent to the face of the riser. Where a visually acceptable finish is required a substantial 'round' between the riser and the tread should be incorporated. Concrete of carefully controlled slump should be used and the removal of the surge can be delayed until just before the final trowelling stage of stair finishing.

As distinct from unwanted recesses caused by panel joints at intersecting surfaces, a defect can arise due to the formation of unwanted excrescences, such as fins or nibs which tend to trap panels. They are generally caused either by differential movement between the panels or by localized damage to the form faces. End grain which is in contact with concrete tends to tear during striking and during subsequent operations the defective areas can become filled with paste or particles of aggregate. As these materials harden and the striking process is again carried out, larger areas of sheathing become damaged.

Where insufficient attention has been given to lead and draw, a failure of the form face which is induced by trapping during striking, allows the form face to key to the subsequent concrete so that the damage becomes progressive. Fibre-reinforced resin pastes can be used to fill formwork defects and early application to the damaged parts as a sound fixing between backing members and sheathing is necessary to ensure that the faces applied in striking do not tear the form to pieces.

Feature formers and the way in which they are attached to the main form face can damage formwork and limit the number of uses. Even after many years of formwork construction there are still many cases in which the basic principles of construction are disregarded, for example timber fillets are often planted onto timber faces, and steel features are spot or plug welded onto steel surfaces. Subsequent leakage of grout then causes trapping, progressive distortion or tearing of the feature and even breakdown. As the series of casting operations continues this damage increases so that a succession of substandard lifts, bays or areas may be formed, before the defect is identified and rectified.

The wrong selection of mould coating as a parting agent or incorrect application will fail to produce the appropriate standard of surface finish or the intended number of re-uses. Over-application of parting agents, may cause staining and scabbing. Boards of different degrees of absorbency or untreated boards used in form sheathing replacement may cause visual defects. All too often a manufacturer's instruction sheet or manual is consulted only after a failure has occured rather than before preparation of the form.

Joints and stopends

One of the main causes of formwork failure is leakage. Cement paste that is subjected to pressure can displace fairly massive forms and can build up during succeeding operations and thus mar and spoil the line and face of the concrete. Any patch on a form which exhibits water staining or local darkening of the concrete indicates that there is some sort of leakage. Gaskets, tongue and groove techniques and the insertion of splines should prevent the leakage. The build-up of paste at joints can cause problems because of the traditional methods used to clean forms. Where cleaning has not been carried out paste build-up can bring about considerable inaccuracy. Evidence of leakage is often seen adjacent to stopends, particularly where continuity reinforcement projects through openings in these stopends. The joints formed at the stopend are seldom detailed and indeed the joints between the stopend and the sheathing face are seldom grout-tight, so that leakage and water staining often occur. Stopend line is usually maintained by cleats or bearers on the form face and for the best visual results this line needs to be straight, and incorporate a joint check or rule to present an accurate line throughout the course of the succeeding lifts. Stopends can be easily displaced inwards, often only being restrained in this direction by the concrete itself and such displacement can cause an expensive failure which will be almost impossible to repair.

Steel formwork and systems

Failure to use system equipment correctly will obviously produce defective work, damaged equipment or even an accident.

With regard to formwork systems the main supporting system must be set up correctly as a suitably prepared base, i.e. either a previously concreted base or a series of plates or sleepers. Props and tubes should fit adequately and be suitably braced to prevent serious accidents. This is particularly important because even props which are founded on what was apparently a carefully prepared blinding have been known to punch through the blinding and thus cause a failure.

Failure of steel forms, even purpose madc formwork may be caused by fatigue induced by handling and striking techniques. In the mountings of vibrators, the damaged bracket or clamp does not allow transfer of compactive effort so that the system breaks down. Ideally, external vibrators should be attached to substantial brackets which are attached to major members of the form carcass.

The wrong selection of a tie arrangement, incorrect spacing of ties or misalignment of ties contribute to a potential failure. Tie spacing and alignment should be key items in the checklist used by those who are responsible for pre-concreting and concreting operations.

Scaffold arrangements which have been incorrectly erected or badly secured can be very dangerous and where these are combined with formwork systems, the problems begin to compound up, because it may be that support is taken from ledgers or scaffold boards, and this spells disaster.

Arrangements between the support members and the carcass of the formwork at the interface are critical for the successful transference of loading into the supports and thence to the bracing points or bearing pads. Crushing of timber due to insufficient bearing areas and rotation of the fittings due to eccentricity of loading are often the causes of failure. By far the greatest number of failures are a direct result of poor workmanship or temporary 'ad hoc' arrangements.

The only positive way to avoid a failure is to accord the appropriate importance to each facet of design, construction and erection and to consider all aspects of striking and handling. A company will always need to utilize the experience and knowledge available either from within or from the specialist supplier. Proper use of the skills of tradesmen and the correct application of formwork technology can overcome many of the problems outlined in this chapter and help with the solution of additional problems as and when they arise.

One of the main factors which helps to govern the avoidance of failures is that in which a critical examination is made of the facts pertaining to the history of some particular failure. Many factors are responsible for a failure and it is very rare to find a situation based on one particular cause.

Figure 21.1 Safety is jeopardized where large areas are only temporarily supported.

21. Formwork safety

General

Inevitably in any kind of design, safety is of prime importance. Formwork arrangements must provide for safe access, and those who have to use equipment must be fully instructed in the safe procedures to be adopted. Work which is carried out in an unplanned and haphazard way during construction spells danger. Even work which has been carefully planned can be made hazardous by 'ad hoc' alterations.

Even the best planned and safest systems can fail because of some breakdown in communication between the designer and the user. For example this could occur where sophisticated equipment is badly used because of a lack of appreciation of the intended method of use. In any situation where equipment is designed in one place for use elsewhere, or where some standard component is to be used in various situations misunderstandings can arise.

Formwork has long been considered a crude area of construction especially where temporary and sometimes makeshift arrangements have been employed. There has always been discussion in contracting circles as to whether formwork comes within the province of the skilled tradesman or that of the semi-skilled operative. While it continues to be viewed in such a way accidents are inevitable.

Safety hazards are rarely highlighted in the press unless the resulting accidents have been particularly dramatic or where there is loss of life. When fatalities are involved these are reported not only in the technical press but also in the national papers.

It is intended here to examine the causes of accidents and in particular those which occur in formwork activities.

Accidents can be categorized as:

Those caused by bad design

These result from a poor appreciation of form loads, stresses, pressures and forces which are set up within the container. Accidents which stem from unsatisfactory design include those where for example a poor assessment has been made of the pressures which result from a particular method of concreting, where sectional sizes have been wrongly calculated or where insufficient bearing area has been allowed between adjacent components that transfer force.

Those caused by unsound construction and mis-use of materials

These are particularly troublesome because they often occur even after a lot of effort has been expended on careful preparation of plan and method. They happen where, inadvertantly or through lack of training, substandard methods have been adopted or where materials have been badly assembled, or where critical components become displaced so that adjacent parts of an arrangement have become overstressed.

Accidents that are caused by poor assembly are those in which adjustable props have been badly located or erected out of plumb; where slender joints rotate and allow massive sheathing deflections and other situations which only become apparent when the formwork is subjected to loading.

Those caused by factors normally considered beyond the control of the formwork designer or supervisor

It could be argued that the skilled formwork designer or supervisor should be able to forsee most of the hazards likely to cause an accident. Obviously with so many activities going on within the construction operation, countless hazards exist or become introduced into the process all the time that the work proceeds. Hazards which are created when the formwork is in use, or after it has been used, are either those in which props have been casually removed to provide access to adjacent works, or where additional excessive loads have been applied to the newly cast concrete.

Those caused by poor workmanship or bad communications and instruction

On first inspection, many of the accidents which come within this category appear to result from one of the previously mentioned groups of hazards. However, there are certain kinds of accident which are related to people or the action carried out by them. As with most situations where the human element is involved, this often boils down to a problem of communication. Unfortunately many accidents are the result of poor labour performance or motivation, and as much of the success of formwork relies on dexterity this is a difficult area to define and control. Accidents coming within this category include those that while everyone concerned know what should be done, either because of trade practice or some established method something goes wrong, because an attempt has been made to achieve faster rates or some economies.

It should be remembered that communication is a two-way process. An accident can occur as a result of breakdown in communication in the designer – supervisor operative chain. Equally one can occur due to poor feedback of information between the operative supervisor and designer.

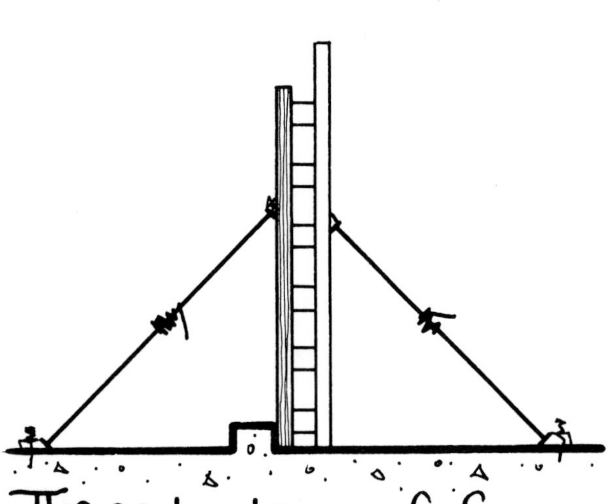

The early stages of form erection are hazardous — wind pressure — dis-lodgement

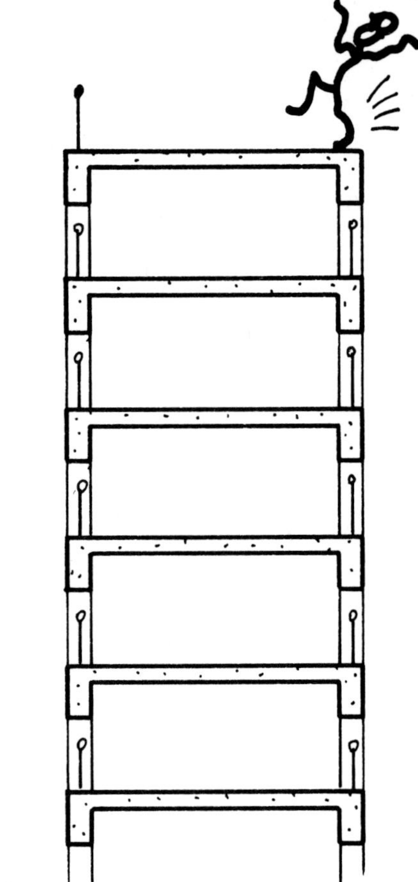

Men who have work on repetitive stages who fall — familiarity causes carelessness.

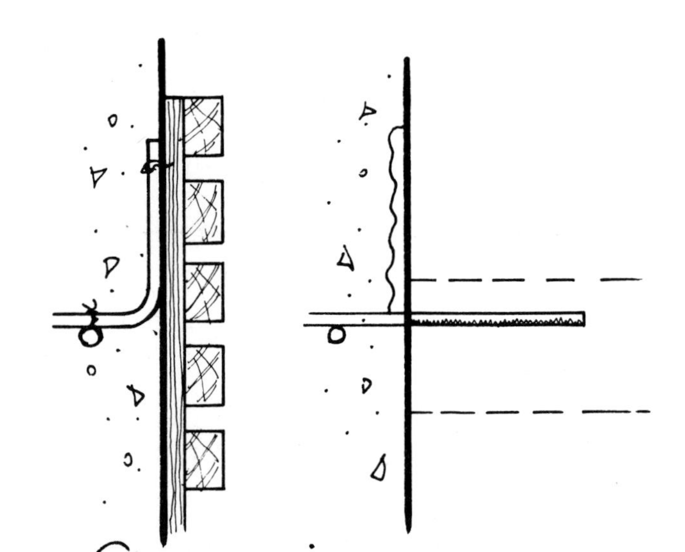

Steel which has been diverted for construction purposes must be re-located and checked.

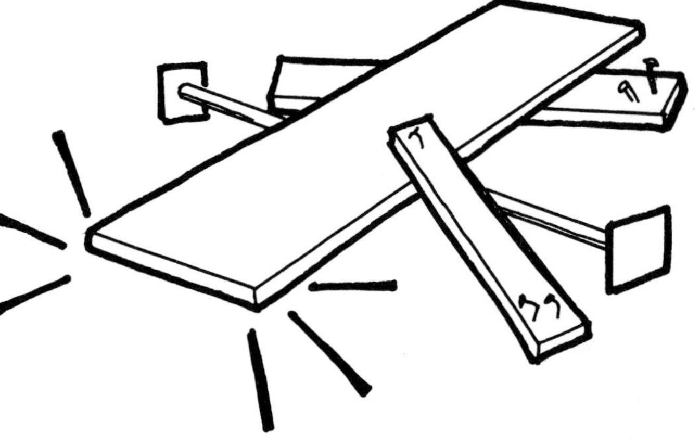

Poor housekeeping kills.

Some causes of accidents.

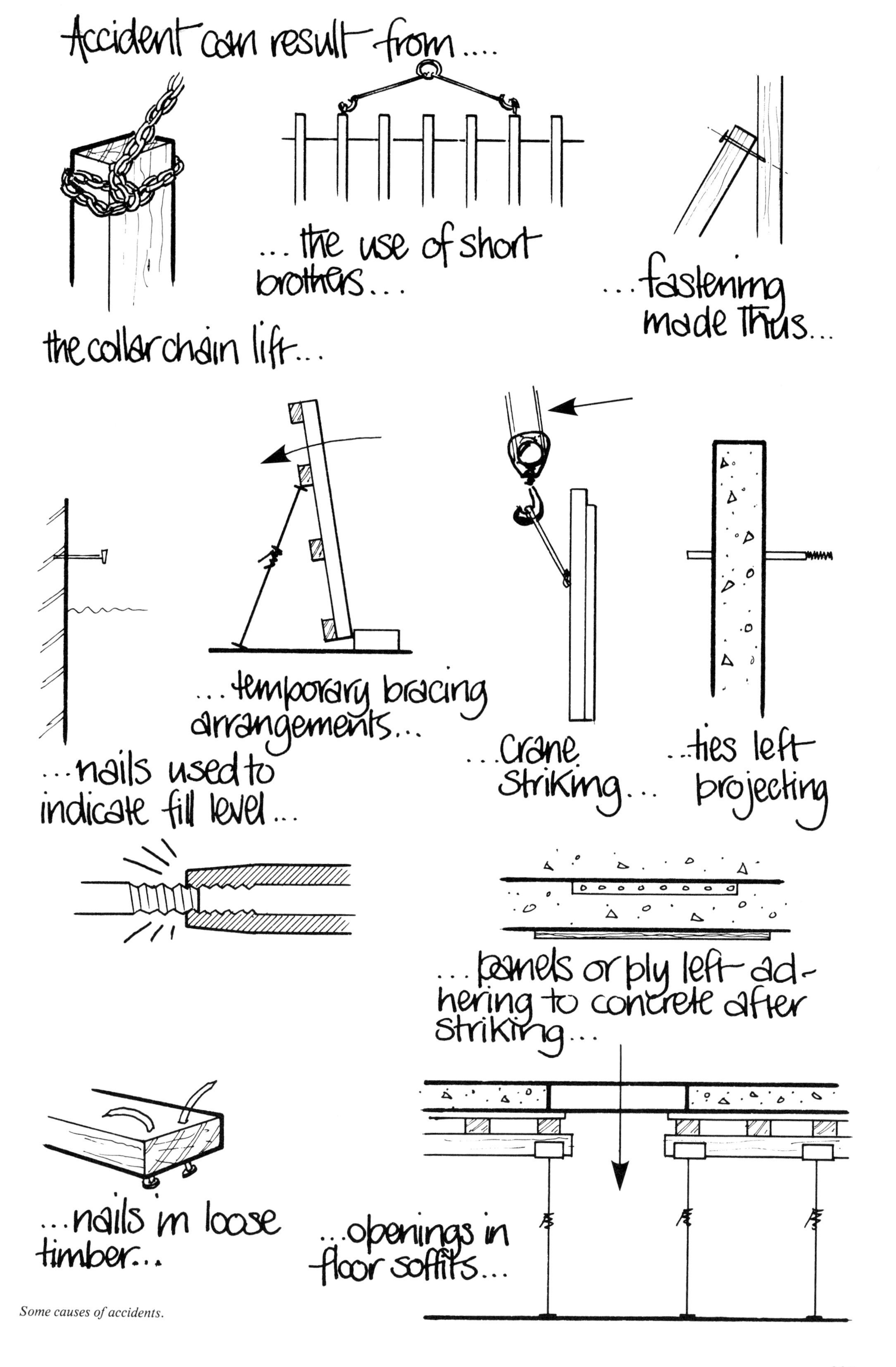

Some causes of accidents.

Responsibility

The designer and supervisor are required to know and understand the statutory requirements which relate to safety and these are clearly set down in the relevant Statutory Instruments of the Construction Regulations. Those who are concerned with lifting equipment and plant and scaffolding are bound to maintain the registers demanded by the Construction Regulations. It is very difficult for the supervisor (and sometimes the designer) to be completely conversant with every point that the Regulations contain with regard to their responsibilities to those with whom they work, although in law this is never accepted as an excuse for default. It is worthwhile to examine the situation in the light of the common law requirements* and the rights of the individual. These can be set out in their simplest terms as follows:

1 A safe place of work

2 A safe method of work

3 Safe fellow workers.

The employer has the right of:

1 Personal service

2 Loyal service

3 Careful and obedient service.

These terms are perhaps rather crude, but it must be remembered that these are historic terms used traditionally in defining laws of employment. The supervisor and designer who remembers them however, when dealing with all matters of method and instruction possesses a sound basis on which to establish the safety of those on site.

Special problems exist in the construction industry. Formwork and falsework operations are often carried out in exposed and detached locations which present particular problems to those who take responsibility. The temporary nature of the work tends towards scant construction, and economy often is sought by reducing the weight of sections, the quantity of supporting members, or by use of secondhand materials. Although research establishes criteria for design, and much is being carried out to devise safe limits on loads and stresses, site conditions can for example rapidly alter pressures, thrusts and loading.

Changes in sequence of construction can quite easily invalidate the original design. Peculiarities of weather can considerably affect methods and materials quite apart from the influence they may have on the performance of those employed on the erection and striking of formwork. Because construction technology continues to advance, there are many more techniques which can present hazards or introduce the possibility of accidents. The greater use of electrical power tools and pneumatic and hydraulic equipment add to the hazards that exist within the formwork field.

Incentive schemes and bonus arrangements have introduced further hazards, because the tempo of operations has increased. While supervisors are concerned with chasing progress they are often hard put to cope with important matters such as the provision of access for scaffolds, or the maintenance of safety barriers.

By the very nature of a construction site with its constantly changing excavations, scaffold arrangements, and other variables the possibility of accident is ever present. Unfortunately those who are most engrossed in their task are likely to be the most vulnerable to accidents. These factors, combined with noise, constant movements and problems of communication can render the formwork operations as being extremely hazardous.

Endless examples can be quoted of accidents that have occurred under the various categories just mentioned and anyone who has worked on a construction site can no doubt recall some particular situation which has resulted in an accident. Causes and hazards are legion, and while some examples are now given the reader can no doubt provide many of this own illustrations. It is essential to establish the cost of accidents and then to identify ways in which those involved can contribute to reducing the occurrence of accident.

While it is possible to determine the cost of accidents to industry in terms of loss of life, injury and lost output, what cannot be calculated so easily is lost opportunity nor indeed the physical suffering caused to those who have been involved and their relatives or dependents. While the direct costs of delays can be established, it is impossible to arrive at the true cost in terms of associated delays that affect other trades.

The cause of an accident can be defined as 'a combination of events creating a dangerous condition subsequently released by some unsafe action'. In the process of design manufacture and construction, it is necessary to be constantly alert to causes of accidents and to design or build in counter measures. Nothing can be absolutely foolproof, but it is the duty of those responsible to assess reasonable margins of safety and incorporate fail-safe arrangements in order to guard against the various possible causes of accident.

Returning to the categories of accident identified earlier in the chapter, the following are courses open to the designer, supervisor or person responsible for the formwork operations.

Accidents caused by bad design

Ensure that design criteria are valid by cross-checking with Codes and Standards, research reports and feedback from work previously carried out. Ensure especially that theoretical approach matches practical information, and that experience confirms the probability of successful operations.

All calculations and details should be checked and where necessary specialists consulted, especially when the formwork method depends on mechanical equipment. For example, the plastics technologist will need to be consulted where special injected moulded or heat-formed moulds are used. Because formwork materials are often used in conditions for which they were not intended, and thus

* The Legal Aspects of Industry and Commerce, W. F. Frank, Published by George G. Harrup & Co. Ltd.

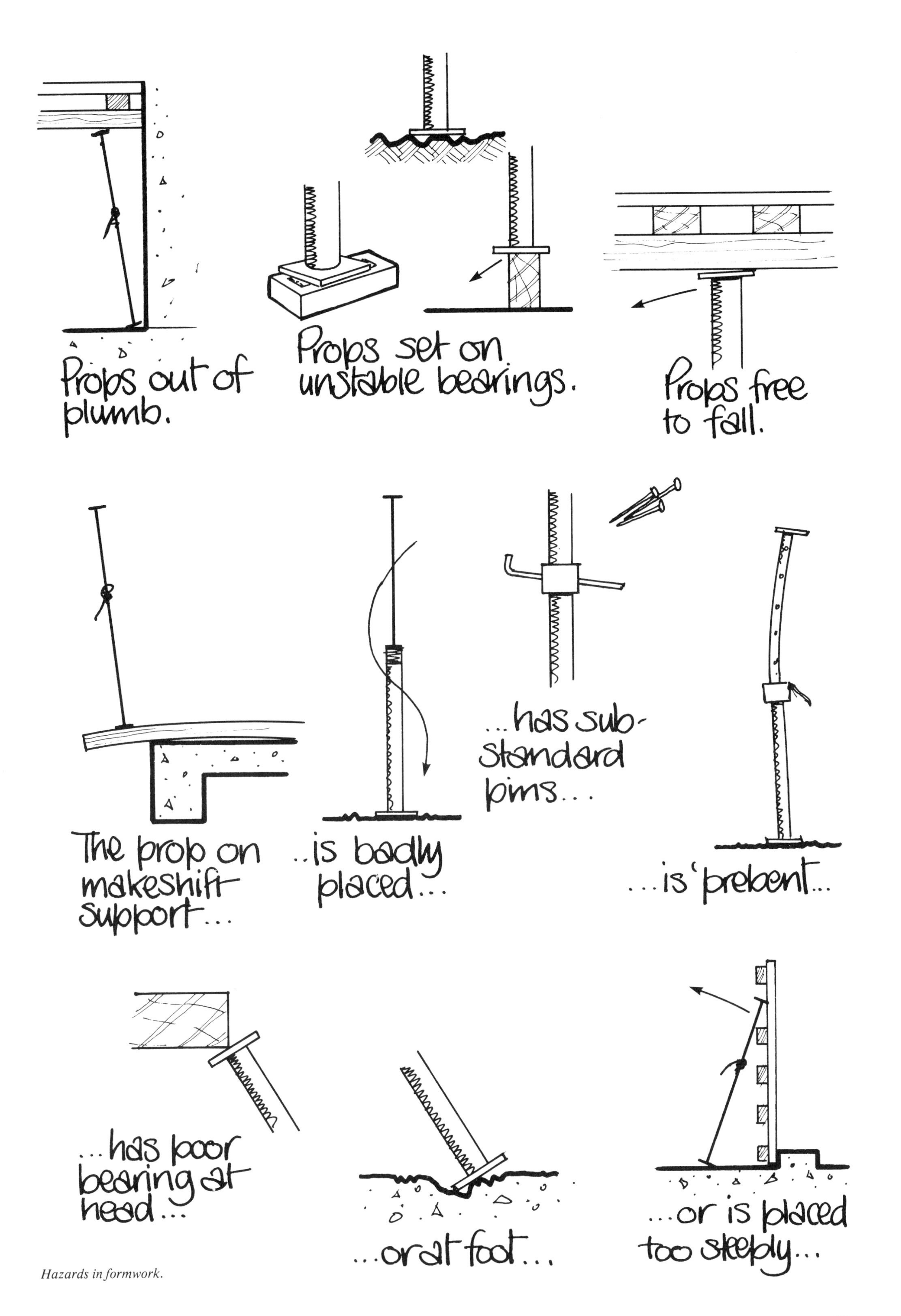

Hazards in formwork.

become inefficient, care must be taken to prevent their misuse. All designs should be carefully drawn up and detailed with respect to steel reinforcement and concrete profile. Form panel positions should be related to other site features. An appreciation of the scale of the various components and their relation to the abilities of men or the capability of the equipment should be made. Always seek a feedback of information on performance, and maintain close contact with the work and those involved in its execution. Always record variations and improvements and maintain records for future use in similar types of work.

Always maintain constant liaison with other designers and those who are concerned with the production and use of formwork. Notes on accidents and hazards should be compared so that all relevant information is passed on throughout the production line. Remember that details and drawings must be readily interpreted by ordinary operatives on site and have to be easily understood by them.

Accidents caused by faulty construction and misuse of materials

Carefully study the drawings and manuals which describe any of the proposed systems. Any variance between the proposed scheme and normal practice should be questioned even though it is probably intentional. However an error may have been made and no harm will come in being cautious.

As soon as materials and equipment are brought onto site they should be checked to see that they meet the specification or formwork details.

Avoid incomplete temporary work, for example where further braces are still to be added someone may make the mistake of loading the work before it is completed. Damaged equipment should never be used, not even as a temporary measure.

Any situation in which it is necessary to apply force during assembly should be checked. Forced fitting or tightening can result in a prestressing state which, when combined with concrete loading and pressures can bring about overloading and failure. Always ensure that bearings are clean and square and of adequate area as given by the drawings. Avoid applying force during the stripping process and do not use chains which could become strained and thus introduce hazards for subsequent operations.

Report any exceptional deflections or movements to a responsible engineer and, if in doubt, stop the erection or concreting process.

Accidents caused by factors normally considered beyond the control of the formwork designer

Accidents which come within this category are the most difficult to prevent and are often caused by all sorts of hazards. However, it should be possible for a skilled designer to forsee potential hazards and deal with them by an extension of his design. For example, a designer should include in his form design some system of handling or strutting which becomes an integral feature rather than leave it to site staff who would normally provide the arrangement.

Brief notes or warnings on the loadbearing requirements of a cast-in fixing would help to prevent some accidents caused by overloading during the early stages of construction. In essence, the designer always needs to remember what requirements are necessary in succeeding operations and allied trades. The basic rules for avoiding accidents under this area are:

Study the activities of all the other trades involved in construction and attempt to visualize their influences on all stages of the formwork construction.

Pay special attention to plant intensive activities and attempt to visualize the movements of all heavy plant and equipment.

Study the contract programme to obtain a picture of the construction process to identify situations where other construction work is to be carried out in close proximity to formwork erection.

Watch out particularly for situations where the demands of other activities, for access reasons, may prompt the removal or adjustment of formwork.

Ensure that the supervisors and managers who handle allied processes are kept informed about formwork arrangements and sequence of operations.

Avoid exposing critical support members which could become displaced by traffic.

Liaise with planning and project engineers to ensure safe continuity of the form processes, special care being taken with work that is being undertaken above or adjacent to other gangs.

Accidents caused by poor labour or bad communications

The simple answer here is to avoid employing unskilled labour. In British law it is required that the employer should provide an operative with suitably skilled fellow operatives, and this must be the constant aim of managers and supervisors. Trade testing of applicants helps in the selection process.

As soon as a new operative has been installed in a gang, it must be the supervisor's prime task to ensure that he is capable of carrying out the task for which he has been engaged and that he can do so in such a way that he does not become a danger to himself or to anybody else on site. Of course many formwork operations are such that if carried out unsatisfactorily, damage to the public or to property may ensue. New men should never work alone, they should receive instruction and be put to work in such a way that they learn the accepted safe methods of working employed on the site. They must be able to understand and receive instructions, particularly those which relate to safety, and it is essential that language barriers are overcome. Operatives should work under the control of somebody who can communicate directly a

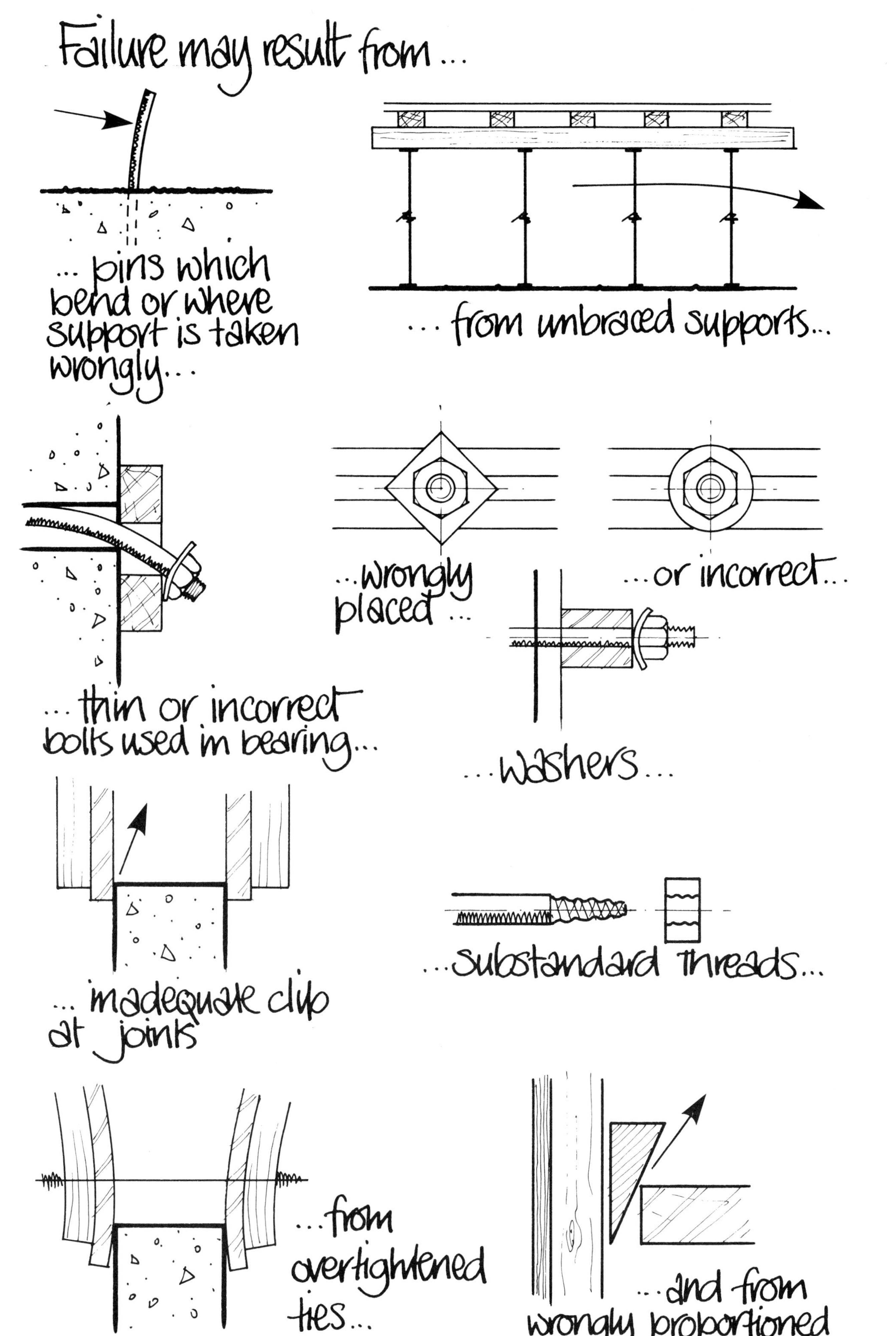

Formwork failure.

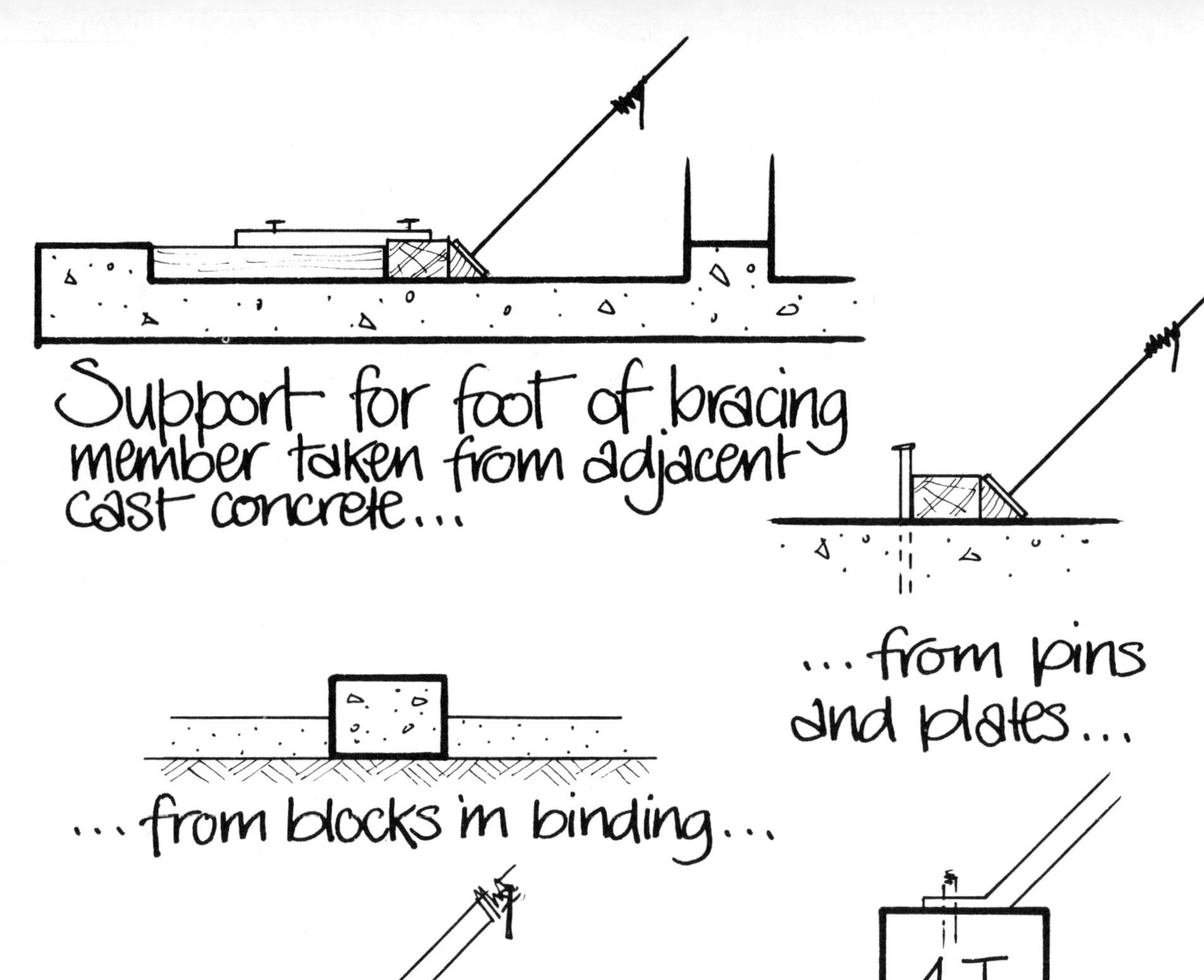

Support for foot of bracing member taken from adjacent cast concrete...
... from pins and plates...
... from blocks in binding...

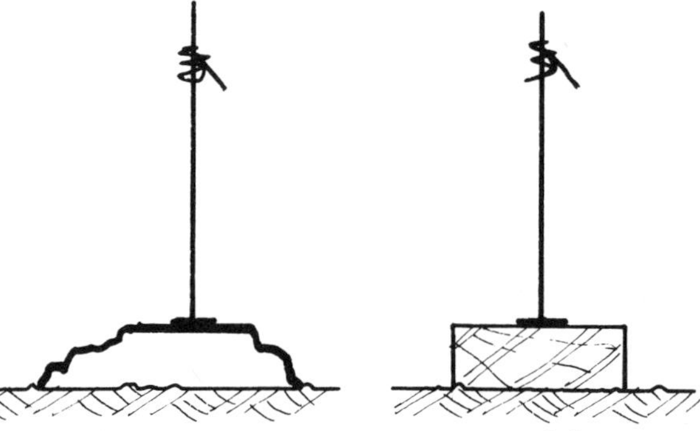

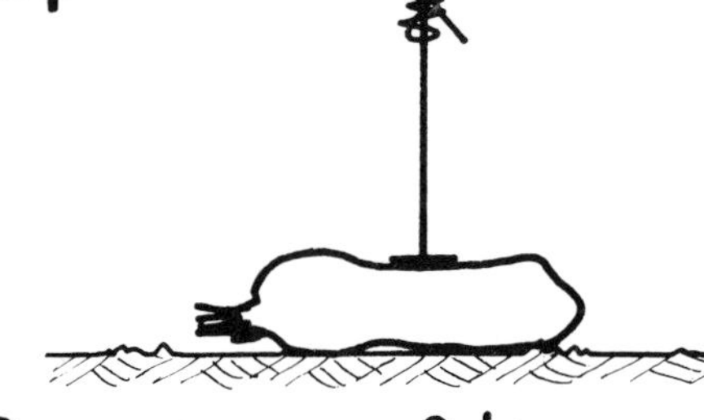

Formwork failure.

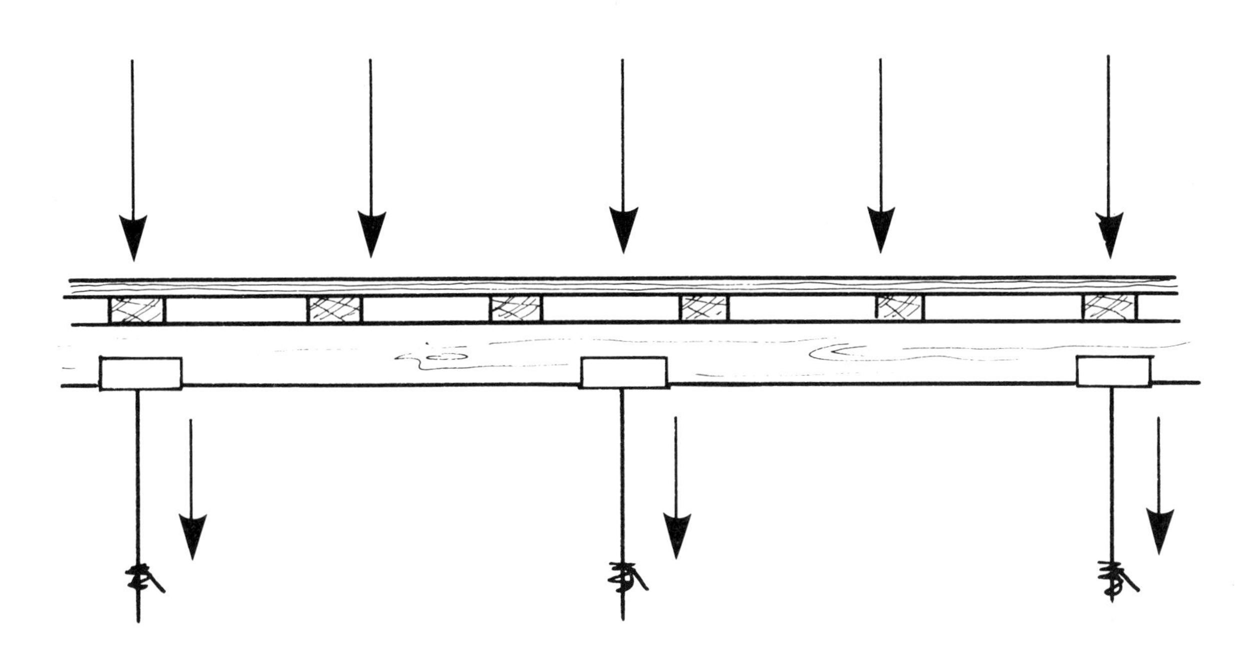
Uniform application of load.

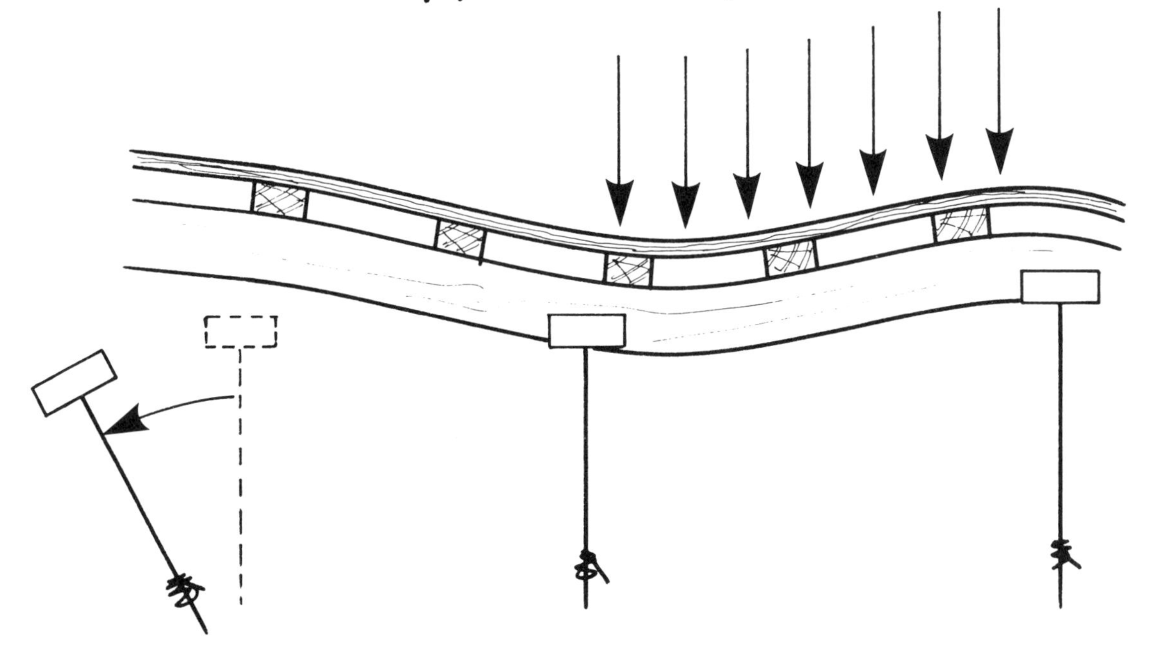
Excessive local loading causing deflection and loosening of props.

Formwork failure.

language that is fully understood. The supervising staff should never accept a man's statement that he can carry out a particular operation until reliable proof or reference is forthcoming.

All site activities should be executed by properly organized gangs or teams through a clearly defined system of control, each man knowing to whom he is responsible and what task he has been allocated.

The designer's responsibility

The designer who takes a keen interest in all the construction activities and who is involved in all phases of the work should be able to design with a view to the safety requirements. A designer who detaches himself from such activities will not appreciate the potential dangers to safety. He must be constantly aware of the problems at all levels and must include in his details and calculations every reasonable safeguard against possible accident. He must seek all the facts that are relevant to the overall construction process and once in possession of these work positively to counter any accident or hazard.

The supervisor's responsibility

In order to successfully supervise work, it is necessary to achieve a sound understanding of the technology and techniques to be employed, and to fully appreciate the designer's intentions and to formulate instructions using this information. These would then be the result of applied thought and planning and would reflect the supervisor's experience from previous work.

The supervisor's task, that of co-ordinating human effort, which is the most fragile factor in production, must include every consideration of the safety for those for whom he is responsible. Even when faced with production demands and economic considerations he must create, install and maintain a safe method of working for those under his control. There is another responsibility – that of maintaining a flow of information such that those who are responsible for the design and serviceability of formwork and form equipment are kept informed about the performance and any points which might affect the safety of those engaged.

Known causes of accident

One of the most common causes of accidents occurs where people fall or where materials are dropped. In many cases the safeguards are obvious, although it should be mentioned that there is always a need for adequate access scaffold and safety harnesses, and that care is required in securing ancillary equipment to the main forms.

Nails or fixings that project from forms are constant hazards as is reinforcement that protrudes from some concrete component.

Crash striking can cause problems both to those on site and to the public. Quite apart from the obvious damage to people and equipment, there is the added danger of structural damage inflicted by the sudden impact of falling equipment.

High winds are often responsible for accidents, especially where large areas of formwork are set up and only temporarily strutted and propped with a few props or raking supports. A cross-wall arranged thus can easily overturn and be carried some distance leaving havoc in its trail. A panel carried by a crane in a high wind can swing about and strike against people or property and cause extensive damage.

Partially struck formwork can prove to be quite a hazard. Panels or individual sheets of ply which have become lodged in the concrete once all the supports and braces have been removed, can be very dangerous because pieces of formwork can easily fall without any warning.

Rapid rates of fill can lead to dangerous overloading of the forms, and sometimes formwork which has been designed to withstand only light loading can effect a bad accident because it may have been overloaded by materials being left on it, which originally had not been there.

Although leakage of concrete through a form can indicate a weakness and also serve as an indication that something is wrong, no such warning exists where a structural failure occurs when sudden buckling and collapse can possibly kill somebody.

Poor footings and foundations and inadequate bracing and support can set off a chain of accidents which could be avoided had a preconcreting check been made.

All lifting and handling equipment such as slings, chains and hooks must be subject to statutory inspection, especially where the equipment involved has been in use for some time.

Safety officers should always ensure that helmets, boots, goggles and protective clothing – often issued free – are, in actual fact, worn, even if operatives object to wearing them.

Electrical leads and compressor air lines often cause accidents by careless use through lack of warning where they trail across the ground or get dragged around equipment. All regulations relating to the use of power tools must be clearly posted and operatives attention directed towards them.

Close liaison with HM Factory Inspectorate is an essential part of the safety aspect of any part of constructions. Factor Inspectors are highly experienced and possess a wealth of information; they are usually only too happy to pass this on. However, they do not approve techniques or suggest systems, their function is to check that sites meet the statutory requirements. They give advice on the interpretation of these requirements and will always welcome enquiries through local branches or even through headquarters, where these are perhaps more specialized.

The safety supervisor or officer is the one responsible for seeing that the safety requirements are being followed. It is therefore important that both design and supervisor have a close working arrangement with him. He will need to know about all special arrangements which concern a formwork system and must have a working knowledge of the way in which drawings and details communicate information to those on site. In other words he must be kept fully informed especially where unusual constructional techniques are being applied.

Figure 22.1 Out of plumb slipforming experiment for the splayed leg of an oil platform.

22. Slipforming

General

The continuous casting of concrete walls using a system of formwork which is jacked, or otherwise moved during the placing and compacting processes while the concrete is 'green', has been carried out for about 50 years. In the late 1920s a number of frame concrete structures were cast by this method which has since been adopted for towers and cores, chimney shafts, pylons and the legs of oil rig platforms. Slipform-type operations which tend towards extrusion have been used for casting walls, safety barriers, kerbs and horizontal structural components. Generally, the techniques have been used where a considerable amount of concrete has had to be formed of similar cross-section. However, slipforming has been employed successfully and economically in situations which have required discontinuity of section.

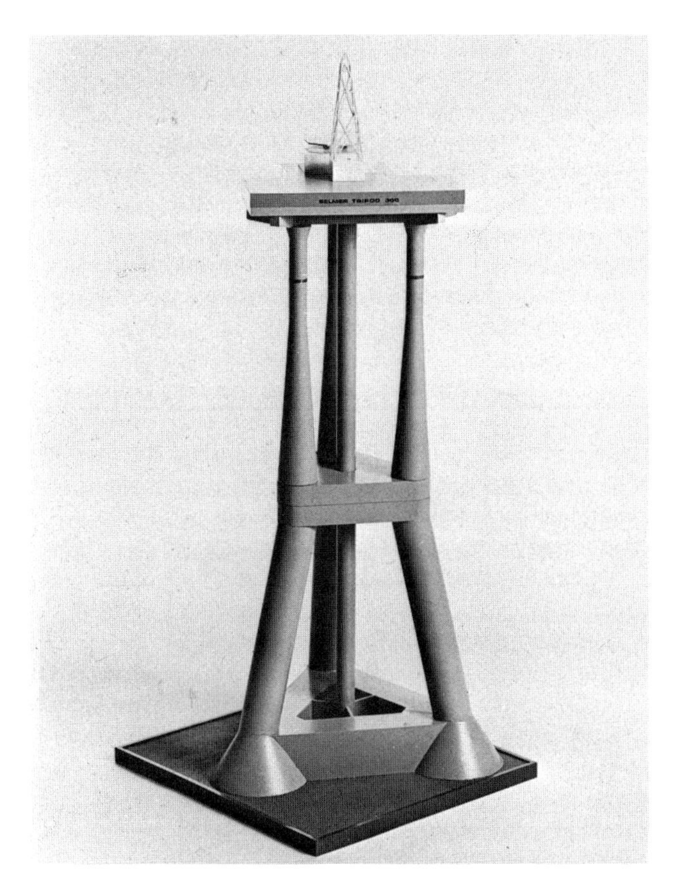

Figure 22.2 Model of an oil platform with splayed legs, the experiment for which is shown in Figure 22.1.

Slipforming techniques

On one contract, slipforming was used to cast a series of five-storey lift and stair walls for the halls of residence in a university. Recently, a major contractor vertically slipformed a series of individual wall bays for the perimeter of a water storage tank. Initially, slipforming was regarded as a solution which could be applied only for continuous vertical runs of walling or columns. However it has been used to cast profiles as diverse as parabolic towers and television masts. Trials are now being conducted, with some degree of success, on the slipforming of skewed or sloping legs for a particular design of platform, to be placed in the North Sea oil fields.

Slipforming is regarded as a somewhat specialist area of construction, and several major national and international companies undertake the operation involved as specialist sub-contractors. Contractors have operated systems that have been devised and manufactured on a one-off basis for shaft manufacture in connection with tunnelling schemes, and while the specialists will use jacks and embedded rods to propel the formwork arrangement, in their simple schemes contractors have used various arrangements of mechanical and hand-operated winches in slipform arrangements.

The main requirements of slipforming are a high degree of uniformity of concrete mix control, maintenance of a suitable degree of workability with cohesion, and early gain of strength. With regard to the formwork arrangement there must be a reliable system for raising or moving an accurate and suitably rigid formwork in such a way that level and plumb can be achieved, maintained and suitably adjusted during the concreting process. The demand on the concrete handling plant requirement is not excessive because concreting, while continuous, is carried out at a relatively slow rate compared with the peak demands met with in lift over lift construction. It is essential that there is constant staffing for plant maintenance; operation and critical engineering control are also necessary.

The actual slipforming technique consists of filling a set of forms which are continuously or intermittently raised or moved to generate the required structural profile. The length of forms, rate of travel and stiffening time of the concrete are so arranged that the concrete which emerges from the bottom opening of the form achieves sufficient strength to support the load of the form and the fresh concrete above as the slide continues.

Figure 22.3 Slipforming an office tower. Note the method employed to form openings.

The steel reinforcement must be inserted at the top of the form and be correctly positioned within the concrete mass; the cohesion of the concrete mix must give a clean face. The jacking or lifting arrangement is used to overcome the friction between the concrete and the form face. Arrangements should be made for differential jacking or lifting efforts to allow the form to be moved vertically and held in a vertical plane.

Some systems depend on 'lost' bars upon which the jacks climb, the lifting effort having been transmitted via yokes to the form. Other systems allow the removal and re-use of bars or tubes. Winch arrangements, used in shaft-linings, require no internal support within the concrete. The upper platforms which generally serve to brace the form to the correct outline serve as steelfixing and concreting platforms. Access ways suspended from this system provide a workplace for the dressers who make good the face as the whole arrangement climbs and the concrete becomes exposed. The steel reinforcement and jacking rods are so designed that the fixing sequence allows continuous work, horizontal steel being inserted below the yoke member and tied as the climb proceeds.

Openings are formed as the concreting operation proceeds by the insertion of a cill member, if required, and jambs at the appropriate positions. These members are suitably braced to prevent distortion and the thickness of the components is such that friction with the form face is minimized. The head member is inserted as required and concrete placing continues above.

Where they are to be transverse beams or connections with subsequent structural members, blocks of polystyrene or timber boxouts are inserted between the forms as sliding proceeds. These allow the subsequent insertion of hairpin-type starter bars, the insertion of threaded sockets and the placing of structural concrete.

Occasionally 'hang-ups' occur where the friction between the concrete and the form face tends to drag the concrete such that it lifts or tears the face. Bridging which results from a lack of compaction can also leave defects. These can be filled from a well made platform, small pieces of static formwork being applied if required. A careful check should be maintained on steel stocks as the shift work proceeds, to make sure that the correct number of horizontal bars are placed for a given rise of form.

Visual checks on line and level need to be maintained. Plumb lines that are suspended within shafts and towers can be damped by suspending the bob in oil, and auto-plumbs can be set up above targets or datum marks at set positions on the base. Instruments are sometimes used at ground level and are sighted vertically onto targets suspended beneath the forms. Care should be taken to measure any sideways displacement and any tendency for the form arrangement to spiral. Available commercial equipment can be used to give a visual display of levels to check the state of the form. Recently, laser beams have been used to provide an available permanent datum at any point throughout the slide.

Where profiles vary, the slide can be broken down into a series of phases, form panels being added to or taken away from the arrangement to provide the required outline. Where walls taper, or where thicknesses vary, i.e. with a parabolic tower, the panels can be arranged to slide or telescope one over the other in such a way that the ranges of sizes generated by the form can be obtained by adjustment.

Setting up a slide form usually takes a long time and requires careful phasing into the overall construction programme. Once a slide operation begins however, considerable progress is made as the form is lifted at a steady rate – 200 mm/h is normal, while exceptional slides have been carried out at a rate of 600 mm/h.

In the early days of slipforming, concrete was always batched and mixed on site and it was not unknown for mixer set-ups to be duplicated when breakdowns occurred. Today many slides are fed from ready-mixed plants with the inbuilt assurance that more than one of the plants are available in case of some breakdown or delay. Slides can be supplied thus, even in cities despite the traffic conditions encountered in the urban areas. Concreting can of course be stopped, and it is usual to elevate the formwork so that it just clips back over the cast concrete. Movement can then be started with the minimum jacking effort.

In connection with work recently carried out on oil rigs and storage tanks, pumped concrete has been extensively used in slipforming, being pumped many metres over jetties and pontoons to the actual point of placing. The designers have to take account of the forces which result from the surge that occurs in some pipelines, depending

Figure 22.4 *An unusual slipforming technique. The massive gantry carried all the equipment for applying the lifting force to the slipform ring. The foundation ring-beam formwork was handled by a tower crane.*

Figure 22.5 *Slipforming the core of a 24-storey office tower. Note the use of a service crane.*

Figure 22.6 *These twin towers were slipformed in seven working weeks.*

Figure 22.7 Slipforming a tower. Note access platforms and safety net.

Figure 22.8 Slipforming a 16 m diameter by 65 m high silo. Note working platforms, distribution points for concreting, steelfixing and platforms for rubbing down. (Cement Limited.)

on what type of concrete pump is used. Fewer operatives are required once a slide has begun when compared with lift over lift construction. However contractors often need to provide an incentive to ensure the attendance of men on shifts. Plant arrangements need to be carefully organized and critical pieces of equipment, such as hoists, carefully maintained. Replacement equipment should be kept on standby if not available on site. In the case of plant breakdown or some delay, the rate of slipforming can be adjusted accordingly and if necessary concrete distributed and placed by hand. Sliding is usually continued by day and night, in which case safety precautions are essential. Lighting has to be provided at all levels and stand-by generators should be on hand.

The ex-mould surface is of an open texture with scour or friction marks. Of course this would not be considered acceptable for visual concrete without some degree of dressing. In Europe architects, to some effect, have expressed the face as cast. It is usual to spend considerable effort on filling and patching however, and it is fair to say that slipform techniques are used where the visual aspects are not critical.

It may be helpful to the reader to examine some details of a slipforming project recently carried out in the construction of twin hollow bridge support towers. The towers were 150 m high and the vertical section varied from 6 m^2 at foundation level to 4.7 m × 3.7 m at the top. The rake of the sides of the towers varied on two of the three raking sides: pitches of 1 : 96 and 1 : 240 being encountered. The adjustment to the plan dimension was achieved by closing up the forms and moving the corner panels. These corner panels were quadrant shaped in plan and were arranged to slide within the side members.

The plane side panels were also arranged to give variable tilt, and consequently a varying throat opening at the bottom of the form. The twin towers were linked by a framework of heavy steel box section which provided the working platform, storage for steel and the control cabin. Climbing was achieved by the use of 40 mm diameter screw connected bars.

The weight of the complete mobile assembly, forms and cabin was about 100 t, and 40 × 6 t jacks were used for the slide. The concreting work and materials handling were carried out by two tower cranes, each located in one of the hollow towers, and two wire-guided hoists; in addition skips were lifted on a hoist which operated through the tower crane frameworks. The slide operated 24 hours a day for 19 weeks. Due to unforeseen circumstances it was stopped only five times.

The 50 mm diameter vertical steel reinforcement was caged and hoisted; all horizontal reinforcement had to be inserted into a 1 m space between the yokes and the top of the concrete. Timber sheathing was used because of its flexibility and steering, although this produced degradation of the face which became apparent at the 100 m mark.

Two 0·5 m mixers located near the foot of the towers were employed to supply the concrete. The two batching plants had a capacity of three times the maximum demand. Total amount of aggregate stored was 2500 t and was handled by boom scrapers. On the mobile platform, which was fully sheeted to provide weather protection for the crew, proprietary control equipment for the jacking system was located. Pumps and clocks were duplicated and the whole of the climbing arrangement set to automatic control with a manual over-ride. Steering of the slide was controlled in several ways including that of adjusting the inclination of the form sides to increase or decrease the friction between the sides and the concrete, and adjusting the pressure of specially sited rams at the bottom of the tower crane leg and varying the jacking rate. The thrust of the bottom members of the tower crane was arranged to be imposed on concrete which was five days old.

During the slide, thousands of sockets were cast in to allow insertion of reinforcement for the tie beams which connected the towers at various points. These sockets were attached to the formers so that recesses were provided at the beam positions over which the form slid. Adjustment to the plan profile was achieved by horizontal screw jacks which were operated proportionally to the jacking action. The speed of the slide averaged 50 mm per hour as the complex steel was being inserted at the intersections with tie beams when some 30 t of steel were fixed in a day.

Initially, the tower cranes were bolted to the concrete foundation but at 20 m they were transferred to the sliding platform and then transferred again to the concrete wall just before completion of concreting at the top of each tower. The whole formwork arrangement was subsequently lowered by specially manufactured wire cables. The tie beams were formed and concreted from heavy structural steel falsework which was erected at the top of the tower and then lowered to other tie beams, the lower beam being 8 m deep and situated 48 m above the foundation.

Figure 23.1 Because costs of materials are continually increasing, efficient site control is essential. There are possibly many valuable pieces of material in this heap.

23. Formwork and the small builder

General

The first question that the thoughtful builder must ask when he considers formwork is 'Is formwork really necessary?'. Formwork in terms of materials usage and labour is expensive, while formwork operations are often such that a lot of wastage can result, even occasionally to the destruction of timber, ply or steel proprietary components.

All too frequently a large number of components brought onto site such as tie arrangements, clips and couplers are lost and have to be replaced before the equipment can be used again. Due to lack of training or skill the items, such as plywood sheets or timber bearers are cut and mutilated in such a way that only a few uses can be achieved and the unit cost of forming concrete rapidly shoots up.

Is formwork necessary?

The quick answer in many situations is 'No'. Simple arrangements of basic materials such as blocks and brick or concrete planks can greatly reduce the demands on the formwork carpenter. The brick, block and plank are types of formwork, but skilled design also renders them an integral part of the construction. The use of these alternatives improves the builder's option on the type of labour that he employs and allows mobility thus removing the emphasis from carpenters or joiners, to bricklayers or simply skilled operatives.

If, on consideration of this basic question, it is decided that formwork must be used the builder must follow a logical series of steps in the selection of materials and technique so that his final selection is not only practical and economical, but also provides a completed concrete component as specified or to the requirement of the client.

Assuming that a component is to be formed on site, the next question will be whether the component is to be cast *in situ*. Precasting of balconies, walls, slabs and beams still requires formwork or mouldwork, but it may provide the builder with options not only on the type of workman to be employed but as to *when* the operations should be carried out. The latter point enables levelling of activities and allocation of resources.

The builder's next concern will be the type of formwork materials to be used, the type of labour to be applied and his policy regarding the standards of accuracy and finish required. Obviously accuracy and finish are tied to the economics of the contract and whether or not the concrete component is to remain exposed as a feature, or if it is to be clad or covered in a completed building.

Concrete offers considerable economies in both exposed and concealed structural work. Because it is 'plastic' it can be squeezed, squirted or dropped into place; it possesses immense potential strength and yet being composed of some of the most beautiful natural aggregates it can provide exotic and interesting surface finishes. The suitability of the concrete for any building purpose be it structural, decorative or a combination of both, depends on the formwork material selected, the available skills and the money to be expended on these items.

Groundwork

Concrete is commonly used for foundation work and the builder should invest in steel forms of the road form-type since these can be handled and placed by skilled operatives. These forms, and a few commercial panels which are used in conjunction with tubular walings, cater for most kinds of foundation work, beams, slab edges and such like where the use of other than secondhand traditional materials proves to be expensive. Even these materials while apparently cheap to provide, make considerable demands on carpenter labour.

In the ground, trench and proprietary props founded on plates or heavy timber baulks are the simplest items for paging or propping. A careful check must be made on all pins and pegs as the loss of just a few of them per base can drastically affect the financial outcome of the work. The largest savings that can be effected are undoubtedly those offered by good housekeeping.

Many foundations and ground beams have openings and crossings which can consume a large amount of timber, ply or expanded polystyrene. Expanded polystyrene provides only one use yet costs approximately 1/6th of the cost of a timber and ply form which is capable of being used 15 or 20 times. As an alternative materials such as damaged blocks or pipe can be used to provide savings, although for slabs the greatest saving can be achieved by the manufacture of one set of concrete opening formers. Concrete formers have a very high re-use value and are virtually indestructible. The smaller sizes can be manhandled and are even easier to handle where cranes are available.

Recently, due to the accelerating cost of form materials, it has proved economical to form slab edges and ground beams by using precast concrete beams which are laid into position and stripped or removed by crane early on in the hardening period. Such simple forms can provide repeated use on site after site and become permanent items of plant.

Suspended slabs

On building sites the next most important components which require formwork is the suspended slab, the corridor, landing balcony or porch slab. For these there can be little doubt that the commercial form and supporting equipment provides the most viable solution for formwork problems. However a further consideration is necessary, that of whether the builder should hire or purchase materials. Where a builder envisages some 40 or 50 uses for the equipment and can of course mobilize sufficient capital he should buy the equipment he needs, although he will have to calculate the costs of cleaning, maintenance and storage between uses, and determine the cost of the space which the equipment will occupy in his yard between contracts.

For economical reasons the builder must ensure that the infill material which makes up the spaces between the proprietary forms and brickwork or bearing walls, is supplied only once, cut-in and fabricated as required and then re-used, throughout the contract without the need for any replacement. The greatest amount of waste occurs where ply or timber sheathing has to be constantly replaced, especially when one takes into account the excessive expenditure of craftsmen's time during the process. Slab edging and beam and lintel forms should be fabricated to a standard that is consistent with the number of uses be taken from the panels during succeeding operations.

Where surfaces are to remain exposed, or for decorative reasons, extra care spent in masking joints and sealing gaps adjacent to the brickwork considerably reduces the succeeding trades' labour and skills necessary to achieve the required standard. Again, the builder must decide where he intends to spend his money, in the initial construction or in the subsequent finishing operations.

The builder is further faced with alternatives when he comes to consider suspended construction, whether to use precast proprietary components, permanent forms or to cast *in situ*. A major factor of the first two is that a supporting skeleton is still required to provide the necessary support while the *in situ* topping is placed and compacted and while the floor assumes its full structural role.

Although it would appear that the cost of formwork has been avoided, it is really only reduced by an amount equivalent to the cost of the sheathing and joists together with perhaps 50% of the supports and runner system.

Precasting

The precasting of components on site using moulds of the substance required for repetitious casting can prove to be expensive during the early stages of making and setting up the moulds. However, when set against all the costs involved in the handling and erection of the forms for the *in situ* work it is possible to achieve savings.

By comparing the numbers of moulds and forms required for the respective processes, overall savings can be effected by the adoption of precasting because the few moulds involved can be spread and phased throughout the construction programme to help reduce the demands required by concrete supply and placing. Initially, the moulds may cost more, but if properly designed and constructed to provide a number of uses for casting the various sizes of component, the whole process can be geared to the demand in such a way that the stocks of cast units which are held on site can be kept to a reasonable minimum. A builder must ensure that sufficient maturity is achieved in his stock components to enable them to be handled and installed without damage.

Formwork arrangements for concrete stairs in maisonettes and blocks of flats are often critical to the whole process of construction. It is often cheaper to translate the formwork item into a mould item, i.e. by precasting the stair flights. A single simple mould which is set up by the delivery point for concrete, or even adjacent to a mixer, can be used to cast on alternate days using normal curing techniques at the rate of two or three flights in a week. Edge-casting can help in achieving surface finishes or the transfer of tiles or non-slip treads.

Lintels and simple beams are the units where greatest savings can be achieved by precasting – structural concrete from other activities being placed into a gang of moulds whenever time allows and a stock being built up over a period of time.

Figure 23.2 Of course, it need not be timber! Concrete blocks form adequate sheathing for ground beams on small works.

Safe loads and propping

Recently published safe load tables for props will be of considerable interest to the builder, since they allow him to choose the optimum safe spacing for his props. A wise builder will pay particular attention on the comments which relate to eccentricity of loading and 'out-of-plumb' propping on load-carrying capacity.

A glance at the props on many building sites will raise doubts on the safety of the formwork arrangements which may well be confirmed by an insepction of the way in which the props take their support from the ground. Supporting systems and their integral bracing arrangements offer the builder a built-in insurance against failure arising from poor erection of formwork – the spacers imposing a discipline on both the spacing and the plumb of the standards. Quickstrip arrangements offer further assurance of successful form operations as well as providing continuity of support when the sheathing has to be removed at three days from initial casting for early re-use.

Pressure and ties

The construction of walls and columns can present a builder with certain problems because the pressures which result from the placing and vibrating of concrete are such that adequate sheathing support must be provided.

From research recently carried out some simple graphs have been published which are designed to assist those who are prepared and able to make simple calculations when designing ply and timber wall forms that will be required to sustain the loads imposed by the concrete under varying conditions of temperature for differing rates of fill.

Whatever form is selected the correct spacing of ties requires urgent consideration for building supervision. The through bolt (setting aside wiring which can still be used in the ground) is the cheapest type of tie arrangement. Through bolts can be used over and over again and are simple to handle. Unfortunately, they need to be supplemented by some kind of spacer which may not always be acceptable in certain circumstances. Recently, a very useful tapered tie bolt has been re-introduced which is simple to strip from the concrete, although it still requires some kind of spacer. Snap ties are also very useful especially where proprietary panels are being manhandled between casts. Careful and correct placing of the required number of ties is necessary.

Exposed concrete

Where concrete surfaces are to be exposed, careful selection of the mould oils, coatings and parting agents is required. They are often selected as the cheapest bulk purchase agreement available and this is rarely successful because the cheapest of oils nearly always fails to meet all the specification requirements. Because particular attention is required for visual concrete, the oil or parting agent should be suited to the material employed for the formwork sheathing.

To ensure adequate savings the builder must see that his operatives apply the appropriate type of agent in the specified way. He will thus be assured that no defects will result from flooding, saturation or the penetration of oil into the concrete during some critical part of construction.

Ancillary components

Modern building techniques often include the rapid provision of refuse chutes, drainage and service facilities. For these purposes, and for any moulded or shaped concrete items, moulds that are made from glass, reinforced plastics or special steel are the cheapest in the long run. One-piece manhole forms can be jacked out of the concrete while the small degree of flexibility inherent within these materials allows simple removal from the concrete.

While the scale of production governs, to a large extent, the method of construction, in many instances formwork in building can be relatively simple. Just as much care is needed for success as in any other operation. The check on the formwork made prior to and during concreting is an essential step in the series of activities of building and the checklist set out in Chapter 18 should be condensed and used in checking work.

Figure 24.1 A first attempt at design and construction by engineering graduates – a lesson in bracing and propping!

24. Formwork instruction

General

From the many visits to technical colleges, colleges of Further Education, Polytechnics and Universities made by the author, and during conversations with architectural, engineering and technical students it is quite evident that only perfunctory instruction is given with regard to form design and construction. When considering the impact that both formwork and concrete technology have on the performance of a concrete structure and the need for skilled designers and informed supervisors, there are quite obviously gaps in constructional education and training. There are a certain number of professional institutions and associations active in the formwork and concrete technology field, while organizations such as the Concrete Society make considerable efforts to emphasize the importance of formwork in the construction process.

Both the Construction Industry Training Board and the Cement and Concrete Association run one week courses on all aspects of formwork from design and construction to planning and design, while the Institution of City and Guilds of London promote a series of one year, part-time courses in colleges throughout the UK which lead to an examination and certificate in *Formwork for Concrete Construction*. An option in the CGLI Carpentry and Joinery Advanced Craft Certificate Course is a module on Formwork, the level set is appropriate to the needs of potential supervisors.

Course content

It is hoped that the following will help tutors in the preparation of courses of instruction on formwork topics. The exercises given in Appendix 2 provide some basic material for group work, and at the same time indicate the kind of material which the tutor can assemble on the activities of constructors who operate within his area.

Course objectives

Generally tutors are well versed in the preparation of course objectives and perhaps it goes without saying that they should first make sure that they have clearly identified the objectives of their courses. The following brief 'objectives' have been prepared for courses organized by the Cement and Concrete Association.

Introduction to formwork

Objective: to provide an understanding of normal site formwork practice and related aspects of concrete technology. A basic knowledge of concrete construction will be assumed.

Intended for: technicians, general foremen, clerks of works and other members of the construction industry who wish to understand formwork practice at a general level.

Formwork construction and practice

Objective: to review current practice and factors affecting the design, manufacture and utilization of formwork, and to provide an introduction to planning and programming techniques.

Intended for: foremen, clerks of works and others who already have considerable experience with formwork.

Formwork design

Objective: to provide an introduction to the design of formwork and to review current practice and recent developments.

Intended for: engineers and higher technicians with some practical experience of formwork. ICE/IMunE/IStructE.

Formwork planning and construction

Objective: to provide a sound basis for planning of formwork construction and to review current trends and new developments.

Intended for: engineers and technicians concerned with the preparation of formwork schemes for construction work. ICE/IMunE/IStructE.

It should be noted that in each case the objectives are carefully related to the needs of the participants.

The syllabus

A topic can be divided into a number of aspects each of which can be treated in different ways to ensure that a student receives the type of training and instruction that fully imparts skills and technical know-how:

Planning: Selection of method
Selection of materials
Staffing – outputs

Construction: Materials
Techniques

Operations: Handling
Erecting
Striking formwork

Technology: Preparation of concrete
Surfaces finishes
Mechanical design
Allied materials

An instructor must of course make sure that all aspects of the topic are covered during the course work.

Current information

One problem experienced by many lecturers is that of maintaining contact with industry and ensuring that information passed on to students is topical. Inevitably lecturers deal with a range of construction or technological topics, and thus formwork can easily become just another topic within a whole variety of allied subjects. Fortunately, there are a number of trade associations who contribute to college courses as part of their general promotional activities. These associations employ members who are suitably qualified to provide useful lectures to supplement the more academic coverage offered by colleges or training centres. The associations through the literature that they publish and the lecturers that they provide, help to disseminate information on specialized topics. The published material is often free and can prove invaluable as part of a technical reference library.

Local contractors, and indeed national contractors who operate in the provinces, are generally prepared to arrange site visits and these can be incorporated within full or part-time courses.

Visiting lecturers

A visiting lecturer should be briefed about the level of appreciation of the course participants and the ground he needs to cover to ensure that he is within the syllabus. It is unwise to combine courses in order to achieve good attendance for a visiting speaker. Generally, an enthusiastic lecturer will be more concerned about holding the interest of his audience than the number of people listening to him. Ideally, a visiting lecturer should be introduced to other speakers who may be contributing to the course so that he can compare notes and thus avoid unnecessary repetition.

The use of visual aids rarely presents problems, because colleges and training establishments are usually well equipped, although bad 'blackout' facilities can often spoil a slide demonstration, Before an invited lecturer speaks, course participants should not only be briefed to some given level but should also receive a handout or some notes which underline the key points covered during the talk. A suggested reading list should also be circulated in advance.

The 'sales' type of lecture should be avoided. Even the most considerate technical representative may unconsciously over-sell and thus tend to bias the participants' thinking. The ideal lecturer is one who speaks for an industry and presents information which covers a broad spectrum of activities. The various trade associations concerned with fringe activities with regard to construction can help to broaden a student's knowledge of allied technologies.

Visits

Visits to local sites and works form useful extensions to the in-college activities. Particular care must be taken over insurance cover and the timing of visits. A lecturer should make a reconnaissance visit to ascertain the scope of the information available and to ensure that a suitably briefed official will be available to communicate essential information to the students as they are shown round. Students should be briefed before a site visit, preferably in groups, so that particular aspects of formwork such as erection, handling, safety and such like can be studied. A report-back session, preferably attended by a site official will help to confirm and conolidate the details gathered.

Projects

When some construction activity or extension to the college is planned or where local jobs are being carried out it is useful to obtain from the contractor details of key points or particular problems or constructional features. Students can then 'pit themselves against the experts' as it were, by devizing schemes and details which can be compared with the way in which the work is actually carried out on site. Contractors are usually only too happy to help and often will explain the reasons for the approach or techniques adopted.

Practical work

The practical work involved in forming and casting concrete is often neglected because of the problems with disposing a cast component, and possibly also because of the difficulties in providing concrete in any quantity. The actual physical handling, i.e. placing and compacting concrete into the forms, helps a student to appreciate the problems of robustness of form, grout tightness and such like.

The cost of 0.5 m^3 of ready-mixed concrete is very little – while form panels can often be borrowed from a local contractor or supplier for little more than the cost of 'a refreshment', thus a class can handle the forms, place and compact concrete and, provided that the timing is monitored, dispose of the concrete in a refuse bin before the hardening process is completed.

Considerable practical work on surface finishes, retarders and mould oils can be simply and cheaply carried out, for just the cost of a bag of ready-mixed concrete and a set of baking tins. Here, students can obtain a feeling for the real thing – even to the extent of appreciating the principles of demoulding and early striking. This sort of contact and appreciation of the properties of the materials used will prove to be extremely valuable.

Models

Instruction in formwork techniques provides an instructor with considerable scope for ingenuity, from the preparation of models of particular formwork items, demountable models of formwork for use in discussing positioning of construction and day joints to the manufacture of moulds and forms with perspex sides for visual examination of techniques such as placing and compacting underwater concreting. Pressure development within formwork can be demonstrated using the normal equipment which is found in the college physics laboratory, and perhaps the simplest illustration of pressure development is that in which water is allowed to escape from holes at different levels in a plastic pipe! Once again formwork can present a challenge to the instructor and can be used to open the way for further investigation of the allied technologies of plastics, adhesives and steel.

Films

Many films have been made on constructional topics although formwork may not be the main subject of many of these films, nonetheless it will always be featured. The instructor should preview as many films as possible so that he can choose examples which explain the importance of formwork with respect to the whole constructional process. Many of the major contractors prepare films of their more prestigious contracts and a study of these will help to provide a basic film library, parts of which can be used to illustrate particular formwork and falsework solutions. Manufacturers of materials and systems can also generally supply, on loan, films which show applications of their particular products which can help with general information. There are some very good films on plywood manufacture, use of cements and other materials which can help to give students some background knowledge and information.

Slides and illustrations

Good visual aids are essential for any kind of instruction. Even well prepared and technically sound lectures can be spoilt by the use of badly prepared or inappropriate slides and vu-foils.

Slides should illustrate some specific point, although slides of general construction work may well contain something which also usefully illustrates other points. They can be taken out on loan from commercial producers of materials and equipment, and there are some prepared slide sets which are actually produced for instructional purposes.

The best kind of slide is the one which illustrates and emphasizes a specific point of instruction. Acceptable transparencies can be produced by using the simplest of equipment provided that the instructor clearly identifies the object taken in the photograph. Good lighting is essential, and an unobstructed view, and in particular, a close-up is necessary.

Photographs or slides should give an appreciation of the general context of use, for example, a general site view. Details of a product or piece of equipment actually in use are necessary and, where possible, some detail of a component taken in a position which is not obstructed or cluttered with unnecessary paraphernalia.

Vu-foils or slides

Vu-foils can be produced from virtually any original printed text, line drawing or half tone illustration. Trade periodicals provide a wealth of useful material, particularly in advertisements in which detailed drawings, often prepared at considerable expense, highlight features of a product or piece of equipment. Attention must be given to copyright, and usually a telephone call to an author or commercial producer brings approval for using printed material provided due acknowledgement is given. Often it is more than likely that as a result of such communication further information, leaflets, display panels and models can be obtained.

Appendix 1 – Prop Selection Chart and Concrete pressure graphs for formwork design

Introduction

This chart is an aid to the selection of telescopic props and their spacings for supporting soffit forms on timber bearers. Calculation of loads is unnecessary. The soffit areas given on the chart will load the props to the safe limit recommended by the CIRIA Report No 27 for props complying with BS 4074 where concentric loading is not guaranteed.

Safety

This data sheet has allowed for minor errors in erection of props, but it is only safe if the following rules are observed :

(1) No prop which is visibly bent or damaged is safe to use. It must be discarded.

(2) Only the high tensile steel pins provided by the manufacturers must be used. Old bolts or pieces of reinforcement are a dangerous substitute.

(3) Props must be placed centrally under the member supported. An eccentricity of more than 38 mm must be corrected.

(4) Props must also be plumb. A prop more than $1\frac{1}{2}°$ or 1 in 60 out of plumb must be corrected. Verticality of props should be checked with a spirit level and errors of more than 4 mm in a 240 mm level or 13 mm in a 800 mm level must be corrected.

The chart overleaf is a rapid aid to the design and checking of props.

How to use the chart

(1) Select the slab thickness along the bottom of the chart.

(2) Select the extended prop height that is required up the side of the chart.

(3) Find the intersection point on the chart of these two dimensions, ignoring for the moment the coloured lines.

(4) Move horizontally to the right from this point to the nearest coloured line (unless the intersection point coincides with a coloured line in which case choose this line). The colour of the line indicates the size of prop, and the area printed on it is the safe area of slab it can support.

(5) In cases where, for the required height, more than one prop may be used, select the most convenient size of prop and use the appropriate coloured line.

Example 1

To support a 300 mm thick slab with props 2.6 m high, the intersection is point A. The nearest line to the right is the yellow one with 1.2 m² written against it. The required prop is a No 1 or 2 size and must not support more than 1.2 m² of formwork. If the bearers are at 1 m centres the props must not be more than 1.2 m apart. If the bearers are at 900 mm then the props can be at 1.3 m but not at 1.4 m spacings, because this would come to 1.26 m² on each prop and they would be overloaded.

A prop set more than 70 mm eccentric (this is not a specially posed photograph)

Example 2

Consider a 560 mm slab with props at 3.2 m high. The intersection is at point B on the graph. To the right there is a choice of magenta, yellow and blue lines, which means that theoretically either props size Nos 2, 3 or 4 can be used. However, it is worth checking that No 4 props can be lowered sufficiently for striking.

As before, coloured lines horizontally to the right of point B are safe so one can safely use either No 4 props supporting not more than 1 m² or Nos 2 or 3 props supporting a maximum of 0.6 m². Here it would be more economical if possible to use No 4 props because fewer would be required.

If bearers are set at 1 m, then No 4 props could be at 1 m intervals, but Nos 2 or 3 props would need to be at 600 mm spacings.

Concrete pressure graphs for formwork design

The graphs given opposite are based, by kind permission of the Director of CIRIA, on the formula given in CIRIA *Research Report RR1, The pressure of concrete on formwork.* This information sheet will give results similar to those obtained from the CIRIA report or the CIRIA *Formwork loading design sheets.* For further information on the effect of the various factors involved, reference should be made to the CIRIA report.

Use of graphs

(1) The graphs are drawn for normal concrete of density 24 kg/m³.

(2) To determine the maximum pressure (kN/m²) on vertical formwork, take the *smallest* of the three values found as follows.

 (a) *Hydrostatic pressure.* Multiply the height of pour, H (m), by 24.

 (b) *Arching.* Where the minimum thickness of the wall or column, d, is 500 mm or less, the pressure may be reduced by arching effects. From graphs, read off the pressure against rate of placing (m/h) using the arching limit line appropriate to d.

 (c) *Stiffening.* This may apply in every case. Read off pressure against rate of placing using radiating line giving appropriate combination of concrete placing temperature (°C) and slump (mm).

Examples

(1) A wall 300 mm thick and 4.5 m high. Rate of placing 1.5 m/h; concrete temperature 10°C; slump 75 mm.

Hydrostatic pressure = 4.5 × 24 = 108 kN/m²
Arching:
for rate of placing = 1.5 m/h, line d = 300 mm
we get pressure = 49 kN/m²
Stiffening:
for rate of placing = 1.5 m/h, line 10°C. s = 75 mm
we get pressure = 69 kN/m²
Therefore
maximum pressure = 49 kN/m²

(2) A wall 600 mm thick and 4.5 m high. Rate of placing 3.0 m/h; concrete temperature 15°C; slump 25 mm.

Hydrostatic pressure = 4.5 × 24 = 108 kN/m²
Arching:
inapplicable, as thickness is 500 mm
Stiffening:
for rate of placing = 3.0 m/h, line 15°C, s = 15 mm
we get pressure = 63 kN/m²
Therefore
maximum pressure = 63 kN/m²

Appendix 2. Formwork exercises and projects

The author wishes to express his appreciation to Dr Ralph P. Andrew, Director of Training, Cement and Concrete Association, for permission to reproduce the exercises and projects contained in this Appendix. These projects have been developed by the author and by Mr. V. Watson and Mr Robin Harold-Barry for use on the Cement and Concrete Association's training courses at Fulmer Grange.

The exercises are arranged in such a way that they can be carried out by groups of participants in periods which can vary from 2 to 24 hours. The work can be carried out during seminars, or during the course of instruction from supervisors, site engineers and similar personnel. The materials can serve as a focus for the instructor who can 'talk through' the construction processes illustrating points of form manufacture and technique.

Although some solutions are offered, it will be obvious that numerous solutions are possible, and in most cases the instructor should be able to suggest method and techniques based on his own experience.

Regarding group work, this is best carried out in groups which consist of six to eight participants. A spokesman should be appointed to present the work during a feedback session after the group work has been completed. The spokesman should be asked to break down the work of the group so that pairs of participants deal with salient points. The production of sketches, drawings and models should be encouraged, indeed in most instances these are essential. The models need not be elaborate, card and tape are adequate in most cases.

AN EXERCISE IN FORMWORK DESIGN

This exercise is based on an actual construction project and contains all the information necessary for engineering students to prepare the calculations involved in the formwork design. It illustrates the actual construction and formwork methods employed, and the author is grateful to Bovis Ltd for their permission to use this work for training purposes.

Assembly hall column exercise

Six similar columns are required for a large town centre assembly hall. The sloping columns are external to the building but the overhanging area at the back of the column containing the rain water pipe recess will, to a certain extent, be shaded by the building.

Ordinary Portland cement concrete with a 28-day characteristic strength of 30 N/mm^2 is to be used. Concrete surfaces must be of uniform colour and remain untouched after the formwork is struck.

Tolerances are to be grade 1 in PD 6440, i.e.

Dimensions on plan, ± 10 mm

Verticality, ± 20 mm

Surface levels, ± fl0 mm

Bow, ± 10 mm

Abrupt changes of a continuous *in situ* surface, 4 mm.

Your brief

List blemishes which might occur and your method of avoiding them.

Design the timber and ply part of the formwork above kicker level; assume that proprietary soldiers and ties of a suitable size are available.

Reinforcement is as shown in the photograph of the kicker, but a satisfactory alternative arrangement will be accepted.

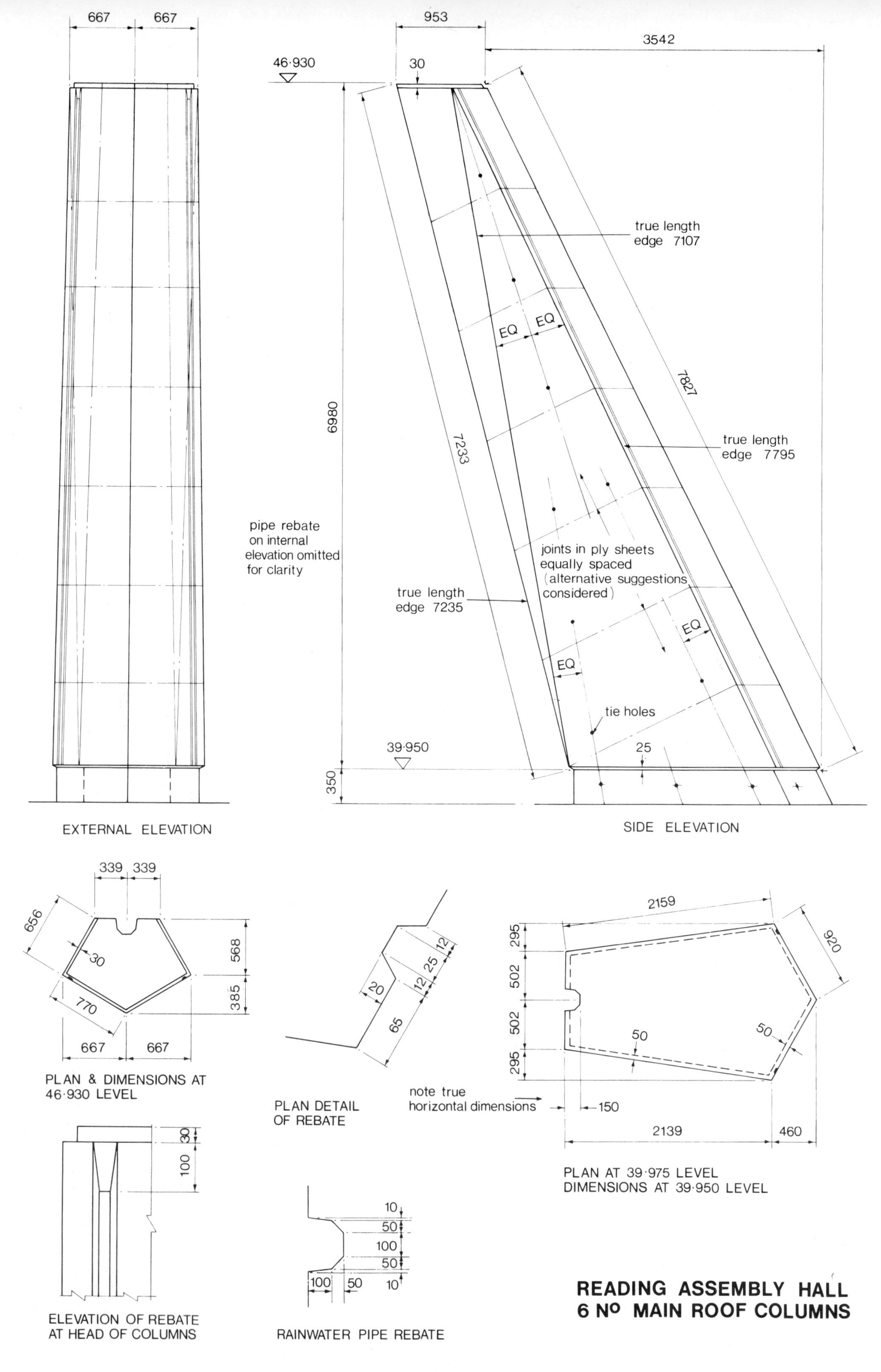

READING ASSEMBLY HALL
6 Nº MAIN ROOF COLUMNS

Try this example yourself

If you have to support a 450 mm thick slab on props which are 3.8 m high, what soffit area would each safely support?

(The answer is given at the bottom of the page.)

Prop lengths

The range of height for each size of prop should be noted. Always allow a margin above the minimum height to allow room for striking. This is a particular problem in the case of No 1 and No 2 props. At the same height these props have the same load-carrying capacity, but different height ranges. Both are covered by the yellow lines on the chart.

Loadings

The chart assumes that traditional timber formwork and normal density concrete will be used (i.e. concrete of density 23.5 kN/m^3). In addition a live load of 2 kN/m^2 has been allowed. If lightweight concrete or heavier live loads are to be used, an equivalent thickness slab of the same weight could be assumed, and the chart used as before.

CIRIA Report No 27

This data sheet has been produced from information given in CIRIA Report No 27 entitled *Effect of Site Factors on the Load Capacities of Adjustable Steel Props.*

The Concrete Society are indebted to the Construction Industry Research and Information Association for sponsoring this research and for permitting the Society to publish their results in this form so that this important data, which concerns the safety of soffit formwork, can be made available to prop users at an early date. The Society is also indebted to John Laing Design Associates Ltd who devised the form of presentation used here.

CIRIA Report No 27 gives recommendations for the use of props in contexts other than that assumed here. In particular, if concentric loading can be guaranteed by jigging of the supported members or by the use of locating heads on the props, then higher loads than those given here may be safely carried.

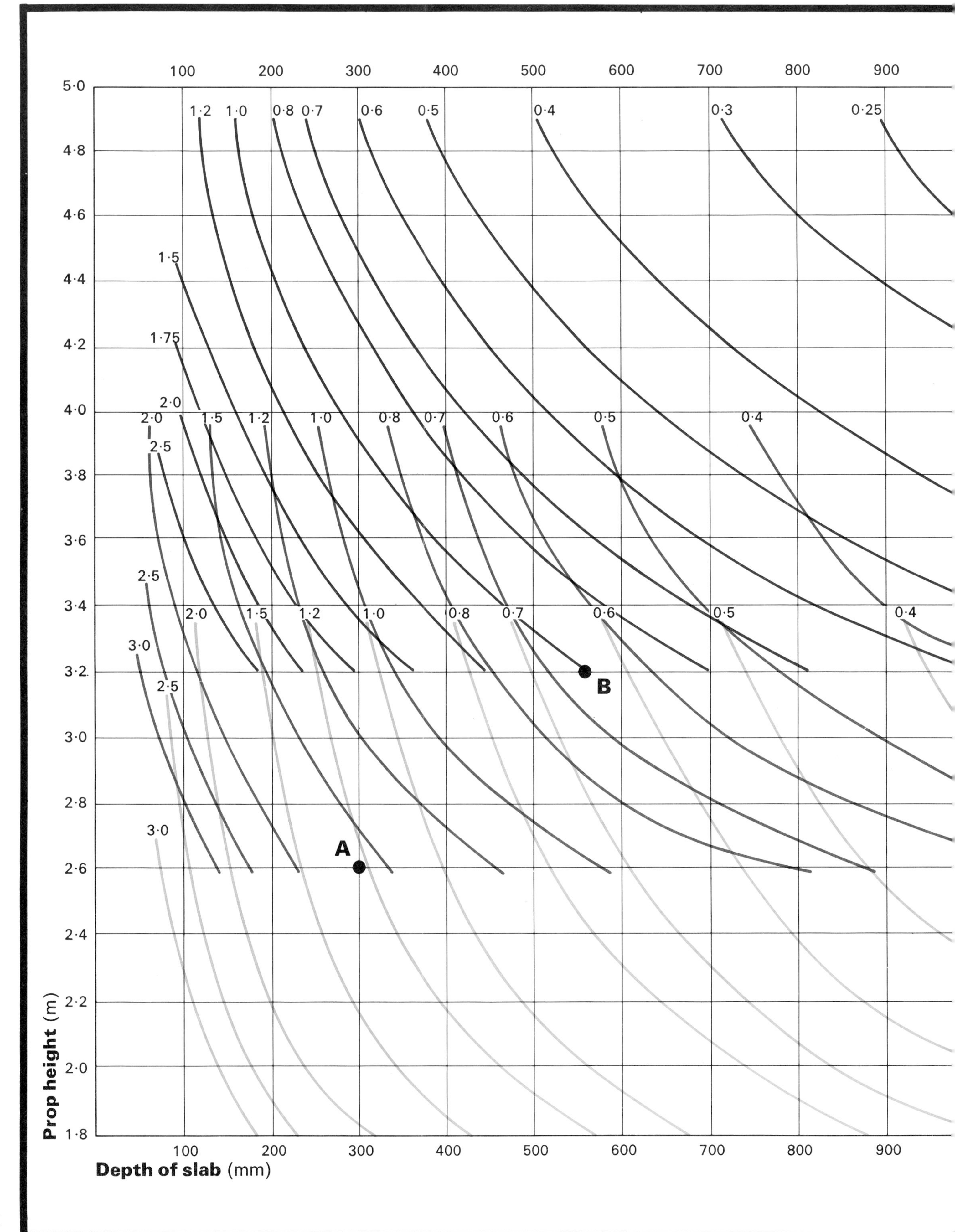

Answer : Either 0.7 m^2 with No 4 props or 0.6 m^2 with No 3 props

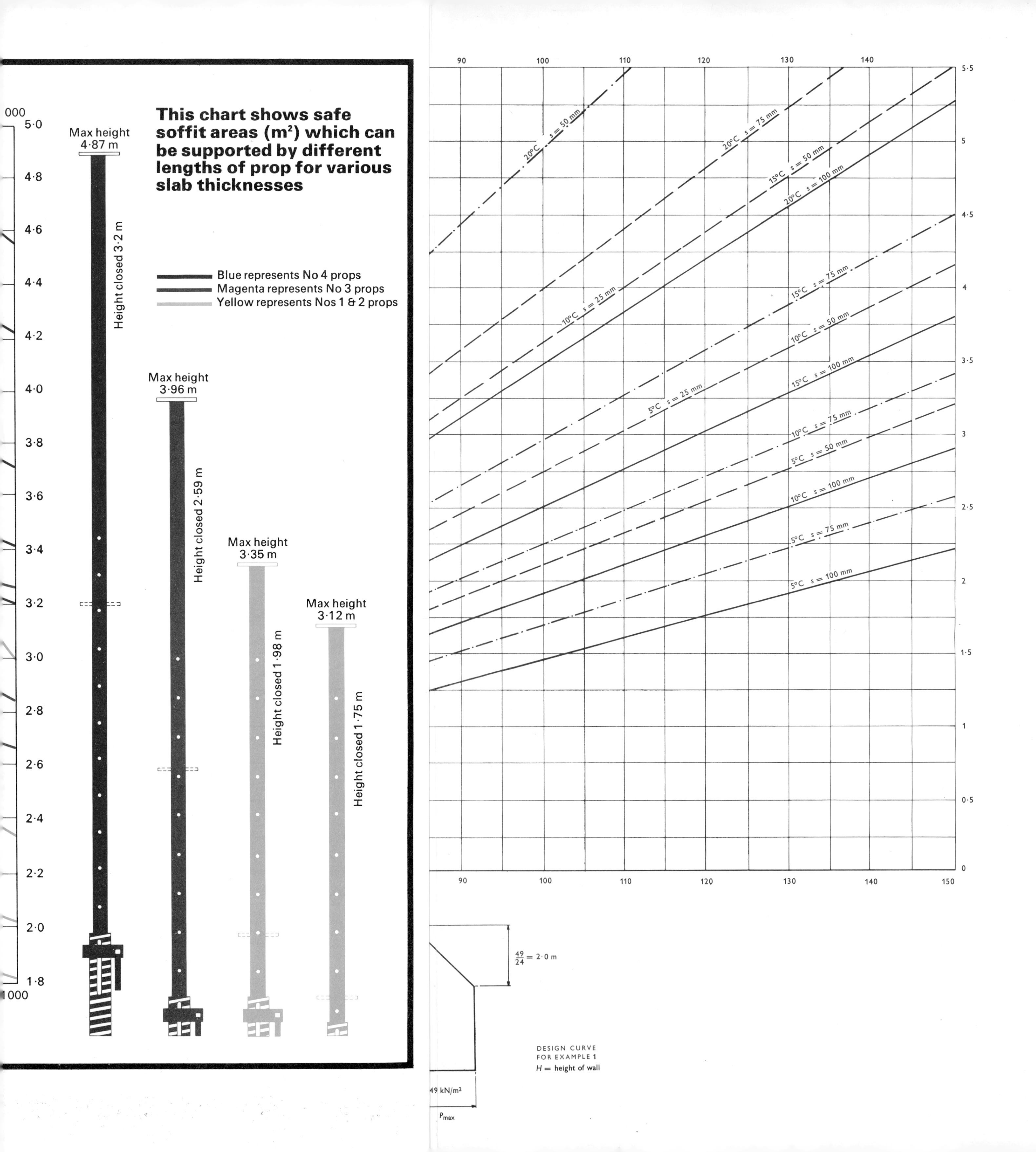

This chart shows safe soffit areas (m²) which can be supported by different lengths of prop for various slab thicknesses
Blue represents No 4 props
Magenta represents No 3 props
Yellow represents Nos 1 & 2 props
Max height 4·87 m
Height closed 3·2 m
Max height 3·96 m
Height closed 2·59 m
Max height 3·35 m
Height closed 1·98 m
Max height 3·12 m
Height closed 1·75 m
000
5·0
4·8
4·6
4·4
4·2
4·0
3·8
3·6
3·4
3·2
3·0
2·8
2·6
2·4
2·2
2·0
1·8
1000
90
100
110
120
130
140
150
5·5
5
4·5
4
3·5
3
2·5
2
1·5
1
0·5
0
20°C s = 50 mm
20°C s = 75 mm
15°C s = 50 mm
20°C s = 100 mm
10°C s = 25 mm
15°C s = 75 mm
10°C s = 50 mm
5°C s = 25 mm
15°C s = 100 mm
10°C s = 75 mm
5°C s = 50 mm
10°C s = 100 mm
5°C s = 75 mm
5°C s = 100 mm
49/24 = 2·0 m
DESIGN CURVE
FOR EXAMPLE 1
H = height of wall
49 kN/m²
Pmax

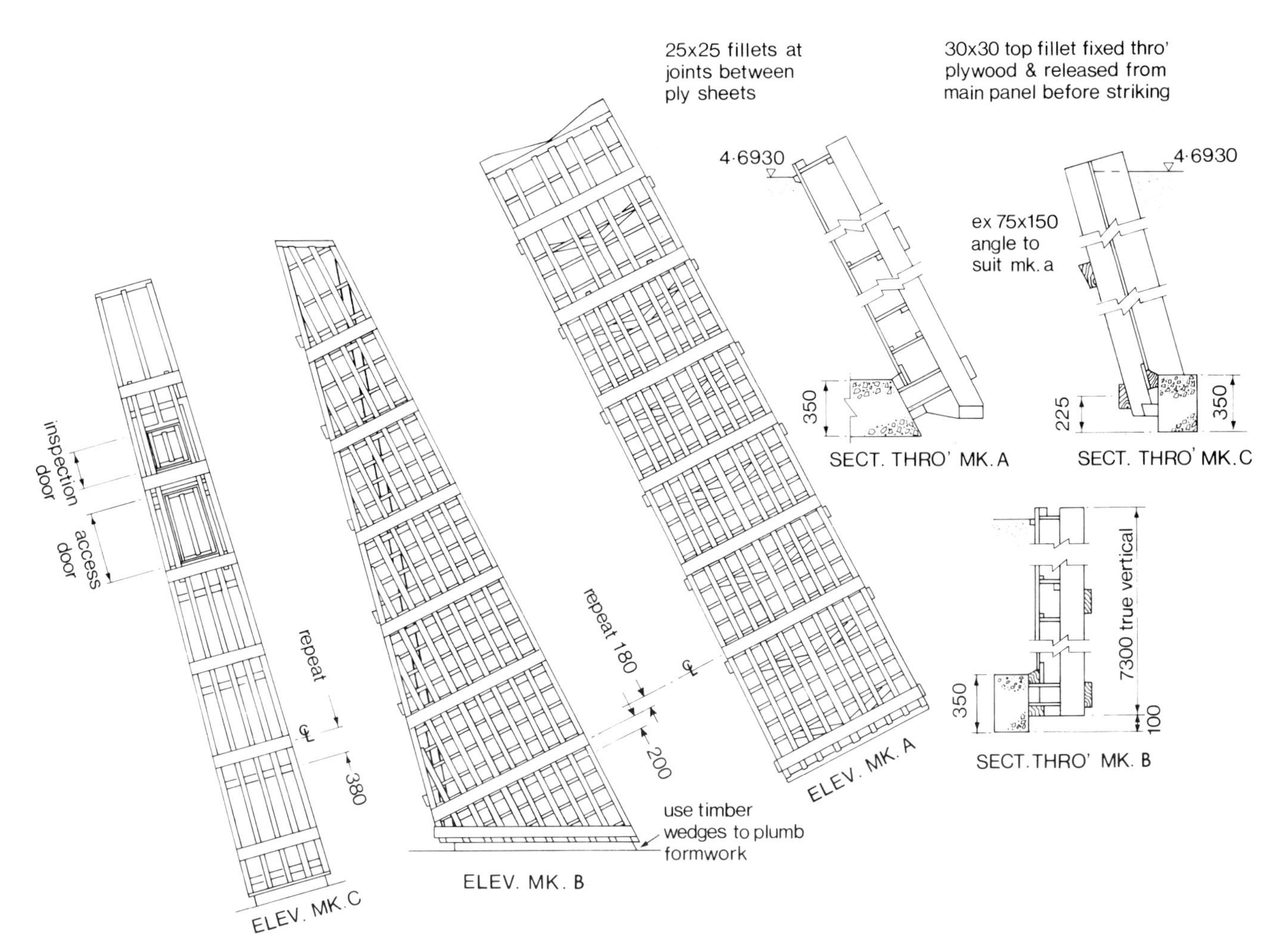

READING ASSEMBLY HALL
MAIN ROOF COLUMN FORMWORK ADOPTED BY BOVIS

A solution to assembly hall column exercise

BLEMISH	REMEDY
Colour variation	Keep mix materials uniform, i.e. from one cement works, from one reliable aggregate source with uniform grading, and from one mixing plant.
Rust marks	Avoid pyrites in aggregates. Cover starter bars and bolts.
Aggregate transparency	Design formwork to have uniform stiffness to reduce variation in vibration characteristics.
Hydration discoloration	Make the form watertight with foam strips and seal edges and face with at least two coats of polyurethane varnish.
Oil discoloration	Spray on chemical release agent and allow to dry before each use.
Lighter colour near the top	Protect the top surface of the concrete, and hence the top edges, from premature drying with damp hessian covered with polythene. Reduce water content of last batch.
Different coloured panels	Strike all exposed surfaces at the same time to allow drying to occur evenly.
Blow holes	Continuous thorough compaction while placing.
Join lines	Place at a uniform rate. Avoid delays.
Bleeding, sand runs, settlement cracks	Use an air entraining agent. Reduce the water content to a minimum for good compaction. Revibrate when the concrete is just stiff enough to support itself without vibration. Vibrate form face with rubber hammer.
Honeycombing air pockets	Avoid a harsh mix.
Incomplete compaction or segregation in the corners	Use access and inspection hatches on the shaded side of the column; the reinforcement would have to be arranged. Deliver the concrete through elephant trunking. Stone content of first batch could be decreased.
Crazing, mottling	Reduce the fines and water content of the mix but not so much as to produce honeycombing. Avoid a polished formwork face.

Area of base $= (2.139 \times 1.004) + (2.139 \times 0.295) + (0.797 \times 0.460) = 3.145\ \text{m}^2$

Area of top $= (0.568 \times 0.667) + (0.568 \times 0.339) + (0.667 \times 0.385) = 0.828\ \text{m}^2$

Volume of concrete $= 6.980 \times (3.145 \times 0.828) \div 2 = 13.9\ \text{m}^3$.

The Bovis design shows a rigid system for distributing the concrete pressure between 18 mm ply diaphragms and 75 × 150 mm timbers both at 275 m centres. The 75 × 150 mm timbers span between a rectangular system of yokes at 1100 mm centres; and from the safe load graph for species group S2 grade 50 wet softwoods the load carried is 16 kN/m run. Since the timbers are at 225 mm centres the safe concrete pressure is $16 \div 0.225 = 71$ kN/m^2.

The plywood size needed for this pressure from the safe load graph for saturated plywoods spanning 225 mm is over 15 mm; the next size up in Finnish birch is 18 mm thick.

Assuming a concrete density of 2400 kg/m^3, a slump of 50 mm, a summer temperature of 20°C and a winter temperature of 10°C; then the maximum rate at which the concrete surface may rise in winter from the concrete pressure graph is 1.9 m/h and in summer is 3.4 m/h.

The time required to pour a column in winter is therefore $6.980 \div 1.9 = 3.7$ hours and in summer $6.980 \div 3.4 = 2$ hours. Two hours may not be enough time to properly place and compact 13.9 m^3 of concrete in a confined space; so plan on 3.7 hours with loads arriving at 1.2 hourly intervals. The loads should contain 6.5 m^3, 4.6 m^3 and 2.8 m^3.

FORMWORK TROUBLE SHOOTING PROJECT

Brief

You have designed the purpose-made steel form for a reservoir wall, as shown in the diagram.

Several tie failures have been reported from the site, and you have been directed to go to site, to determine the cause of these failures, and decide what investigations are necessary to convince all parties that your diagnosis is correct: finally, you are to recommend remedial action to prevent future tie failures.

The tutor will circulate amongst the groups, and answer direct questions on matters of ascertainable fact.

Design

On about 20 previous contracts, forms of the same general design have proved quite satisfactory. Hence, one would not look for a design error in the first instance.

Information to be supplied to students only on request for specific information

Formwork situation to date

Two similar forms were supplied, and five casts have been taken from each.

Three ties have broken, in each case at the water bar, and in one case two during the same pour.

In each case, the failed tie was in the next to bottom row.

Extra holding-down fixings have been added to the forms, to overcome uplift problems which arose with the early pours.

Tie quality

Two portions of failed ties were taken and subjected to independent tensile tests, after having threads rolled onto them to provide a grip. In addition, one of these had a water bar welded onto it in the normal way. The ends of the rods which had been damaged in the initial break were analysed chemically. Results of these tests showed that

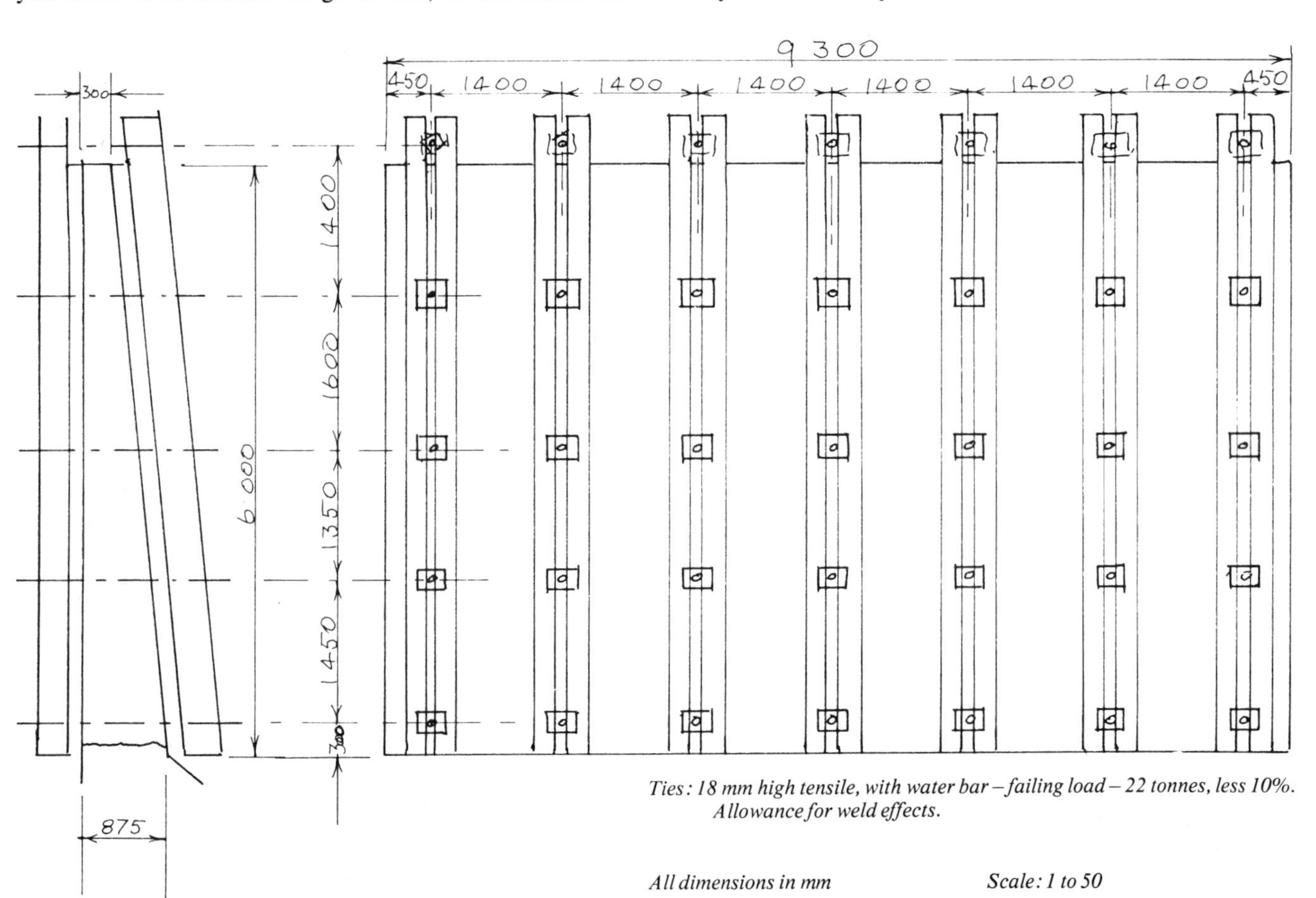

Ties: 18 mm high tensile, with water bar – failing load – 22 tonnes, less 10%. Allowance for weld effects.

All dimensions in mm *Scale: 1 to 50*

the tensile strength of the rods, and their carbon and manganese content were satisfactory. In each case, the test piece broke at the thread, showing that the welding of the water bar had not produced a serious strength reduction in this case.

Tests were also carried out on other ties taken from the site stock, as a further check, and these too performed satisfactorily. In the light of this evidence, tie quality is unlikely to be at fault.

Workmanship in tie assembly

A common form of error in this direction arises through the she-bolt threads not being properly cleaned out, so that the tie thread is only partly gripped within the she-bolt: in such a case, however, the break would occur within the length of the threaded portion of the tie rod. In this instance, the breaks occurred at the water bars, thus eliminating this cause.

One could conceivably produce excessive or uneven loading of ties by over-tightening some of the she-bolt nuts, but this would be indicated by an irregular profile to the wall, which was not observed. It is likely that, if such a problem had arisen, an experienced contractor would in any case have sorted this out for himself.

Excessive loading

There were several indications that excessive loading, brought about by the longer than usual stiffening time of the concrete, was responsible for the breakages.

(a) The early difficulties with the lifting of the forms suggested that the concrete remained fluid longer than expected.

(b) The mix contained a fine powder admixture, which would in itself account for delayed stiffening, and this effect would be augmented by the admixture (A).

(c) It was found that the external Acme thread on a she-bolt had been partially stripped within a nut; another sure sign of excessive loading.

(d) Many of the washers attached to the she-bolt nuts were dished, the corners being bent slightly outwards. This was also a sign of excessive loadings.

(e) The face of the concrete showed some horizontal settlement cracks below the tie positions. This hinted that bleeding and settlement of the concrete had been protracted, which again would be consistent with delayed stiffening.

(f) An extended she-bolt was fitted with a load cell during the casting of a subsequent bay, and this indicated loads rising to 15 tonnes during a three-hour filling operation. In this case, the admixture was omitted, and one can assume that maximum supervision and control were exercised.

The real-life conclusions

The four possible causes of tie failure are:

1. Formwork design
2. Quality of tie
3. Workmanship in tie assembly operation
4. Excess applied loads to tie

Remedial action taken

The contractor overcame the problem by omitting the admixture (A) from the concrete, and by inserting an extra row of ties.

Details of concrete

Grade 30

Materials per cubic metre

	kg
Ordinary Portland cement	230
A fine powder admixture (A)	100
Fine aggregate	790
10 mm aggregate	308
20 mm aggregate	717
Water	225
Total	2370

Originally a proprietary admixture was used, primarily as a plasticizer, but it also had some retarding and thickening properties. The admixture dosage had been calculated on the basis of the total weight of cement and fine powder admixture (A).

Tests on tie rods

Two new ties, with water bars, were taken from stock, and subjected to a tensile test. The loads carried were:

23.8 tonnes

23.65 tonnes

In each case, the failure occurred at the thread.

Measurement of tie rod load

A load recording cell was attached to the she-bolt on the second tie from the right-hand end of the inside face of the form, on the second row from the bottom.

The pour commenced at 09.00, and was completed at 12.00, the cell indicating a load of 15 tonnes had been applied to the tie by concrete pressure. Two points were noted as follows:

1. At one stage during the pour the vibrator was removed from the concrete one end and moved to the other end where the recording cell was positioned.

   Immediately the vibrator was plunged into the concrete, the recording cell showed an increase of 2 tonnes, from 10 tonnes to 12 tonnes.

2. Approximately 20 hours after the completion of the pour the recording cell showed a decrease in load of 5 tonnes, from 15 tonnes to 10 tonnes. This decrease can only be attributed to the concrete shrinkage.

Tensile tests on failed tie rods

Two halves of broken tie rods were cut short at the broken ends, and re-threaded. In addition, in one case a water bar

Figure 1. General view, showing both forms. Note site batching plant in background.

Figure 2. Leading form.

Figure 3. Infill form.

Figure 4. Close-up of infill form.

was welded onto the re-threaded rod, using the normal works production procedure.

These two ties were tested in tension, and the ends which had been cut off were examined and analysed chemically.

The results were as follows:

Failing loads	plain rod 21.9 tonnes water specimen 21.65 tonnes (Both rods failed at the thread)
Chemical analysis	plain rod carbon 0.47% manganese 0.60% water bar specimen: carbon 0.50% manganese 0.72%

The fracture faces showed some signs of brittle failure. This could be due to the welding of the medium carbon steel without any pre-heating, forming acicular brittle structures: to ascertain if this is so, would require metallographic examination of several components.

Acknowledgements

The assistance of both the contractor involved, and also of Rapid Metal Developments Ltd, who designed and supplied the formwork are gratefully acknowledged.

Further information available during progress of exercise.

Appendix 3: Recent examination questions

The questions in this Appendix have been reproduced from past examination papers set by the City and Guilds of London Institute and the author wishes to express his appreciation for permission to reproduce them.

The questions illustrate typical formwork and mould problems which tradesmen and supervisors meet in current construction practice and are published here to provide exercises for students of formwork. A fully illustrated answer is required for each of seven such questions within the four-hour period of the examination. Questions under Section A require sketches and descriptions, while those for Section B require scale drawings for the answers.

The CGLI *Formwork for Concrete Construction* examination follows a course of instruction which is conducted in a number of Technical Colleges and Schools of Building in the UK and overseas. The course runs for one year and is intended for tradesmen and supervisors who wish to improve their knowledge for formwork methods and techniques.

During the past few years the CGLI have been updating and revising certain of the schemes for part time further education courses. Students on the Advanced Craft Certificate course are invited to study formwork as a module within the course. The final examination is intended to test the achievement of certain objectives that are related to formwork. For associated practical activities these are as follows:

1. Develop and extend machine and hand tool skills beyond Craft Certificate level.
2. Apply the necessary knowledge to assess the requirements of a job and to devise realistic solutions.
3. To demonstrate the interdependency of craft practice and theoretical principles.
4. Engage in workshop and laboratory investigations.
5. Develop the abilities required in job planning and organization.
6. Apply safe working procedures.

Syllabuses and copies of past examination papers can be obtained from:

City and Guilds of London Institute
Sales Department
76 Portland Place
London W1N 4AA

The questions

1. One small contract consists of a wall 60 m long as shown in Fig. 1 to be cast *in situ* using ready-mixed concrete. The rate of construction is your decision.

(a) Indicate with the aid of a sketch but without detailed calculation the formwork to be used.

(b) Prepare a cost analysis to determine the speed at which the job shall be constructed and the amount of formwork you would use.

(c) Show your method of construction and explain in detail, with the aid of sketches, how any construction joints would be formed.

(d) Produce a programme for the job including labour and plant requirements.

(e) Estimate the cost of labour, plant and formwork (including consumable items) for your scheme. Reasonable estimates of costs for all items are acceptable.

It may be assumed that the reinforcing steel for foundation and wall will be made up into cages of any length as required.

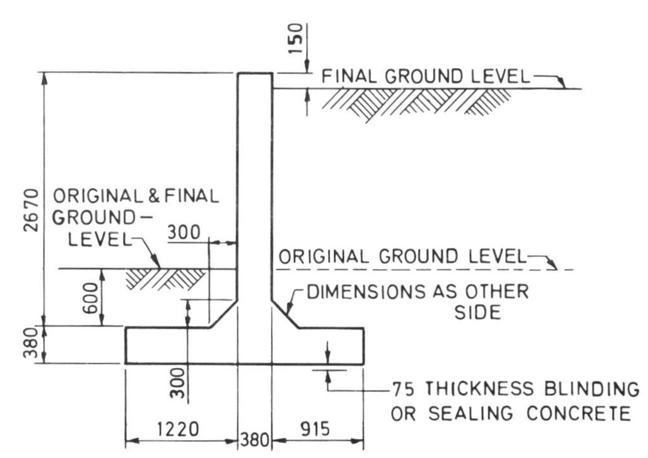

Figure 1

Formwork Planning and Design, Section A
May 1971

2. Produce sketches to illustrate typical formwork arrangements for the following items. Note on the sketches the materials and fixings to be used, and indicate methods of fixing and obtaining support

   (a) Arrangements to provide monolithic kickers 75 mm high for a reinforced concrete lift shaft which occur at 12 floor levels in a multi-storey building. The inside dimensions of the shaft are 2200 mm by 2700 mm and the walls are to have a finished thickness of 200 mm. The structural concrete floor slab is 200 mm thick.

   (b) Formwork to be used in forming a circular roof-light opening in a reinforced concrete slab 200 mm thick. The inside diameter of the opening will be 1200 mm. The upstand kerb surround will project above the slab 350 mm and will be 120 mm thick. The formwork will be used eight times.

   (c) A mould for use in precasting a series of 12 duct cover slabs in reinforced concrete. The cover slabs are to be 3200 mm by 1000 mm by 200 mm thick and will have a central recess 2000 mm by 600 mm by 75 mm deep placed centrally in the upper face. The edges of the recess can have a 1 in 4 splay to assist in the removal of the former.

3. State the treatment that should be applied to the following items of formwork prior to their use in casting the concrete described. Include for each brief notes on precautions which should be taken on site to ensure the required number of re-uses being obtained from the formwork and moulds described, mentioning storage and handling arrangements for them.

   (a) Timber formwork used to cast pile caps and ground beams. There is a possibility that the forms may be used many times.

   (b) Plywood-faced, timber-framed panels for repeated use in casting lift walls and stair walls in a multi-storey block. At least 12 uses are to be obtained from the forms.

   (c) Steel-framed and steel-faced special formwork used to cast a length of wall around a reservoir. Over 100 uses are to be obtained.

   (d) Glass-reinforced plastic sheathing panels used to mould the face of precast featured panels cast horizontally in a casting yard on site; 30 uses are expected.

Formwork for Concrete Construction, Section A June 1973

4. The panel sketched in Fig. 2 is to be used in casting reinforced concrete walling. Produce a material list stating the sizes of the materials required to manufacture the panel. Include in the list, all blocks and suitable woodscrews for framing and lining. As more than 12 uses are to be achieved, the supporting framing and sheathing will be fixed by woodscrews throughout.

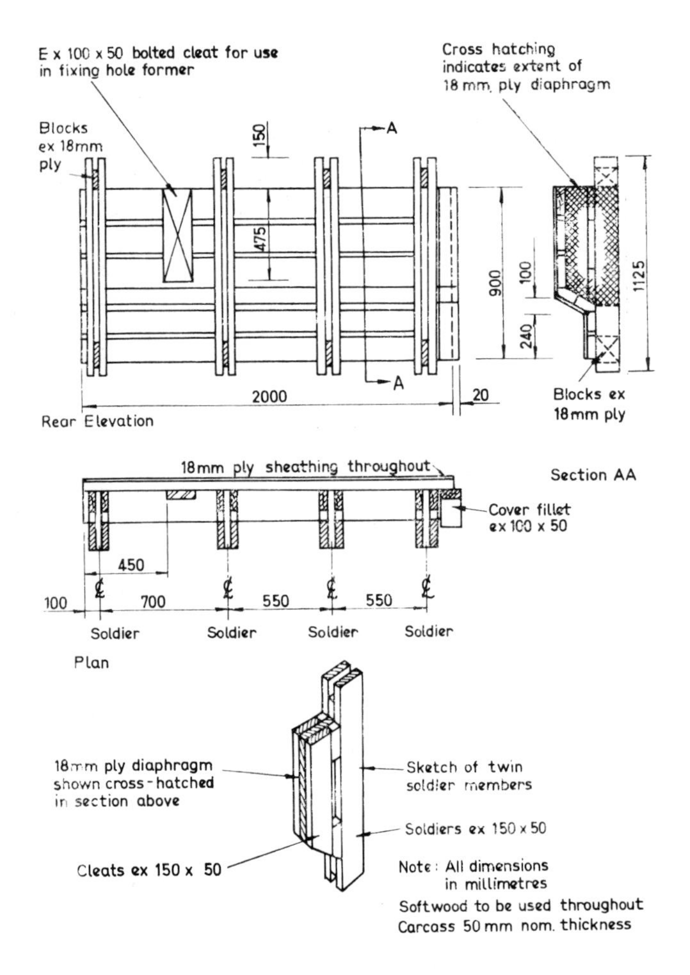

Figure 2

Formwork for Concrete Construction, Section A June 1973

5. The stair flight illustrated in Fig. 3 is to be cast using timber formwork. The wall, loading dock and column have already been cast.

   Sketch the form panels which will be required, indicating the method of obtaining fixing and support.

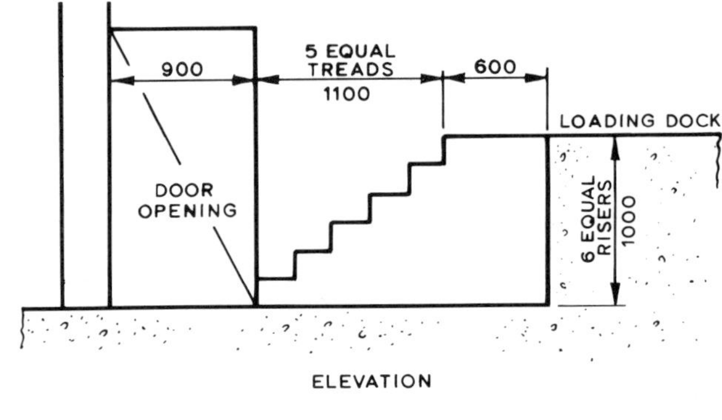

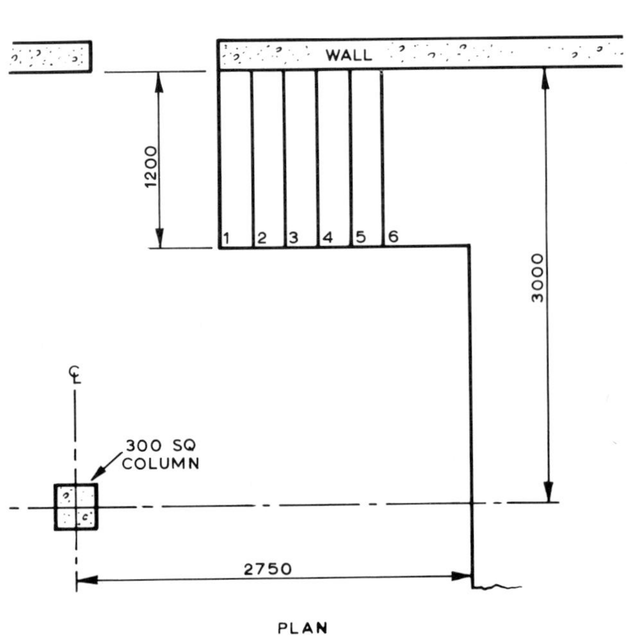

Figure 3

Formwork for Concrete Construction, Section B, June 1974

6. The 20 panels indicated in Fig. 4 are to be precast using one timber mould.

The mould carcass will be constructed of timber and the sheathing will be ply. The mould will be set up on a level concrete casting deck.

The units are to be cast face up, the exposed aggregate face being achieved by washing and brushing.

Prepare a detailed drawing from which the mould can be constructed noting all sectional sizes and indicating sheathing joint arrangements.

7. Give an example of the special provisions which may have to be made in the construction of formwork for EACH of the following:

(a) The steelfixer.
(b) The electrician.
(c) The heating and ventilating engineer.

8. Describe with the aid of sketches the use of THREE of the following items of proprietary formwork:

(a) Column clamps.
(b) Coil ties.
(c) Adjustable steel props.
(d) Strongbacks.
(e) Telescopic centres.

9. Describe with the aid of sketches the mould required to cast on site the boot lintol illustrated in Fig. 5. The mould is to provide 10 uses and will be constructed on site using timber and plywood. The reinforcement is such that the unit may be cast with any face down.

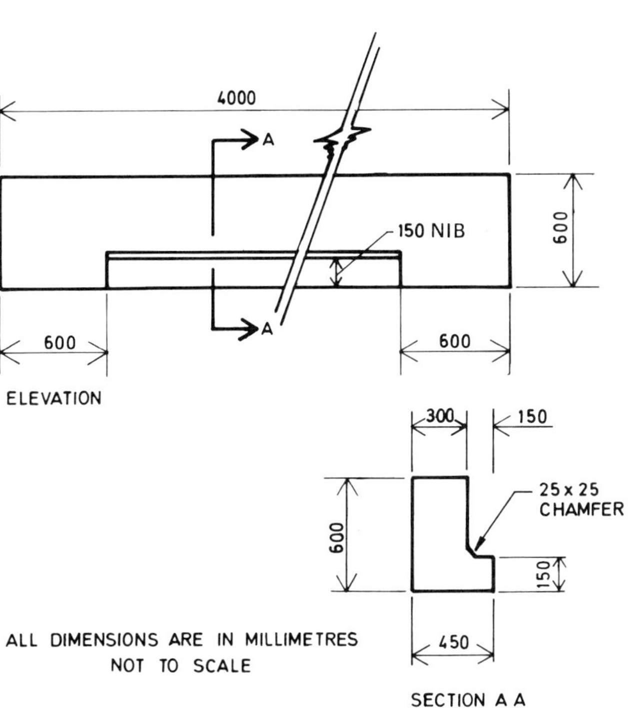

Figure 4

Carpentry and Joinery Advanced Craft Certificate – New Scheme, June 1976

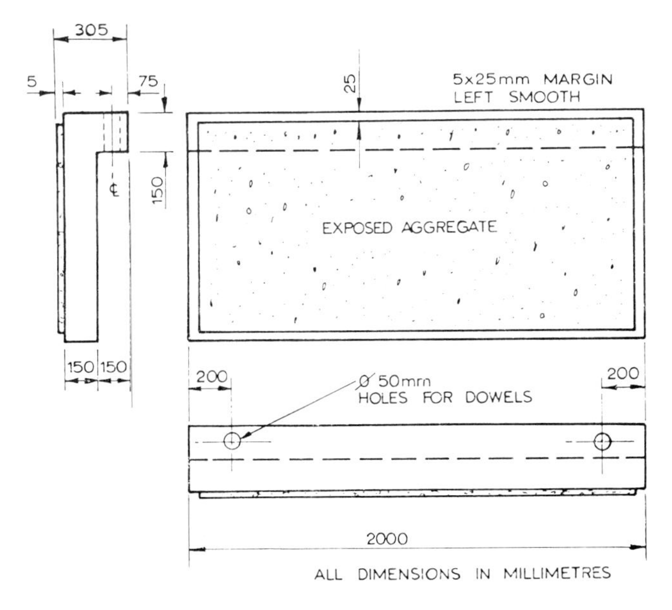

Figure 5

Formwork for Concrete Construction, Section B, June 1975

10. Describe how the beam formwork and supports illustrated in Fig. 6 should be made and erected.

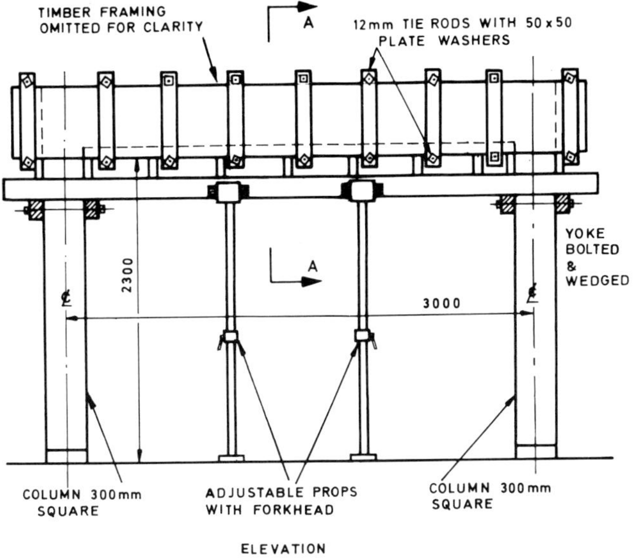

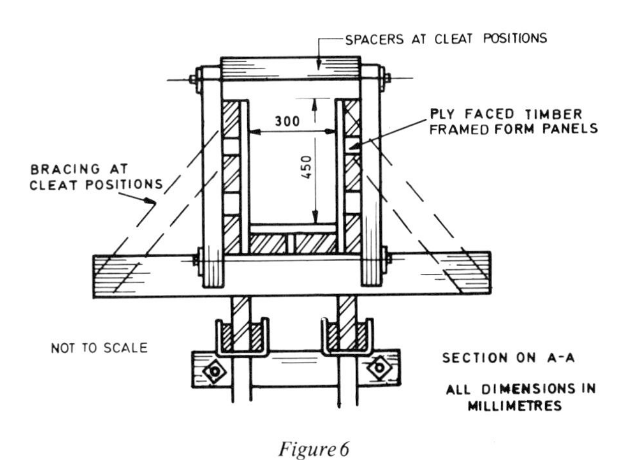

Figure 6

Formwork for Concrete Construction, Section A, June 1976

Appendix 4. Bibliography

Books, Reports and Articles

AUSTIN, C. K. *Formwork to concrete.* First edition 1960; Second edition 1966. London, Cleaver-Hume and Macmillan. pp. 283 and pp. 308.

RICHARDSON, J. G. *Formwork notebook.* London, 1972. Cement and Concrete Association. pp. 98. Publication 12.047.

HURD, M. K. *Formwork for concrete.* Third edition, Detroit, 1973. American Concrete Institute. pp. 364. ACI Publication SP–4.

RICHARDSON, J. G. *Practical formwork and mould construction.* Second edition, Barking, 1976. Applied Science Publishers. pp. 294.

GAGE, M. *A guide to exposed concrete finishes.* London, 1970. Architectural Press in collaboration with Cement and Concrete Association. pp. 164. Publication 15.334.

SNOW, Sir FREDERICK. *Formwork for modern structures.* London 1965. Chapman and Hall Ltd. pp. 128.

WYNN, A. E. and MANNING, G. P. *Design and construction of formwork for concrete structures.* Sixth edition, London, 1974. Cement and Concrete Association. pp. 428. Publication 13.013.

RICHARDSON, J. G. *Concrete notebook.* Wexham Springs, 1974. Cement and Concrete Association. pp. 92. Viewpoint Publication 12.063.

BLAKE, L. S. Recommendations for the production of high quality concrete surfaces. London, 1967. Cement and Concrete Association. pp. 38. Advisory Publication 47.019.

THE CONCRETE SOCIETY/INSTITUTION OF STRUCTURAL ENGINEERS. *Falsework.* Report of the Joint Committee, London, July 1971. pp. 52.

THE CONCRETE SOCIETY/INSTITUTION OF STRUCTURAL ENGINEERS. *Formwork.* Report of the Joint Committee, London, April, 1977. pp. 76.

BIRCH, N., WALKER, M. B. A. and LEE, C. T. *Effect of site factors on the load capacities of adjustable steel props.* CIRIA Report 27. London, Construction Industry Research and Information Association, 1971.

KINNEAR, R. G. *et al. The pressure of concrete on formwork.* CERA Report No. 1. 1965. pp. 44.

MAYNARD, D. F. and MACLEOD, G. *Factors influencing the deflection of plywood sheeting in formwork.* CIRIA Report 37. London, Construction Industry Research and Information Association. 1971. pp. 33.

WEAVER, J. and SADGROVE, B. M. *Striking times of formwork-tables of curing periods to achieve given strengths.* CIRIA Report 36. London, Construction Industry Research and Information Association. 1971. pp. 76.

LEE, I. D. G. Film faced plywood for concrete formwork. *Civil Engineering and Public Works Review.* Vol. 62, No. 728. March 1967. pp. 339–342.

ADDISON, D. W. Travelling and standard tunnel formwork, *The Consulting Engineer.* Vol. 31, No. 11, 1967. pp. 38–40.

STEIN, J. and DONALDSON, P. K. Vertical slip-forming in Europe. *Concrete Construction.* Vol. 13, No. 1, 1968. pp. 11–13.

CHAMPION, S. *Falsework with tubes and scaffold fittings, props and proprietary systems.* Paper presented at a Symposium on Falsework arranged by the Institution of Structural Engineers and the Concrete Society, London, 3 February 1969. pp. 9. Concrete Society Technical Paper 34.

JESSOP, K. G. Steel formwork: accuracy, money or myth? *Civil Engineering and Public Works Review.* Vol. 65, No. 773, December 1970. pp. 1439–1442, and 1468.

DAFFY, J. *Formwork design and detailing.* Sydney, 1971. Cement and Concrete Association of Australia. pp 29.

British Standards and Codes of Practice

BS 4: Structural steel sections
  Part 1: 1972 Hot-rolled sections
  Part 2: 1969 Hot-rolled hollow sections

BS 302: 1968 Wire ropes for cranes, excavators and general engineering purposes

BS 308: 1964 Engineering drawing practice

BS 405: 1945 Expanded metal (steel) for general purposes

BS 449: The use of structural steel in building
  449: Part 1: 1970 Imperial units
  449: Part 2: 1969 Metric units

BS 565: 1963 Glossary of terms relating to timber and woodwork

BS 647: 1969 Methods of sampling and testing glues (bone, skin and fish glues)

BS 648: 1964 Schedule of weights of building materials

BS 693: 1960 General requirements for oxy-acetylene welding of mild steel

BS 881, 589: 1955 Nomenclature of commercial timbers, including sources of supply

BS 916: 1953 Black bolts, screws and nuts

BS 1088: & 4079: Plywood for marine craft

BS 1105: 1972 Wood wool slabs up to 102 mm thick

BS 1139: 1964 Metal scaffolding

BS 1192: 1969 Building drawing practice

BS 1202: Nails
  1202: Part 1: 1966 Steel nails
  1202: Part 2: 1966 Copper nails
  1202: Part 3: 1962 Aluminium nails

BS 1203: 1963 Synthetic resin adhesives (phenolic and aminoplastic) for plywood

BS 1204: Synthetic resin adhesives (phenolic and aminoplastic) for wood
  1204: Part 1: 1964 Gap-filling adhesives
  1204: Part 2: 1965 Close-contact adhesives

BS1210: 1963 Wood screws

BS 1449: Steel plate, sheet and strip
  1449: Part 1: 1972 Carbon steel plate, sheet and strip

BS 1455: 1972 Plywood manufactured from tropical hardwoods

BS 1494: 1951 Fixing accessories for building purposes
  1494: Part 2: 1967 Sundry fixings

BS 1579: 1960 Connectors for timber

BS 1856: 1964 General requirements for the metal-arc welding of mild steel

BS 2094: Glossary of terms relating to iron and steel
  2094: Part 3: 1954 Hot rolled steel products (excluding sheet, strip and tubes)
  2094: Part 4: 1954 Steel sheet and strip
  2094: Part 5: 1954 Bright steel bar and steel wire
  2094: Part 6: 1954 Forgings and drop forgings

BS 2539: 1954 Preferred dimensions of reinforced concrete structural members

BS 2604: Resin-bonded wood chipboard
2604: Part 2: 1970 Metric units

BS 2787: 1956 Glossary of terms for concrete and reinforced concrete

BS 2900: 1970 Recommendations for the co-ordination of dimensions in building, Glossary of terms

BS 3012: 1970 Low and intermediate density polythene sheet for general purposes

BS 3444· 1961 Blockboard and laminboard

BS 3534: Epoxide resin systems for glass fibre reinforced plastics
3534: Part 1: 1962 Wet laying systems
3534: Part 2: 1964 Pre-impregnating systems

BS 3583: 1963 Information about blockboard and laminboard

BS 3589: 1963 Glossary of general building terms

BS 3809: 1971 Wood wool permanent formwork and infill units for reinforced concrete floors and roofs

BS 4011: 1966 Recommendations for the co-ordination of dimensions in building. Basic sizes for building components and assemblies

BS 4047: 1966 Grading rules for sawn home grown hardwood

BS 4071: 1966 Polyvinyl acetate (PVA) emulsion adhesives for wood

BS 4074: 1966 Metal props and struts

BS 4318: 1968 Recommendations for preferred metric basic sizes for engineering

BS 4320: 1968 Metal washers for general engineering purposes

BS 4330: 1968 Recommendations for the co-ordination of dimensions in building. Controlling dimensions

BS 4340: 1968 Glossary of formwork terms

BS 4345: 1968 Slotted angles

BS 4360: 1972 Weldable structural steels

BS 4471: Dimensions for softwood
4471: Part 1: 1969 Basic sections
4471: Part 2: 1971 Small resawn sections

BS 4549: Guide to quality control requirements for reinforced plastics mouldings
4549: Part 1: 1970 Polyester resin mouldings reinforced with chopped strand mat or randomly deposited glass fibres

BS 4606: 1970 Recommendations for the co-ordination of dimensions in building. Co-ordinating sizes for rigid flat sheet materials used in building

BS 4643: 1970 Glossary of terms relating to joints and jointing in building

BS 4646: 1970 High density polythene sheet for general purposes

CP 3: Code of basic data for the design of buildings
CP 3: Chapter V: Part 1: 1967 Dead and imposed loads
CP 3: Chapter V: Part 2: 1972 Wind loads

CP 97: Metal scaffolding
CP 97: Part 1: 1967 Common scaffolds in steel
CP 97: Part 2: 1970 Suspended scaffolds

CP 110: The structural use of concrete
CP 110: Part 1: 1972 Design, materials and workmanship

CP 111: Structural recommendations for loadbearing walls
CP 111: Part 1: 1964 Imperial units
CP 111: Part 2: 1970 Metric units

CP 112: The structural use of timber
CP 112: Part 2: 1971 Metric units

CP 114: Structural use of reinforced concrete in buildings
CP 114: Part 2: 1969 Metric units

CP 115: Structural use of prestressed concrete in buildings
CP 115: Part 2: 1969 Metric units

CP 116: The structural use of precast concrete
CP 116: Part 2: 1969 Metric units

CP 117: Composite construction in structural steel and concrete
CP 117: Part 1: 1965 Simply-supported beams in building
CP 117: Part 2: 1967 Beams for bridges

CP 123.101: 1951 Dense concrete walls

Legislation

Statutory regulations which affect the formwork operation and allied works include the following:

Construction (General Provisions) Regulations 1961.

Construction (Health and Welfare) Regulations 1966.

Construction (Lifting Operations) Certificates Order 1962.

Construction (Lifting Operations) Regulations 1961.

Construction (Working Races) Regulations 1966.

Engineering Construction (Extension of Definition) Regulations 1960.

Factories Act 1961. Section 176 (Definition of building operations).

Health and Safety at Work etc. Act 1974.

Lead paint regulations 1927.

Industrial Relations Act 1971.

Contracts of Employment Act 1963.

The Diving Operations Special Regulations 1960.

The Electricity (Factories Act) Regulations 1908.

The Electricity (Factories Act) Special Regulations 1944.

The Woodworking Machinery Special Regulations 1922–1945.

The Petroleum (Consolidation Act) 1928.

The Petroleum Regulations 1929.

The formwork designer should possess copies of the relevant statutory instruments and is well advised also to obtain copies of the following publications.

Guide to the construction regulations 1961 and 1966. (Published by The Federation of Civil Engineering Contractors and The National Federation of Building Trades Employers).

Safety, Health and Welfare booklets. New series numbers 1, 2, 45.6A. 6B. 6C. 6D. 6E. 8 10. (Published by HMSO.)

Industrial Relations.

A guide to the Industrial Relations Act 1971. (Published by Department of Employment.)

Health and Safety at Work etc. Act 1974.

Advice to employers leaflet HSC3.

The Act outlined leaflet HSC2. (Published by HMSO.)